Lecture Notes in Computer Scie

Commenced Publication in 1973
Founding and Former Series Editors:
Gerhard Goos, Juris Hartmanis, and Jan van Leeuwen

Kenjiro Cho   Philippe Jacquet (Eds.)

Technologies for Advanced Heterogeneous Networks

First Asian Internet Engineering Conference, AINTEC 2005
Bangkok, Thailand, December 13-15, 2005
Proceedings

 Springer

Volume Editors

Kenjiro Cho
Internet Initiative Japan, Inc.
1-105 Kanda Jinbo-cho, Chiyoda-ku, Tokyo 1010051 Tokyo, Japan
E-mail: kjc@iijlab.net

Philippe Jacquet
INRIA, Campus of Rocquencourt
Domaine de Voluceau, B.P. 105, 78153 Le Chesnay Cedex, France
E-mail: philippe.jacquet@inria.fr

Library of Congress Control Number: 2005936809

CR Subject Classification (1998): C.2.4, C.2, C.3, F.1, F.2.2, K.6

ISSN	0302-9743
ISBN-10	3-540-30884-9 Springer Berlin Heidelberg New York
ISBN-13	978-3-540-30884-3 Springer Berlin Heidelberg New York

Springer is a part of Springer Science+Business Media

springeronline.com

© Springer-Verlag Berlin Heidelberg 2005
Printed in Germany

Typesetting: Camera-ready by author, data conversion by Scientific Publishing Services, Chennai, India
Printed on acid-free paper   SPIN: 11599593   06/3142   5 4 3 2 1 0

Preface

The Asian Internet Engineering Conference (AINTEC) brings together researchers and engineers interested in practical and theoretical problems in Internet technologies. The conference aims at addressing issues pertinent to the Asian region with vast diversities of socio-economic and networking conditions while inviting high-quality and recent research results from the global international research community. The first event was jointly organized by the Internet Education and Research Laboratory of the Asian Institute of Technology (AIT) and the WIDE Project with support from the APAN-TH community.

In response to the recent natural disaster in Asia, AINTEC 2005 solicited papers, among other things, on the survival of the Internet in order to provide alternative means of communication in emergency and chaotic situations. The main topics include: Mobile IP
Mobile Ad Hoc and Emergency Networks
Multimedia or Multi-Services IP-Based Networks
Peer-to-Peer
Measurement and Performance Analysis
Internet over Satellite Communications

There were 52 submissions to the Technical Program, and we selected the 18 papers presented in these proceedings. In addition, we have three invited papers and one invited position paper by leading experts in the field.

Finally, we would like to acknowledge the conference General Chair, Kanchana Kanchanasut of AIT, and the Local Organizers team from AIT, namely, Pensri Arunwatanamongkol, Withmone Tin Latt and Yasuo Tsuchimoto, for organizing and arranging this conference. We are also grateful to the French Ministry of Foreign Affairs through its French Regional Cooperation and the ICT Asia project (STIC-ASIE) for providing travel support.

December 2005   Kenjiro Cho and Philippe Jacquet

Organization

General Chair

Kanchana Kanchanasut (AIT, Thailand)

Program Committee Co-chairs

Kenjiro Cho, WIDE Project
Philippe Jacquet, INRIA, France

Program Committee

Kazi Ahmed (AIT, Thailand)
Patcharee Basu (SOI/ASIA, Japan)
Randy Bush (IIJ, USA)
Thomas Clausen (Polytechnique, France)
Noel Crespi (INT, France)
Tapio Erke (AIT, Thailand)
Thierry Ernst (Keio University, Japan)
Chalermek Intanagowiwat (Chulalongkorn U, Thailand)
Alain Jean-Marie (LIRMM/INRIA, France)
T.J. Kniveton (NOKIA Research Center, USA)
Youngseok Lee (CNU, Korea)
Bill Manning (USC/ISI, USA)
Thomas Noel (University Louis Pasteur, France)
Alexandru Petrescu (Motorola)
Anan Phonphoem (Kasetsart U, Thailand)
Poompat Saengudomlert (AIT, Thailand)
Teerapat Sa-nguankotchakorn (AIT, Thailand)
Kazunori Sugiura (WIDE Project, Japan)
Jun Takei (JCSAT, Japan)
C.W. Tan (USM, Malaysia)
Takamichi Tateoka (University of Electro-Communications, Japan)
Antti Tuominen (Helsinki University of Technology, Finland)
Ryuji Wakikawa (WIDE Project, Japan)

Local Organization

Pensri Arunwatanamongkol (AIT, Thailand)
Withmone Tin Latt (AIT, Thailand)
Yasuo Tsuchimoto (AIT, Thailand)

Table of Contents

Measurement and Performance Analysis

Efficient Blacklisting and Pollution-Level Estimation in P2P File-Sharing Systems

Jian Liang, Naoum Naoumov, and Keith W. Ross

Department of Computer and Information Science,
Polytechnic University, Brooklyn NY 11201, USA
{jliang, naoum, ross}@poly.edu

Abstract. P2P file-sharing systems are susceptible to pollution attacks, whereby corrupted copies of content are aggressively introduced into the system. Recent research indicates that pollution is extensive in several file sharing systems. In this paper we propose an efficient measurement methodology for identifying the sources of pollution and estimating the levels of polluted content. The methodology can be used to efficiently blacklist polluters, evaluate the success of a pollution campaign, to reduce wasted bandwidth due to the transmission of polluted content, and to remove the noise from content measurement data. The proposed methodology is efficient in that it does not involve the downloading and analysis of binary content, which would be expensive in bandwidth and in computation/human resources. The methodology is based on harvesting metadata from the file sharing system and then processing off-line the harvested meta-data. We apply the technique to the FastTrack/Kazaa file-sharing network. Analyzing the false positives and false negatives, we conclude that the methodology is efficient and accurate.

1 Introduction

By many measures, P2P file sharing is the most important application in the Internet today. There are more than 8 million concurrent users that are connected to either FastTrack/Kazaa, eDonkey and eMule. These users share terabytes of content. In the days of Napster (circa 2000), most of the shared files were MP3 files. Today the content includes MP3 songs, entire albums, television shows, entire movies, documents, images, software, and games. P2P traffic accounts for more than 60% of tier-1 ISP traffic in the USA and more than 80% of tier-1 traffic in Asia [1].

Because of the their decentralized and non-authenticated nature, P2P file sharing systems are highly susceptible to "pollution attacks". In a pollution attack, a polluter first tampers with targeted content, rendering the content unusable. It then deposits the tampered content, or only the metadata for that content, in large volumes in the P2P file sharing system. Unable to distinguish polluted files from unpolluted files, unsuspecting users download the files into their own file-sharing folders, from which other users may then later download the polluted files. In this manner, the polluted copies of a title spread through the

K. Cho and P. Jacquet (Eds.): AINTEC 2005, LNCS 3837, pp. 1–21, 2005.
© Springer-Verlag Berlin Heidelberg 2005

file-sharing system, and the number copies of the polluted title may eventually exceed the number of clean copies. The goal of the polluter is to trick users into repeatedly downloading polluted copies of the targeted title; users may then become frustrated and abandon trying to obtain the title from the file-sharing system. As a side effect, however, the polluted content becomes a persistent "noise" in the P2P system that interferes with research measurement work. Pollution is currently highly prevalent in file-sharing systems, with as many as 50% to 80% of the copies of popular titles being polluted [2].

In this paper we study mechanisms to measure the effectiveness of a pollution attack. We emphasize, however, that we do not take a side in the P2P file-sharing debate, neither condoning nor condemning the pollution attacks that are commissioned by the music, television and film industries. But given that P2P file sharing traffic is currently the dominant traffic type in the Internet, and that the files being transferred are frequently polluted, a significant fraction of Internet bandwidth is clearly being wasted by transporting large, corrupted files. It is therefore important to gain a deep understanding of the pollution attack and develop effective mechanisms to measure it.

In this paper we explore two techniques for countering the pollution attack:

- **Identifying pollution source IP ranges:** The goal is to identify IP address ranges that are broad and complete enough to cover the hosts providing polluted content, yet narrow enough to exclude the vast majority of ordinary users.
- **Identifying the pollution level of titles:** With knowledge of which titles are being polluted and to what extent users and researchers can use the file sharing system accordingly.

In developing methodologies we have not only aimed for accuracy but also for efficiency. For source identification, one approach would be to download copies of titles from a vast number of IP addresses and then manually check the copies for pollution; the IP addresses that consistently supply polluted content could then be marked. Such an approach would be highly inefficient, requiring enormous bandwidth, computing and human resources, and would also introduce significant "probing" traffic into the Internet.

Our methodologies do not involve the downloading of any files. Instead, they identify polluting IP address ranges and targeted titles by collecting and analyzing metadata from the file sharing system. The metadata is harvested by crawling the nodes in the P2P system and sending tailored queries to each of the crawled nodes. The harvested metadata can then be analyzed to obtain detailed information about the numbers of versions and copies, and the IP subnets containing the versions and copies, for a large number of investigated titles. From this detailed information, our methodology constructs the blacklisted IP ranges and the estimated pollution levels for the targeted titles. The methodology is efficient in that it collects metadata (text) rather than content (which is typically 3MB to several GB per file for music and video) and that a large number of titles and virtually all the file-sharing nodes can be investigated in one crawl.

Our contribution is as follows:

– We developed a methodology for creating a blacklist set. The methodology is based on identifying high-density prefixes, which are prefixes in which the nodes that have a copy of a particular title have, on average, a large number of copies. We provide a heuristic for separating the low density prefixes from the high density prefixes, and a mechanism to merge prefixes that are topologically close. The set of resulting merged prefixes constitutes the pollution source set. We then developed several metrics for measuring the accuracy of the set. The two principal metrics are probability of false positive and false negative. We also examine secondary metrics, including comparing the download times and last-hop RTTs at nodes we have identified and regular nodes.
– We developed a methodology for estimating the pollution level of a title, which is defined as the ratio of polluted copies in the network to the total number of copies in the network. This estimate does not involve the downloading of any files and is solely based on the harvested metadata. We then evaluate our estimate by measuring the actual pollution levels of selected titles.
– We crawled FastTrack for 170 titles, including songs and movies. We then applied the methodologies to the FastTrack metadata harvested during the crawling procedure. Our resulting pollution source set contains 112 prefixes. Our evaluation metrics indicate that the set is accurate, with low probabilities of false positives and false negatives. We also find that the estimates for pollution levels in examined titles is accurate.

This paper is organized as follows. In Section 2 we describe the pollution attack in detail and introduce important terminology. In Section 3 we describe in detail the metholodogies and the evaluation procedures for creating the pollution source set and the pollution-level estimates. In Section 4 we describe the experimental setup, including the crawler and PlanetLab experiments. Section 5 provides the results of our experiment, including evaluation results for the methodologies. Section 6 describes previous work related to this paper. We conclude in Section 7.

2   Overview of Pollution

2.1   File Sharing Terminology

We first provide an overview of a generic P2P file-sharing application. This will allow us to introduce some important terminology that is used throughout the paper. In this paper we are primarily concerned with the sharing of music and video. We shall refer to a specific song or video as a **title**. A given title can have many different **versions** (in fact, tens of thousands). These versions primarily result from a large number of rippers/compressors, each of which can produce slightly different files when created by different users. Modifications of metadata can also create different versions. Users download different versions of titles from each other, thereby creating multiple **copies** of identical file versions in the P2P file sharing system. At any given time, a P2P file-sharing system may make available thousands of copies of the same version of a particular title.

A file in a P2P file sharing system typically has **metadata** associated with it. There are two types of metadata: metadata that is actually included in in file itself and is often created during the ripping process (e.g. ID3 Tags in mp3 files); and metadata that is stored in the file-sharing system but not within the shared files themselves. This "outside-file" metadata may initially be derived from the "inside-file" metadata, but is often modified by the users of the file-sharing systems. It is the outside-file metadata that is employed during P2P searches. In this paper, when using the term metadata, we are referring to the outside-file metadata. Because different copies of a version of a title may be stored on different user nodes, the different copies can actually have different metadata.

When a user wants to obtain a copy of a specific title, the user performs a keyword search, using keywords that relate to the title (for example, artist name and song title). The keywords are sent within a query into the file-sharing network. The query will visit one or more nodes in the file sharing network, and these nodes will respond if they know of files with metadata that match the keywords. The response will include the metadata for the file, the IP address of the node that is sharing the file, and the username at that node. For many file sharing systems, the response will also include a **hash**, which is taken over the entire version. To download a copy of a version, one sends a request message (often within an HTTP request message) to the sharing user. In this request message, the version is identified by its hash. Many file sharing systems employ parallel downloading, in which case requests for different portions of the version are sent to different users sharing that file.

Many nodes in P2P file sharing systems are behind Network Address Translators (NATs). When crawling a NATed node's private IP address and private port number may be provided rather than its public IP address and public port number. Since the range of private IP address is relatively narrow, different NATed users may have the same private IP address. Thus, from the crawling data, we cannot distinguish between different users solely by their IP addresses. In order to distinguish between different users, including NATed users, we define a **user** as the triple (IP address, port number, username).

2.2   Intentional Pollution

Naturally pollution occurs in P2P systems when users share corrupted versions of some titles. However, the amount of such pollution is negligible. Other users of the system, however, may intentionally introduce a large number of corrupted files. They create numerous versions of their targeted title by tampering with it in one or more ways with the binary content of the file. Then, they connect one or more nodes to the P2P file-sharing system and places the tampered versions into its shared folders on these nodes. Users query for the title and learn about the locations of versions of the title, including the polluted versions and download one or more polluted versions. The P2P software then automatically places the file in the shared folders of those users and the pollutions spreads further. That kind of pollution is prevalent in modern P2P file sharing systems such as FastTrack/Kazaa [2]

Most of the pollution today emanates from "professional" polluters that work on the behalf of copyright owners, including the record labels and the motion-picture companies. From this economic context and from our own testing and usage experience, we conclude that the professional polluters tend to pollute popular content, such as recently-released hit-songs and films. In [2] a random sample of recent, popular songs were shown to be heavily polluted whereas a random sample of songs for the 70s were shown to be mostly clean.

In order to facilitate the spread of the polluted content polluters have high-bandwidth Internet connections ,have high availability, and they are not behind firewalls or NATed routers.

In our methodology for detecting polluted content and blacklisting polluters, we will make the following assumptions about polluters. Many of these assumptions will be corroborated in Section 5, where our measurement results are presented.

- Because polluters share popular titles at attractive file-transfer rates, there is a high demand for their content from unsuspecting users. To meet the demand, the polluter often uses a server farm at one or more polluter sites. The nodes in a server farm are concentrated in a narrow IP address range.
- Whereas regular P2P users run one client instance per host, polluters often run many clients in each of their nodes, with each instance having a different username and sharing its own set of copies and versions for the targeted titles. This is done to improve placement of search results in the users' GUIs.
- A polluter distributes multiple polluted versions of the same title. This also improves the placement of search result in the users' GUIs. As we will show in Section 5, an ordinary user typically has a small number of versions of any title. To compete with all the clean versions in the display of the search results, a polluter needs to provide many different versions (each with a different hash) to increase the chances that its versions are selected from the users' GUIs.

3 Methodology

In this paper we develop methodologies for two tasks. The first task, which we refer to as **blacklisting**, is to find the IP address ranges that include the large majority of the polluters. The second task, referred to as **pollution level estimation**, is to determine the extent of pollution for specified titles. For both of these tasks, the first step is to crawl the file sharing system, as we now discuss.

3.1 Crawling

Crawling a P2P file sharing system is the process of visiting a large number of nodes to gather information about the copies of files being shared in the system. The crawler might gather, for example, the IP addresses and hashes of all copies of files being shared in the network for a set of specific titles over a given period of time. Several independent research groups have developed crawlers for P2P file sharing systems. A crawler for the original single-tier Gnutella system is described in [3]. A crawler for the current two-tier Gnutella system (with "ultrapeers") is described in [4]. A crawler for eDonkey is described in [15]. A crawler for the FastTrack P2P file sharing is described in [2]. Since P2P networks are dynamic, with nodes frequently joining and leaving, a good crawler needs to rapidly crawl the entire network to obtain an accurate snapshot.

The first step in our methodologies is to crawl the P2P file sharing system and obtain the following information for each title of interest: the number of versions in the file sharing system for the title; the hash values for each of the versions; the number of copies of each version available in the file sharing system; for each copy, the IP address of the node that is sharing it; the port number of the application instance at that node (many modern P2P systems vary the port number across nodes to bypass firewalls); the username at that node; and, for each copy, some copy details (e.g., playtime, file size, description, etc). For each title of interest, the crawler deposits this information in a **crawling database**, which can then be analyzed off-line. We will describe a crawler for the FastTrack network in Section 4.

3.2   Identifying Pollution Sources

Polluters typically control blocks of IP addresses and can easily move their nodes from one IP address to another within the block. Thus, rather than idnetifying individual IP addresses, we should find ranges or IP addresses that are likely to include the polluters in the near future as well as the present. Our methodology has the following steps:

1. Crawl the P2P file sharing system as described above.
2. From the data in the crawling database, identify the /24 prefixes that are likely operated by polluters.
3. Merge groups of /24 prefixes that are topologically close and don't cross BGP prefixes. The set of merged prefixes becomes our blacklist set.

We now describe the second and third steps in more detail.

The second step is to identify /24 prefixes that are likely operated by polluters. A polluter typically leases from a data center a set of server nodes in a narrow IP address range. Data centers do not normally include ordinary P2P users, which typically access the Internet from residences and universities. A /24 prefix is small enough so that both polluters and ordinary users do not operate from within the same prefix; and it is large enough to cover multiple polluting servers in most subnets. In the third step, we search for larger subnets.

Let N denote the number of titles investigated and T_n denote the nth title. For each title T_n, we determine from the crawling database the /24 prefixes that contain at least one copy of title T_n. Suppose there are $I^{(n)}$ such /24 prefixes; denote the set of these prefixes by $\mathcal{P}^{(n)} = \{p_1^{(n)}, p_2^{(n)}, \ldots, p_{I^{(n)}}^{(n)}\}$.

We now introduce the important concept of the "density of a prefix," which will be used repeatedly in this paper. For each such prefix $p_i^{(n)}$, define $x_i^{(n)}$ to be the number of IP addresses in the prefix with at least one copy of the title and $y_i^{(n)}$ to be the number of copies (included repeated copies across nodes) of the title stored in the prefix. Finally, define the **density** of prefix $p_i^{(n)}$ as $d_i^{(n)} = y_i^{(n)}/x_i^{(n)}$.

From our assumptions about how polluters operate (see Section 2), we expect the prefixes with high density values to be operated by polluters and prefixes with low densities to contain only "innocent" users. We consider prefixes with a density higher than a threshold $d_{thresh}^{(n)}$ to be operated by polluters. There are many possible heuristics that can be used to determine this threshold. We now describe a simple heuristic that gives good performance. It is based on the median value of the distinct density values in $\{d_1^{(n)}, d_2^{(n)}, \ldots, d_{I^{(n)}}\}$ denoted by $d_{median}^{(n)}$. Of course, different prefixes have different numbers of users and different densities, so in order to allow for a variance in user behavior we set a threshold to a multiple of the median. Specifically, our heuristic sets the threshold to

$$d_{thresh}^{(n)} = kd_{median}^{(n)} \tag{1}$$

where k is an appropriately chosen scaling factor (see Section 5). We say that a prefix $p_i^{(n)}$ is a **polluting prefix** if $d_i^{(n)} \geq d_{thresh}^{(n)}$. Let $\mathcal{Q}$ be the union of all the polluting /24 prefixes over all N titles.

A polluter may actually operate within a network that is larger than a /24 prefix. The third step of our methodology is to create larger prefixes which encompass neighboring /24 prefixes in $\mathcal{Q}$. For this, we merge adjacent prefixes in the IP space. We also merge some non-adjacent prefixes. To this end, we perform a traceroute from each of 20 PlanetLab nodes to one IP address in each of the prefixes in $\mathcal{Q}$. Prefixes which have the same last router become candidates for merging. In doing this we need to account for the possibility that some of the traceroutes passing through the same last router may actually pass through the router via different interfaces (and thus IP addresses) [9]. Suppose there are J groups of prefixes, with each prefix in a group sharing the same last router. (Some groups may contain a single prefix.) Let $\mathcal{G}_j$, $j = 1, \ldots, J$, denote the groups. For each group of prefixes $\mathcal{G}_j$, denote p_j as the longest prefix that covers all the prefixes in $\mathcal{G}_j$. For each such p_j we verify that it does not cross prefixes found in a BGP table. If it does, we decompose p_j back into its original /24 prefixes. Let $\mathcal{P}$ be the resulting set of prefixes. $\mathcal{P}$ is our final pollutions source set, and consists of all the p_j's that pass the BGP test and all of the decomposed /24 prefixes as just described.

Note that this methodology for creating a pollution source set does not involve the downloading of content. Indeed, any download-based methodology would require the downloading of an excessively large number of files as well as an automated procedure to determine whether a downloaded file is polluted. Our approach is instead based on the metadata that is gathered by the crawler. This approach is efficient in that crawling a large-scale P2P file-sharing system can be quickly done with modest resources.

3.3   Evaluation Procedure for Pollution Source Sets

The pollution source set $\mathcal{P}$ may not be completely accurate in that it may not contain all polluting nodes (false negatives) and it may contain some active nodes that are innocent users (false positives). We evaluate a blacklisting methodology by estimating the probability of false positives and false negatives.

To this end, we need a procedure to determine whether a downloaded version of any given title is polluted. This can be done by downloading the version and manually watching or listening to it. Such a manual procedure would require an excessive amount of human resources. Instead we use a simple automated procedure which has been shown to give accurate results [2]. Specifically, we download the version into RAM and declare the version to be clean (unpolluted) if the following three criteria are met:

1. Rehashing the file results in the same hash value as the one that was used to request the file;
2. The title is decodable according to the media format specifications; for example, an mp3 file fully decodes as a valid mp3 file [24].
3. The title's playback duration is within 10% of the one specified in release information for that title.

In any one of the three criteria is violated, we consider the version to be polluted. We refer to this procedure as the **automated version-checking procedure**.

Having described our procedure to determine whether a downloaded version is polluted, we can now state our false-negative and false-positive evaluation procedure. To evaluate the false-negative probability, we randomly select 1,000 users having IPs outside of $\mathcal{P}$ and having a copy of at least one of the investigated titles. For each randomly selected node, we randomly download 5 versions stored at that node. We declare a node to be a **false negative** if all of the following conditions are satisfied: (i) it has at least 5 versions of any one of the investigated titles; (ii) its upload throughput is greater than a given threshold. (In Section 4 we describe how we estimate a node's upload throughput); (iii) its Last Hop RTT is less than a threshold (we define Last Hop RTT in Section 4) ; and (iv) all of the randomly selected versions are polluted. Thus a randomly selected node is declared a false negative if that node has the main characteristics of a polluting node. (See Section 2.) The false-negative probability is simply the the number of randomly selected nodes declared to be false negatives divided by the total number of randomly selected nodes.

A false positive occurs when an "innocent" non-polluting node is blacklisted as a polluter by our methodology. This can happen when a /24 prefix is labelled a polluting prefix but contains non-polluting users, or when innocent users are added to $\mathcal{P}$ during the merging process. To evaluate the false-positive probability, we randomly select 1,000 nodes in $\mathcal{P}$ containing a copy of at least one of the titles. For each randomly selected node, we randomly download five versions stored at that node. We declare a randomly selected node to be a **false positive** if any of the following criteria are satisfied: (i) its throughput is smaller than the threshold; (ii) its last hop RTT is larger than the threshold; and (iii) at least one of the randomly selected versions is clean. The false-positive probability is simply the the number of randomly selected nodes declared to be false positives divided by the total number of randomly selected nodes.

3.4   Estimating Content Pollution Levels

In this subsection we provide our methodology for estimating the pollution level of any arbitrary title T_n. The methodology builds on the blacklisting methodology. We define the **pollution level** of a title as the fraction of copies of the title available in the P2P file sharing system that are polluted. The pollution level of a title can be estimated by randomly selecting a large number of copies of the title, downloading each of the copies, and then testing the copies for pollution (either by listening to them or through some automated procedure). This requires an exorbitant amount of resources, particularly if we wish to accurately determine the pollution levels of many titles. We instead estimate the pollution levels of titles directly from the metadata available in the crawling database. To this end, we make the following assumptions:

1. All copies of the title that are stored in a blacklisted node (that is, in a node in $\mathcal{P}$) are polluted.
2. For each node outside of $\mathcal{P}$ with at least one copy of T_n, all copies stored at that node are polluted except for one copy.

With these assumptions, we now derive an expression for $E^{(n)}$, the **estimated pollution level** of title T_n. For a title Recall that $y^{(n)}$ is the total number of copies of the title available in the crawling database. Also let $z^{(n)}$ be the number of nodes outside of the blacklist set $\mathcal{P}$ that have at least one copy of T_n. The above two assumptions imply that the number of copies of T_n that are polluted is $y^{(n)} - z^{(n)}$; thus, our estimate of the pollution level for title T_n is

$$E^{(n)} = \frac{y^{(n)} - z^{(n)}}{y^{(n)}} \tag{2}$$

3.5   Evaluation Procedure for Pollution-Level Estimation

$E^{(n)}$ is an estimate of the pollution level of title T_n, derived solely from the metadata in the crawling database. To evaluate the accuracy of this estimate, we compare it with a measured value, which is obtained by actually downloading content. Specifically, for a given title T_n we do the following:

1. We download the most popular versions of the title. The number of versions downloaded is such that the downloaded versions covers at least 80% of all copies of the title in the file-sharing system. For title T_n, let J_n be the number of versions that meet this 80% criterion.
2. For each of these versions, we determine if the version is polluted or not using the automated version checking procedure described in Section 3.3. Let $\delta_i^{(n)}$ be equal to 1 if version i is determined polluted and be equal to 0 otherwise.

3. The crawling database provides the number of copies that each version contributes. Let $c_i^{(n)}$ be the number of copies of version i in the database. We calculate the fraction of polluted copies $L^{(n)}$ as

$$L^{(n)} = \frac{\sum_{i=1}^{J_n} c_i^{(n)} \delta_i^{(n)}}{\sum_{i=1}^{J_n} c_i^{(n)}} \qquad (3)$$

We then define the error in the pollution-level estimate as:

$$Error = \frac{|E^{(n)} - L^{(n)}|}{L^{(n)}} \qquad (4)$$

We present the resulting error and its implications in Section 5.

4   Experimental Setup

We evaluated our methodologies from data collected in the FastTrack file-sharing network. We first briefly describe the FastTrack network and the FastTrack crawler.

4.1   FastTrack Crawler

With more than two million simultaneous active nodes (in Nov 2004), FastTrack is one of the largest P2P file sharing systems. It has at least an order of magnitude more users than Gnutella. It is used by several FastTrack clients including Kazaa, kazaa-lite, Grokster, and iMesh. It is also known to be the target of the pollution attack. For these reasons, we chose to test our pollution source set and pollution-level methodologies on FastTrack.

In [2] we already described the FastTrack Network structure and a crawling platfrom that we developped for it. In thise paper we use our crawler to collect metadata and analyse it.

4.2   PlanetLab Measurement

As part of our evaluation procedures, we also determine the download throughput and the last-hop Round Trip Time (RTT) at various nodes in the FastTrack network. We now describe how we measure those metrics. Download throughput and last-hop RTT depend on the measurement host from which the measurement is being initiated. To reduce the observer bias, we distributed our measuring hosts on a number of PlanetLab nodes.

Throughput. We used 20 well connected PlanetLab hosts that downloaded data from each measured host. The measurement was performed in the following way. The PlanetLab nodes sequentially establish TCP connections to each measured FastTrack node. For each connection, the PlanetLab host requests 500KB of data. If t denotes the time from when the PlanetLab host begins to receive the 500KB until when it has received all of the 500 KB, then the throughput of the connection is defined to be $(500KB)/t$. The throughput of the node is then obtained by averaging the throughput over all successful connections. A custom program was developed to download and report those measurements from the different nodes.

Estimated Last-Hop RTT. To verify our blacklisting methodology, we also measured the Last-Hop Round Trip Time (RTT). The last-hop RTT is the time it takes a small packet to travel from this last router to the destination host and back to the last router [22]. We can only estimate the last-hop RTT with indirect measurement from our sources, since we don't have access to the routers. To estimate the last-hop RTT, we measure the minimum out of 3 RTTs from the source to the destination from which we subtract the minimum RTT from the source to the last router, as shown in Figure 1. For each PlanetLab source, we found the difference of those values and took the minimum one as an estimate of the last-hop RTT for that target IP address.

In both experiments we used 20 PlanetLab nodes and we selected them from different parts of North America, Europe, Asia and the Pacific.

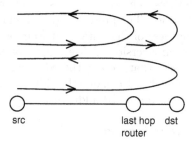

Fig. 1. Computing Estimated Last-Hop RTT: The estimated last-hop RTT (grey) is the difference between the RTT times to the destination host and the last-hop router

5   Results

In this section we present the results of our experiments. We first describe the raw data. We then provide the results pertaining to the blacklisting and pollution-level methodologies.

5.1   Raw Data

The crawling platform captured 124GB of metadata on the KaZaA network from Nov 21 to Nov 27. The crawler queried supernodes around the world for 170 titles, consisting of 133 songs and 37 movies. We choose these title as follows. As discussed in Section 2, popular, newly released titles are often targets for pollution. Even though this paper mostly focuses on music, we include some movie titles to illustrate that the technique is universal. Most of the chosen songs are new popular songs, with their titles obtained from the listing available

Table 1. Raw Data for Representative Titles

title	versions	users	copies	total IPs	Public IPs
040	225341	135102	2120160	8627	7577
008	155642	112542	1575686	6188	5298
060	91447	172879	300865	57681	52252
052	48607	126226	301075	46129	40419
097	9648	28583	37173	17468	15289
009	3795	41405	56215	24689	22388
111	5503	14519	17562	9801	8437
005	5856	56351	67945	37051	32162

at itunes.com [6]. Some of the chosen songs are "oldies" obtained from about.com 70s charts [7]. The remaining chosen content is newly released DVDs from the list of top rentals at blockbuster.com [8]. This selection gave us a varied list of popular titles without particular taste or style preference bias in our results.

Table 1 includes some of the data that we gathered from the crawler for a few of the representative titles. The titles in this table are chosen to represent a diversity of data distributions. The presented data includes the title numbers; the number of versions (hashes) of that title that were observed; the number of users who possess a copy of any version of title; the number of copies of the title; the total number IP addresses that were gathered (Because multiple FastTrack clients can be present on one IP address, one IP address can represent more than one user.); and the number of public IP addresses.

5.2   Pollution Source Identification

We now present and analyze the results obtained from our blacklisting methodology. We clustered all public IP addresses from our database into /24 prefixes and calculated the density of each prefix. Figures 2(a) and 2(b) show the density distribution for the title *"Pain" by "Jimmy Eat World"* (040). Figures 2(c) and 2(d) show the same plots but for the title *"Let's Get It Started" by "Black Eyed Peas"* (005). For both titles, most of the prefixes have a density of 1. For "Pain", there are 5,012 prefixes with density 1 and for "Let's Get It Started", there are 28,506 such prefixes. It is clear from these figures that for some titles there are prefixes with extremely high densities (there are prefixes with over density of over 10,000 in title 040), while for other titles, all prefixes have low densities(all prefixes have density less than 15 in title 005). It is also clear from these figures that title 005 has not been targeted by the pollution attack whereas the title 040 has.

We now turn our attention to determining the blacklist set using the blacklisting methodology developed in Section 3. As defined in Equation 1, we use a blacklisting threshold of $kd_{median}^{(n)}$. In that formula k is an experimental constant; through many trials we found that $k = 8$ consistently achieves good separation of ordinary users and polluters. Figures 2(b) and 2(d) include the thresholds for the two titles. In the case of the polluted title "Pain," with median of 17

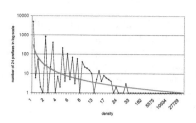

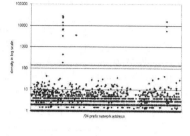

(a) Number of /24 subnets per density value of one title 040

(b) Density distributions of the /24 subnets of the entire IPv4 range and blacklist threshold of title 040

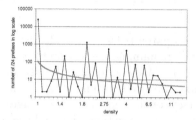

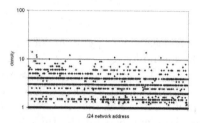

(c) Number of /24 subnets per density value of title 005

(d) Density distributions of the /24 subnets of the entire IPv4 range and blacklist threshold of title 005

Fig. 2.

and resulting threshold of $d_{thresh}^{(040)} = 136$, it successfully manages to separate the majority users from the outstanding few with high density, while in the case of the clean song "Let's Get It Started," with median of 4 and thesold of 32, the threshold is above all prefix densities and thus does not blacklist any prefix.

For the prefixes with densities larger than the threshold for each title, there are 114 /24 prefixes, containing 1,218 IP address (with one of the titles), 70,224,279 title copies, and 10,518,683 versions of 154 of the 170 titles that we had in our crawling database. Note that a very small fraction of the /24 prefixes in FastTrack are responsible for pollution.

The next step of our methodology is the merging of the prefixes that are topologically close. Merging the consecutive /24 prefixes and those that have the same last hop router resulted in decreasing the number of clusters to 101 prefixes, with the masks ranging from /24 to /16. We then performed the BGP prefix verification. The BGP prefixes are from [10] obtained on 12/06/04. We had information about 17,037,611 prefixes. Some of the prefixes that resulted from merging were part of different BGP prefixes. Those merges had to be abandoned. After the BGP verification we had a final list of identified polluter IP ranges, details for which we present in Table 2. We see from the table that the resulting blacklist set $\mathcal{P}$ has 112 prefixes. These prefixes contain 1,218 IP addresses that

Table 2. Pollution Source Results

Nodes	Number of IPs	Number of prefixes	Number of BGP prefixes	Number of BGP ASs
Blacklisted	1218	112	79	59
Non-Blacklisted	1,303,954	325,075	15,747	4,296

Table 3. Node Statistics

Nodes	Avg. copies per title	Variance of # of copies	Avg. users per IP
Blacklisted	11.67	20.0	269
Non-blacklisted	1.56	1.88	1.01

contain at least one copy of one of the investigated titles. The table also shows that the methodology does not blacklist the remaining 325,075 prefixes, which contain over 1.3 million IP addresses with the investigated content. To understand better the distribution of those prefixes we also determined the number of BGP prefixes and ASs that the were supersets of the the blacklisted ranges. We present those numbers also in Table 2. The 112 prefixes that we found are parts of 79 BGP prefixes, or 59 BGP ASs - a very limited set of prefixes compared to the total number of prefixes and ASs found in the BGP tables.

Table 3 provides important insights into the characteristics of the nodes blacklisted by the methodology. A non-blacklisted node, when it has at least one copy of a particular title, has on average 1.56 copies of that title. On the other hand, a blacklisted node, when it has at least one copy of a particular title, has on average more than 11 copies of the title (each of a different version)! The variance of the number of copies per title is also reported in Table 3. Finally, the number of users (client instances) per node is also reported. It is interesting to note that a blacklisted node has on average a remarkable 269 user instances per IP address. In contrast, non-blacklisted nodes have essentially just one instance per IP. The exact value of 1.01 can be explained by the use of SOCKS proxies that allow different users to connect to the P2P network with the same public IP address but different username and port number.

5.3   Evaluating the Accuracy of the Pollution Source Estimation

The last-hop RTT experiment was described in Section 4. Fig 5.3 shows the results of the experiment. We compare the last-hop RTTs of 3,120 randomly chosen non-blacklisted nodes with all 1,218 blacklisted nodes. Since not all of these nodes were up and not all routers replied to the traceroutes, we were able to successfully measure 401 non-blacklisted nodes and 523 blacklisted nodes. Fig 5.3 shows that the vast majority of the non-blacklisted nodes have a last hop RTT in the 5-15 ms range with average value of 15.27 ms and median of 5.32ms. Over 45% of non-blacklisted nodes have a last-hop RTT below 5ms, while for less then 10% it is over 45 ms. This diversity is quite reasonable and in

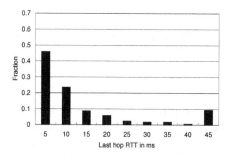

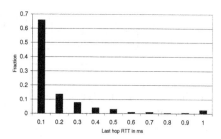

(e) Last-hop RTTs for non-blacklisted nodes

(f) Last-hop RTTs for blacklisted nodes

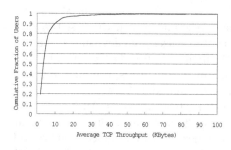

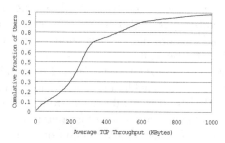

(g) Average throughput of non-blacklisted nodes

(h) Average throughput of blacklisted nodes

Fig. 2. *(continued)*

agreement with previous research [22]. The different values match the different Internet access links that users typically have (< 1ms for LAN, >5m for cable, >15ms for ADSL and >150ms for dial up modem). In contrast, the average estimated last-hop RTT for the blacklisted nodes is 0.67ms, and the median is 0.1ms (typical for LAN connections). Thus, the last-hop RTTs provide evidence that the nodes in our blacklist set are polluters whereas the nodes outside the blacklist set are ordinary users. The average values for both blacklisted and non-blacklisted nodes are listed in Table 4.

We now turn to our TCP throughput experiment as described in Section 4. We used a distributed approach to avoid any limits imposed on our campus connection and obtain an average throughput from different geographic locations. We again compare blacklisted nodes with non-blacklisted nodes. Figure 5.3 shows the CDFs for these two classes of nodes. In these CDFs, the nodes are re-ordered from lowest throughput to highest throughput. We see that more than 95% of the measured non-blacklisted nodes had a throughput less than 20 KBytes/sec. At the same time, more than 95% of the blacklisted nodes have a throughput of more than 20 KBytes/sec. Thus, the TCP throughput provides further evidence that the nodes in our blacklist set are polluters whereas the nodes outside the blacklist set are ordinary users. This observation made us

Table 4. Average Throughput and Last-Hop RTT

Nodes	TCP Throughput	Last Hop RTT
Blacklisted	2,478 kbps	0.67ms
Non-Blacklisted	75.8 kbps	15.27ms

chose the value of 20KBytes/s as a threshold in our false positive and false negative evaluation. The average values for both types of nodes are presented in Table 4.

In order to evaluate the false negatives and the false positives, we use the methodology described in Section 3. We set the threshold for the estimated last hop RTT to 1ms and the threshold for throughput to 20KBps.

We tested 1,000 users (with unique IP addresses) from outside of $\mathcal{P}$ for false negatives. 28 had more than 5 versions of a title and passed the first test. We randomly downloaded 5 versions from each of those users and determined that for 4 of the users all versions were corrupted. We finally applied the TCP throughput and last hop RTT tests. Only 2 users failed all 4 tests. Thus, the false negative ratio of the blacklisting methodology can be estimated to 0.2%.

We also tested 1,000 users from inside of $\mathcal{P}$ for false positives. After randomly downloading 5 versions for content testing, we found that 46 users provide at least one non-polluted version of a title. The next test determined that 38 and 25 users failed the throughput and last hop RTT tests respectively. Overall, 71 users failed at least one of the 3 tests causing a false positive ratio of 7.1%. We suspect those to be regular KaZaA clients that the polluters use to study the network and their targets. Details on the false positive and negative results are shown in Table 5.

Table 5. Pollution Source Evaluation Tests

	Versions	Polluted	Throughput	RTT	Total
False neg	2.8%	0.40%	7.5%	12%	0.20%
False pos	N/A	4.60%	3.8%	2.5%	7.10%

We also evaluate the effect of the blacklisting methodology by comparing the average number of copies per user in FastTrack without blacklisting and with blacklisting. Figure 2(i) shows a plot of the average number of copies and the variance for the 122 titles that we analyzed when blacklisting is employed. Fig 2(j) shows the same graph without the blacklisting. Note that the scale of the graph changes by a factor of 20 and becomes much more uniform. After blacklisting, the average number of copies of a title dropped down to less than 3 for all titles with a maximum variance of about 3.

5.4 Estimating Pollution Levels

We used our methodology described in Section 3 to determine the pollution levels of the 122 investigated titles.

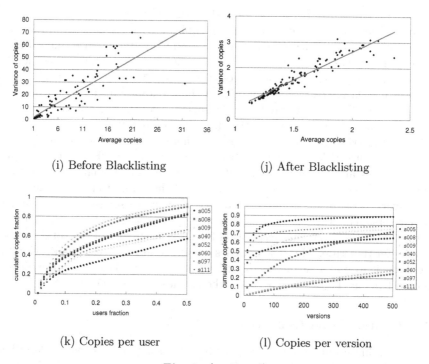

(i) Before Blacklisting   (j) After Blacklisting

(k) Copies per user   (l) Copies per version

Fig. 2. *(continued)*

We present the pollution levels, obtained by the formulas in Section 3, for all 122 titles. Since this would take up too much space, in Figure 3 we present a bar graph showing the number of titles with their pollution level in different intervals. The picture shows that 58 of the titles have estimated pollution level of 90% or more, while 51 titles have pollution level of less than 50%. There are no titles with pollution level of 0% because every title has some number of polluted versions out there.

To give further insight into pollution levels, we plotted the CDFs of the fraction of copies versus the fraction of users for the titles under investigation. Due

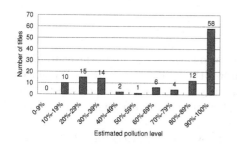

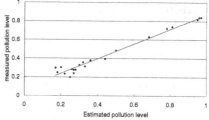

Fig. 3. Estimated Pollution Level of all 122 analized titles

Fig. 4. Measured vs. estimated pollution levels for 24 titles

to space constraints, in Fig 2(k) we only show the plots for 8 representative titles. For some titles the large percentage of all copies is concentrated within a small number of users; those titles have a very skewed CDF. Other titles have a more uniform distribution for the number of versions per user. The CDF of a title gives a visual representation of its pollution level. A very skewed CDF shows that just a few users have copies of most of the versions of a title. A regular user does not normally have hundreds or thousands of versions of the same song; so those users must be polluters and the title must be polluted.

Another interesting result that we present here is the distribution of the number of copies of a title and the number of versions. The CDFs for the 8 representative titles are shown in Figure 2(l). The figure shows that for some titles the top 100 versions account for 80% or more of the copies that are available on the network. This result is indeed expected since clean songs usually have few popular versions. Other titles, however, are highly scattered, having as many as the top 500 of their versions accounting for a less than 30% of the total number of copies for that title. This also matches our expectations and explains why it is very difficult to find a clean version of a highly-polluted title. Thus, the highest polluted title on the plot is 008, while the cleanest one is 005.

5.5   Pollution Level Evaluation

We now evaluate our pollution level estimates, using the procedure described in Section 3. Recall that this procedure compares our estimated pollution level $E^{(n)}$ with the measured pollution level $L^{(n)}$. We selected 24 songs for which the top 200 versions represented 80% or more of the total number of copies. We then downloaded the top 200 versions and used our automatic testing procedure to determine if they are polluted or not. Figure 4 shows the correlation of the measured and estimated pollution level for those titles. The plot is consistently linear and indicates that our procedure for estimating the pollution levels of titles is quite accurate. The 10% difference in the correlation can be explained by the fact that we don't actually download all the versions but just the top 80% or so (the top 200 versions sometimes correspond to more than 80%). We then computed the value of the error for the estimated pollution level as discussed in our Methodology (Equation 4) and its value was 6.8%.

6   Related Work

Although a relatively new Internet application, there are many measurement studies on P2P file-sharing systems. Bandwidth, availability, and TCP connection duration for popular filesharing systems such as Gnutella and Napster are examined in [11] [12] [14] [3]. P2P user behavior and content characteristics are studied in [13] [15].

Several independent research groups have developed crawlers for P2P file sharing systems. A crawler for the original single-tier Gnutella system is described in [3]. A crawler for the current two-tier Gnutella system (with "ultrapeers") is described in [4]. A crawler for e-donkey is described in [15]. A crawler for the FastTrack P2P file sharing is described in [2].

There is related work on attacks and shortcomings of P2P systems. The freerider problem, potential attacks to and from P2P systems, and the DRM are considered in [16], [17] and [18]. DoS attackes in P2P systems are investigated in [16] [19]. Viruses are addressed in [16] and [20].

There are also relevant related studies of CDNs and peer selection. RTT bandwidth, and TCP throughput are examined in [21] for server selection purpose and RTT, throughput probing and bandwidth measurement is considered in [22] for heterogenous P2P environments. In particular, our throughput and last-hop RTT techniques were derived from [22].

In [2] it was established that pollution is widespread for popular titles. The methodology in [2] to determine pollution levels is inefficient in that it requires downloading and binary content analysis of hundreds of versions for every investigated title. Although the current paper makes use of the FastTrack crawler in [2], the contribution is very different. It is the first paper to develop a black-listing methodology for file-sharing systems. Furthermore, all the methodologies in this paper do not require any downloading and are solely based on meta-data gathered from the crawler. A user-supported antip2p banlist is available in [23].

7   Conclusion and Future Work

In this paper we considered two related problems: creating blacklists for polluting IP address ranges, and estimating the pollution level of targeted titles. Adequate solutions to both of these problems require accuracy and efficiency. Extensive tests with data collected from the FastTrack P2P file-sharing system have shown that our methodologies meet both of these goals.

Our methodologies do not involve the downloading of any files. Instead, they identify polluting IP address ranges and targeted titles by collecting and analyzing metadata from the file sharing system. The metadata is harvested by crawling the nodes in the file sharing system. From this harvested metadata, our methodology constructs the blacklisted IP address ranges and the estimated pollution levels for the targeted titles. The methodology is efficient in that it collects metadata (text) rather than binary content and that a large number of titles, and virtually all the file-sharing nodes can be investigated in one crawl.

To address accuracy, we developed several criteria to evaluate the methodologies. We then applied these criteria to a comprehensive test case for the FastTrack file-sharing system. For blacklisting, we found the probability of false negative and false positive to both be low, namely, 0.2% and 7.1%, respectively. After applying the blacklist, the average number of versions per user for polluted titles decreases dramatically. These results testify to the overall accuracy of our black-listing methodology. For estimating pollution-levels, we compared our estimated pollution-levels with measured estimates, which involved the downloading and binary analysis of titles. For our comprehensive test case, we found the percentage error to quite low, less than 7%.

In a real deployment, it is important that the blacklist set and the pollution-level estimates adapt as polluters change hosts and target new content. Our methodology is naturally suited for such a dynamic environment. The crawler can operate continuously, collecting metadata for fresh titles as they become released. The fresh titles can be obtained directly from on-line billboard charts. Similarly, our methodology can continuously be applied to the data in the crawling database, thereby dynamically adjusting the blacklist set and the pollution-level estimates.

References

1. CacheLogic Research: The True Picture of P2P File Sharing, http://www.cachelogic. com/research/
2. J. Liang, R. Kumar, Y. Xi, K. Ross. Pollution in P2P File Sharing Systems, *IEEE Infocom 2005, Miami, FL*
3. M. Ripeanu, I. Foster, and A. Iamnitchi,"Mapping the Gnutella network: Properties of large-scale peer-to-peer systems and implications for system design," *IEEE Internet Computing Journal*, vol. 6, no. 1, 2002.
4. D. Stutzbach, R. Rejaie. Characterizating Today's Gnutella Topology, *submitted*.
5. R. Schemers, fping utility, http://www.fping.com/
6. Apple iTunes Top 100, http://www.apple.com/itunes/
7. Top 100 songs from 1970 to 1979, http://top40.about.com/cs/70shits/
8. Blockbuster's Top 100 Online Rentals, http://www.blockbuster.com/
9. Z. M. Mao, J. Rexford, J. Wang, and R. Katz, Towards an Accurate AS-Level Traceroute Tool, *Proceedings of ACM SIGCOMM*, Karlsruhe, Germany, August 2003
10. CIDR Report, http://www.cidr-report.org
11. S. Sen, J. Wang. Analyzing Peer-to-Peer Traffic Across Large Networks, *ACM/ IEEE Transactions on Networking, Vol. 12, No. 2, April 2004*
12. S. Saroiu, P. K. Gummadi, S. D. Gribble, A Measurement Study of Peer-to-Peer File Sharing Systems, *Multimedia Computing and Networking (MMCN'02)*, San Jose, January 2002
13. K. P. Gummadi, R. J. Dunn, S. Saroiu, S. D. Gribble, H. M. Levy, J. Zahorjan. Measurement, Modeling, and Analysis of a Peer-to-Peer File-Sharing Workload. *Proceedings of the 19th ACM Symposium on Operating Systems Principles (SOSP-19), October 2003*
14. V. Aggarwal, S. Bender, A. Feldmann, A. Wichmann. Methodology for Estimating Network Distances of Gnutella Neighbors. *Proceedings of the Workshop on Algorithms and Protocols for Efficient Peer-to-Peer Applications at Informatik, 2004*
15. F. Le Fessant, S. Handurukande, A.-M. Kermarrec, L. Massouli. Clustering in Peer-to-Peer File Sharing Workloads. *IPTPS'04*
16. D. Katabi, B. Krishanmurthy. Unwanted Traffic: Attacks, Detection, and Potential Solutions. *ACM SIGCOMM'04 Tutorial*
17. P. Biddle, P. England, M. Peinado, and B. Willman. The Darknet and the Future of Content Distribution. *ACM DRM 2002*

18. M. Feldman, C. Papadimitriou, J. Chuang, I Stoica. Free-Riding and Whitewashing in Peer-to-Peer Systems. *ACM SIGCOMM'04 Workshop on Practice and Theory of Incentives in Networked Systems (PINS), August 2004*
19. S. Dropsho. Denial of Service Resilience in Peer to Peer File Sharing Systems.*EPFL Tech Report*
20. http://www.bullguard.com
21. Sandra G. Dykes, Kay A. Robbins, Clinton L. Jeffery. An Empirical Evaluation of Client-side Server Selection Algorithms. *Infocom 00*
22. T.S.E Ng, Y. Chu, S.G. Rao, K. Sripanidkulchai, H. Zhang. Measurement-Based Optimization Techniques for Bandwidth-Demanding Peer-to-Peer Systems. *IEEE INFOCOM'03*
23. Bluetack Internet Security Solutions, http://www.bluetack.co.uk/
24. The FFmpeg project, http://ffmpeg.sourceforge.net/index.php

Building Tailored Wireless Sensor Networks

Yoshito Tobe

Ubiquitous Networking Laboratory,
Tokyo Denki University,
2-2 Kanda-nishiki-cho, Chiyoda-ku, Tokyo 101-8457, Japan
ytobe@acm.org,
http://www.unl.im.dendai.ac.jp

Abstract. Since sensor networks became a challenge in the network research community, the basic concept of sensor networks has been widely understood and many field experiments have been conducted. Currently, we see the full spectrum of sensor networks ranging from personal area networks to the entire earth network. In this talk, two specific applications, human-body monitoring and disaster-area surveillance, are introduced and the generality and the customization of wireless sensor networks are discussed. In addition, the details of a project on multi-robot sensor networks are explained.

1 Introduction

Sensor networks (SNs) has brought new challenges in the network research community [Estrin99]. After abundant investigation on sensor networks, we have now common understanding of SNs although many research issues such as security remain. The hitherto typical applications of SNs have been monitoring of environment, climate, and structural health of buildings. There is assumptions that a huge number of sensor nodes are used in these applications. However, SNs can also be applied to a variety of environments ranging from the living space and the entire earth. Therefore we need to identify the fundamental design principle and per-application customization. In some cases, a routing protocol can be very simple if we limit the traffic patterns [Tobe05-1]. In this talk, I will introduce two specific applications, human-body monitoring and disaster-area surveillance by multiple robots. In particular, the details about a project of multi-robot sensor networks called WISER are explained. The rest of this paper is organized as follows: Section 2 gives the overview of WISER. Section 3 describes the details about the design of WISER system and finally section 4 summarizes this talk.

2 Wiser

Multiple cooperated robots are being paid much attention to provide difficult tasks that cannot be done by human beings. We call such a network Multi-Robot Sensor Network (MRSN). MRSNs will have a potential to perform sensing adaptive to important events for disaster areas. MRSNs have unique characteristics

K. Cho and P. Jacquet (Eds.): AINTEC 2005, LNCS 3837, pp. 22–29, 2005.

that are not observed in MANETs: intentional mobility and disconnection/delay tolerance[Fall03].

- intentional mobility: In MANETs, mobility is considered spontaneous; a node cannot control mobility of other nodes. In contrast, in MRSNs, a node can move toward an wireless communication area that is covered by the destination node of data or instruct another node to move to a specific location to transfer data. This means that movement is a means of conveying data.
- disconnection tolerance: In MANETs, a path from a source to a destination is established when data is transmitted at the source. Otherwise, the source will receive a "message unreachable" error; the reachability is guaranteed by an underlying routing protocol. However, in MRSNs, when data is transmitted at a source node, a path to the destination may not be guaranteed due to disconnection in intermediate nodes. This is allowable since disconnection of links can be compensated by movement of nodes.

Our group is designing a routing architecture which determines the balance between movement and wireless communications. We call this architecture WISER [Tobe05-2]. WISER allows two policies of routing, delay minimization and energy efficiency, and changes the behavior of mobile nodes dynamically. When determining which means of movement and wireless communications is selected, wireless link quality and easiness of movement are taken into account.

3   Basic Design

Let us begin with classifying MRSN nodes into three types: peer, management, and sink nodes. Peer nodes have sensing functions and can move to any location in a sensing area. Management node is responsible for managing the whole sensing area and control the policy inside the area. Sink node is an entity of collecting data from peer nodes. Since peer nodes play a key role in an MRSN, the design is focused on peer nodes. Figure 1 depicts the whole architecture of a peer node.

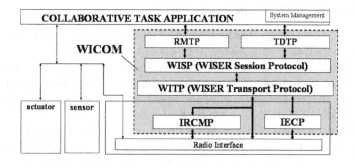

Fig. 1. System architecture of peer node

As wireless communication medium, we assume that any kind of device can be used; IEEE 802.11b, IEEE 802.15.4, and short-range wireless radios.

In designing cooperative actions by multiple robots, a common format of messages exchanged among the robots should be defined. These messages have a wide spectrum ranging from direct actions, values of sensed data, and parameters for operations. Thus a protocol which does not depend on underlying communication channels is desired. We design a protocol named Robot Message Transfer Protocol (RMTP) based on ACTRESS to satisfy such conditions.

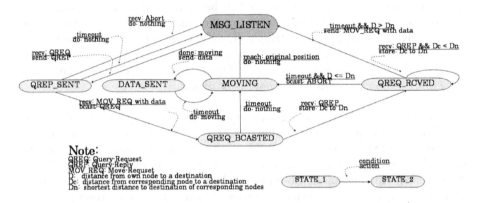

Fig. 2. IRCMP state transision

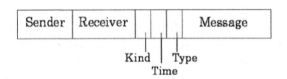

Fig. 3. RMTP format

3.1   Robot Message Transfer Protocol

In WISER, RMTP is defined as a protocol for exchanging messages among robots to provide cooperating operations. A Protocol Data Unit (PDU) of RMTP includes the destination and the source address, the size of the message in addition to the message body which is either an instruction, sensing data, status of the robot such as the left battery energy and operating time. Figure 3 shows the format of the PDU of RMTP. The sender field is set to the robot ID of the source, whereas the receiver field is the robot ID of the destination, the position, or the geographical area. If the destination field is not specified, the PDU is delivered as a broadcast PDU. The kind field in this figure is set to the used communication medium if necessary. The time and type fields include the lifetime of this PDU and the distinction between request and response, respectively.

3.2 Transparent Data Transfer Protocol

Messages exchanged in an MRSN is not only dedicated to control other robots. We introduce Transparent Data Transfer Protocol (TDTP) to transfer pure data among nodes which may include sensed data. TDTP is an end-to-end protocol mainly for various kinds of sensed data between cooperative task applications. The data includes the source of sensor and required delivery time.

3.3 Wiser Session Protocol

In an MRSN, data can be transmitted at a source even when a path to the destination of the data is not established due to partitioned networks. Therefore, if the size of the data is large, a session-level control is required to determine the following.

- how the data is fragmented
- what kind of formation nodes should construct
- how acknowledgment is returned to the source

WISER Session Protocol (WISP) deals with the protocol at this level. In WISP, transferring data that are segmented into several segments is considered a transaction and an operation mode is specified. Depending on the policy of whether energy efficiency or delay minimization, the above decision is made.

3.4 Wiser Transport Protocol

WISER Transport Protocol is defined to ensure the reliability fo Application Data Unit (ADU) in an end-to-end fashion. If the underlying communication accompanies an existing reliable transport protocol, the protocol is used at the transport. For instance, when IEEE 802.11b is used, Transmission Control Protocol (TCP) is executed. On the other hand, a tiny device communication such as a wireless link of Mica Mote2 is used, a light-weight transport protocol is applied merely to guarantee the reliability.

3.5 Inter-robot Control Message Protocol

Since each node can move, the topology of the network frequently changes. Unlike a MANET,an MSRN allows a node to control the mobility of other nodes. Inter-Robot Control Message Protocol (IRCMP) is a protocol to decide routing actions of nodes. In IRCMP, the next-hop of a node is calculated based on the relative location among neighbor node and the policy, and if a suitable next-hop node cannot be found, the node itself moved towards the destination.

Since the emergency level of data and the topology of the network may frequently change, a reactive type of routing is suitable. To obtain the information about neighbor nodes, a node of holding the data sends a broadcast of query and decides the next-hop by examining the location information contained in reply messages. Figure 4 shows the PDU format of IRCMP.

If a node is reserved for a specific data transfer, the node needs to hold required entries in its routing table. In another case in which the topology seldom changes, each node is desired to hold entries in routing table. IRCMP assumes both cases of whether or not routing information is retained. Figure 2 shows the state transition diagram of IRCMP.

In Figure 2, the default state when a node standstills is Message Listening (MSG_LIS). If the node receives Query Request (QREQ), the state moves to QREP_SENT and returns Query Reply (QREP) to the node of originating the QREQ. When a node of waiting for QREP receives multiple QREPs, it selects one node which is the nearest to the destination. If no other nodes are nearer to the destination than the node of holding data, the state moves to MOVING. A node at QREP_SENT may receive the data with move instruction (MOV_REQ). In such a case, the node looks for its next-hop node by broadcasting QREQ.

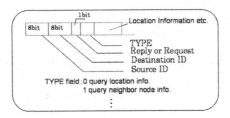

Fig. 4. IRCMP PDU format

3.6   Inter-robot Exception Control Protocol

Inter-robot Exception Control Protocol (IECP) is a protocol to be used among nodes to notify each other of exceptions and errors in a MRSN. We define two categories for IECP messages: destination unreachable and destination node failure.

The destination unreachable error is notified from a node that detects un-reachability to the source node of data within a lifetime specified by the source. In some case when obstacle objects exist in a moving path, a node of holding the data can set detour allowance time. If the data eventually reaches the desti-nation of the data by extending the lifetime with this allowance time, the node send the source the detour information.

When a node around the destination node detects the physical failure of the destination node, it send the source a message about the failure. This failure include battery shortage and abnormal operation. of wireless links.

3.7   Flow of Messages

Assume that one robot node sends sensed data to another robot node. The sensed data is encapsulated into a TDTP PDU. When the TDTP data can be fit into a single PDU, no session control is done and the TDTP PDU is encapsulated into a WITP PDU, which will have an IRCMP header required for routing. At the destination node, this process goes in the reverse way and a message contained in the TDTP is extracted. When a TDTP data will have multiple PDUs, session control is applied.

3.8 Prototype

We are developing a prototype of WISER. Figure 6 shows the prototype robot. A control system using a board computer is mounted on the omni-directional mobile platform in order to realize an omni-directional motion in two dimensions for flexible mobility. Batteries are also mounted on the robot for electrical devices and actuators. The robot can behave autonomously and independently without any cables. As seen in Figure 5, the robot has an encoder which is attached to actuators, and the location is calculated using the total rotation of the tires. The main software is installed in a board computer with Linux kernel and this board computer communicates with a Mica2 Mote with a serial line. IRCMP and IECP in the protocol stack shown in Figure 1 are processed in the Mica Mote, but other protocols are executed inside the board computer.

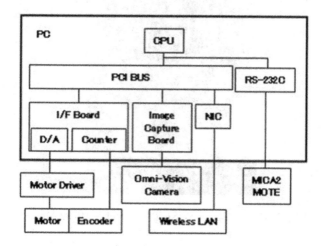

Fig. 5. Hardware configuration of WISER robot

We conducted a preliminary experiment of conveying data from one sensor network to another using a robot. The packet to which moving commands were input in the Message field of RMTP was generated and transmitted via the network. Mica Mote on the robot received the packet from it, and then communicated to the control system through the serial link. The control system read the content on the message field, and controlled the actuators of the robot.

The experiment of data delivery was done by using the robot as the gateway between sensor networks to communicate the sensor data between two indepen-

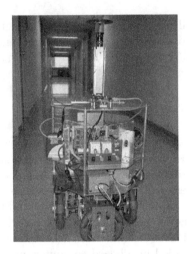

Fig. 6. WISER robot

Fig. 7. Upper side of WISER robot with Mica2 Mote

dent sensor networks not connected mutually. In the experiment, the communication between different sensor networks was confirmed by using the mobile robot as a gateway for delivering the sensor data in one sensor network to another disconnected sensor network.

4   Summary

In this talk, two special types of sensor networks are shown: human-body monitoring and multi-robot surveillance. After several trials of application to environmental monitoring, sensor networks have already shifted to diverse application areas. Although each application area is associated with its own problem, we need to distinguish between a common structure of sensor networks and per-area customization. In our WISER project, we have finished prototyping a system. For our future work, we will investigate the coordination of actuation and sensing in the multi-robot wireless sensor networks.

References

[Estrin99]   Estrin, D., Govindan, R., Heidemann, J., Kumar, S.: Next Century Challenges: Scalable Coordination in Sensor Networks. Proc. ACM/IEEE Mobicom, (1999), 249–255.

[Fall03]   Fall, K.: A delay-tolerant network architecture for challenged Internet. Proc. of ACM SIGCOMM'03, (2003) 27–34.

[Tobe05-1]   Tobe, Y., Thepvilojanapong, Sezaki, K.: Autonomous Configuration in Wireless Sensor Networks. IEICE Trans. Fundamentals of Electronics, Communications and Computer Science, (2005).

[Tobe05-2]   Tobe, Y., Suzuki, T.: WISER: Cooperative Sensing Using Mobile Robots. Proc. HWISE 2005, (2005).

Users and Services in Intelligent Networks

Erol Gelenbe

Memb. Acad. Europ., FIEE FIEEE FACM,
Dennis Gabor Chair, Intelligent Systems and Networks Group,
Electrical & Electronic Engineering Dept.,
Imperial College, London SW7 2BT
e.gelenbe@imperial.ac.uk

Abstract. We present a vision of an Intelligent Network in which users dynamically indicate their requests for services, and formulate needs in terms of Quality of Service (QoS) and price. Users can also monitor on-line the extent to which their requests are being satisfied. In turn the services will dynamically try to satisfy the user as best as they can, and inform the user of the level at which the requests are being satisfied, and at what cost. The network will provide guidelines and constraints to users and services, to avoid that they impede each others' progress. This intelligent and sensible dialogue between users, services and the network can proceed constantly based on mutual observation, network and user self-observation, and on-line adaptive and locally distributed feedback control which proceeds at the same speed as the traffic flows and events being controlled. We review issues such as network "situational awareness", self-organisation, and structure, and relate these concepts to the ongoing research on autonomic communication systems. We relate the search for services in the network to the question of QoS and routing. We examine the need to dynamically protect the networked system from denial of service (DoS) attacks, and propose an approch to DoS defence which uses the detection of violations of QoS constraints and the automatic throttling or dropping of traffic to protect critacl nodes. We also discuss how this vision of an Intelligent Network can benefit from techniques that have been experimented in the Cognitive Packet Network (CPN) test-bed at Imperial College, thanks to "smart packets" and reinforcement learning, which offers routing that is dynamically modified using on on-line sensing and monitoring, based on users' QoS needs and overall network objectives.

1 Introduction

Sheer *technological capabilities and intelligence*, on their own, are of limited value if they do not lead to enhanced and cost-effective capabilities that improve and add value to human beings – or even beyond humans to other living beings.

In the field of telecommunications, fixed and then mobile telephony and the Internet have been enablers for major new developments that improve human existence. However advances in telecommunications have also had some undesirable and unexpected outcomes during the past century. A case in point is television broadcasting. It was initially thought that television broadcasting would become a wonderful medium for education. Unfortunately in many instances it has lowered public standards for entertainment by forcing a limited number of programs upon the public; it has often displaced reading, sophisticated cinema, theatrical and musical forms by the introduction

K. Cho and P. Jacquet (Eds.): AINTEC 2005, LNCS 3837, pp. 30–45, 2005.

of facile talk shows and soap operas. This is a great example of a tremendous success in technology which has not always been applied in the most broadly intelligent manner. The "one-to-very-many" broadcast nature of television does not give users, or communities of users, the possibility to significantly influence the system that they use. Other models of communications, such as the peer-to-peer concept which was born in the Internet, can offer a greater degree of user choice. Thus we suggest that through intelligent organisation of networks, which should include a just compensation for services and intellectual property ownership, one can achieve improved communications for the sake of an enhanced cultural and humanistic environment.

We envision Intelligent Networks (INs) to which users can ubiquitously and harmoniously connect to offer or receive services. We imagine an unlimited peer-to-peer world in which services, including current television broadcasts, voice or video telephony, messaging, libraries and documentation, live theater and entertainment, and services which are based on content, data and information, are available at an affordable cost. In these networks the technical principles that support both the "users" and the "services" will be very similar if they are framed within an autonomic self-managing and self-regulating system. In fact This network will be accessible via open but secure interfaces that are compatible with a wide set of communication standards, including the IP protocol.

We imagine an IN in which users and services play a symmetric role: users of some services can be services of other users, and services can be users of some other services. Users and services can express their requests dynamically to the network in terms of the services that they seek, together with Quality-of-Service (QoS) criteria that they need, their estimate of the quantity or duration of the requested service and the price that they are willing to pay. The users could also have the capability to monitor on-line to what extent their requests are being satisfied. In turn the services and the network would dynamically try to satisfy the user as best as they could, and inform the user of the level at which their requests are being satisfied, and at what cost. The network would also provide guidelines to users to avoid that the latter impede each others' progress. Similarly, network entities and services would also conduct a dialogue, so that they can collectively and autonomously provide a stable, evolving and cost effective network infrastructure. We will sometimes find it useful to distinguish between users and services, merely to indicate the relationship that exists between a specific user requesting a specific service. But we wish to stress that at a certain level of abstraction, these two entities are indeed equivalent.

The IN will offer the facilities for an intelligent and sensible dialogue between all users, including services, and it will adapt to users' needs based on mutual observation, network and user self-observation, and on-line distributed feedback control which acts in response to the events that are being controlled.

2   Towards Autonomic Communications

The growth in both communication needs and communication-intensive technologies and services through which we are living is a trend that is bound to continue and to intensify. In a not-so-distant future, everyday activities will be supported by a ubiquitous

ITC environment or "networked service" that will cater dynamically to our needs in a situation-aware manner.

The Internet is becoming an immense organism of composite, highly distributed, pervasive, communication intensive services; in order to operate effectively, these services need to [27, 46] autonomously detect and organize the knowledge necessary to understand the physical, as well as human user based and social context. Services will need to autonomously adapt to the human, social and technical environment. They should improve our ability to interact with the world by providing us with any needed information about our surrounding physical environment. Imagine a traveller arriving at some location with given objectives (e.g. go to a meeting at hotel X, then lunch with A and restaurant Y with stop-over to do some work at coffee shop Z)) [23] and imagine how the relevant information can be collected by the communication intensive service via local networks, web sites, as well as the user's office system, possibly the ITC system related to the local or the organisation that is hosting the meeting and of person B, and then displaying all the situational information to the traveller on his laptop, mobile phone or PDA. Imagine also that this is done by using the best and cheapest network service, i.e. at the lowest cost and the best possible QoS [26, 38].

Such future services raise major challenges for today's networks, including protocol mismatches and differences that may exist between networks in terms of operational and behavioural semantics at the service level. The technical premises for these communication intensive services will require the integration into the networking world of technologies borrowed from other application domains and from research in ITC, including:

- *Situational awareness* which is a concept borrowed from defence applications. It represents the ability to understand, represent and evaluate the detailed context in which an operation will be conducted. In the military domain, situational awareness is designed to enhance and support the human in the loop, and is also designed in a manner which encompasses the role of human factors. In the present context the human factors aspects can also be important if we wish to provide a certain level of human control and decision making in the context of communications. Technology that can support situational awareness in communications includes sensors, location systems, user profiles, and tools for system and network monitoring [30], which can be utilised for the provisioning of adaptive services [49, 45].
- *Self-organization* will be required of such networks to be able to automatically exploit to their ability for situational awareness. However, this will need to go beyond the structural self-organisation that has already been addressed in communications in the context of Peer-to-Peer networks and computing [11], with techniques inspired from ant-based optimisation, reinforcement learning, and social systems [4, 5, 12, 21]. The systems we consider should create a bridge between robust but conceptually simple self-organising models and the far more complex semantics [18, 43] that context-aware systems will require.
- *Structure and Components*. The traditional top-down hierarchical protocol stack that adopts a vertical hierarchy, going down layer by layer from the application to the physical connection at the bottom, that has dominated our thinking about networks for the last thirty years, must probably be abandoned in favour of a model

composed of a collection of agents or "autonomic components". Such components could use common templates [27, 37] to provide both a conceptual and software framework for programmable network elements and their interactions. Although these ideas are still in their formative stage, they can benefit from research in other areas such as multi-agent systems [28] and programmable networks [7, 22].

3   A Simple Architecture for the Intelligent Network

A sketch of the IN architecture is shown in Figure 1. For reasons of backward compatibility, the IN will have to offer at least the appearance of a standard communication interface inspired by the Internet Protocol (IP). End users U (shown with small purple rectangles as U1, U2, etc.) will generally be mobile; they can be recognised via their ID and password. Users may have a "credit value" with the network and with certain network services, as represented by a credit allocation or via a "pay as you go" scheme (e.g. with a credit card), or they can access certain free services or services that may be paid for by the service provider (e.g. advertisements). Users can have a user terminal which may be as simple as a Personal Digital Assistant or mobile phone, or as complex as intelligent network routers (INRs) shown as blue octagons in Figure 1. Users are connected to the IN via INRs or directly to a network cloud (shown as clouds of different colours). Services S (shown as S1, S2, ..) are very similar to users in that they have an ID and they may have a credit allocation; they can also receive credit when their services are used by users, just as users may be reimbursed by services or by other users. Services can also be mobile. However:

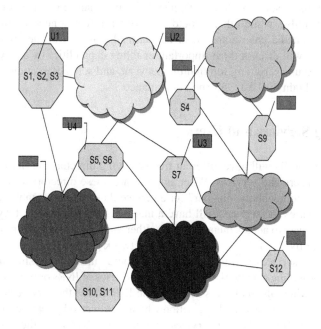

Fig. 1. Architecture of the Intelligent Network

- Most end users will in general be light-weight (a mobile phone, a PDA, or just a user ID and password).
- Other end users and services will be much more complex and may often be resident on one or more INRs. They may own one or more INRs for their needs.

When some other user or service asks something of a user, the chances are that there will be an automatic answer saying "sorry no; I am just a simple user". On the other hand, services will often be equipped with authentication schemes to recognise the party who is making a request, billing schemes that allow for payment to be collected, schemes allowing a service to be used simultaneously by many users, and so on, depending on the complexity of the service being considered.

INRs are machines or clusters which can be identified by the community of users and services. Network clouds on the other hand are collections of routers internally interconnected by wire or wireless and which are only identified as far as the users and services are concerned via the ports of INRs which are linked to a cloud; in other words, users and services do not actually know who and what is inside a network cloud. However INRs, and hence users and services, can observe the QoS related to traversing a network cloud; this may include billing of the transport service by the cloud. Also, clouds may refuse traffic, or control and shape the traffic that wishes to access them, depending on the clouds own perception of the traffic.

The IN architecture we have described can be viewed as an overlay network composed of INRs with advanced search, QoS (including pricing and billing), that links different communities of users and services. The networked environment of the future will include numerous INs, and there may be specific INs whose role is to find the best IN for a given user. Some of the se INs may be quite small (e.g. a network for a single extended family), while others would be very large (e.g. a network that provides sources of multimedia entertainment, or educational content). In the three following sub-sections we will discuss three important enabling capabilities of the system: finding services and users, routing through the network, and self-observation and network monitoring to obtain the best QoS and performance.

4   Finding Services and Users

We expect that the IN will have different free or paying directory services that will be used to locate users and services. When appropriate, these directories may provide a "street address and telephone number" for a service that is being sought out; however, since in many cases the services will have a major virtual component, they will especially provide a way to access them virtually, either via an IP address, or more probably via one or more INR addresses or one or more network paths.

The directory services will offer "how to get there" information similar to a street map service, providing a network path in terms of a series of INRs or of network clouds, from the point where the request is made, to the INR where the service can be found. Directory services may have a billing option which is activated by services to reward the directory for being up-to-date, or services or users can subscribe to them, or they may be paid for via advertisement information, and so on. These directories will be updated pro-actively by the services or by the directories themselves, or on demand when the need occurs. Updates would also occur when INR or network cloud landmarks change. Directories can be "smart" in the sense that they offer information about faster or less congested paths to services that are requested, or paths to less expensive services, or paths that are better in some broader sense. If the user does not know how to find a service, in the worst case it can broadcast its request which will be relayed by INRs and directories. However som of the "smart probing" techniques described below can offer a more efficient solution.

4.1   Smart Search and Routing

Let us describe how search can be "smart", by extending some ideas from the Cognitive Packet Network (CPN) algorithm [5, 42]. In CPN, the purpose of the search is to find a network destination (rather than a service), and SPs and ACKs in CPN play a role that is identical the one we described earlier, except that we are looking for a path to some destination which optimises a QoS requirement. Thus we can imagine that the CPN algorithm which runs at the packet transport level and finds destination nodes, can be abstracted to a higher level where it searches for services.

In order to provide a practical grounding for the preceding discussion, we will discuss how the CPN protocol currently runs; more detail can be found in the collection of papers that are stored in the web site at $http://san.ee.ic.ac.uk$. In CPN dumb packets (DPs) carry the payload traffic, while CPN routers are similar to INRs, and are interconnected either via portions of the Internet which plays the same role as the network clouds that we have described in Figure 1, or via point-to-point Ethernet or other (e.g. ATM) connections. SPs find routes and collect measurements, but do not carry payload.

DPs are source routed, using paths which best match the users' QoS requirements. On the other hand, SPs are routed using a Reinforcement Learning (RL) algorithm that uses the observed outcome of previous decisions to "reward" or "punish" the mechanism that lead to the previous choice, so that its future decisions are more likely to meet the QoS goal.

When a SP arrives to its destination, an acknowledgement (ACK) packet is generated; the ACK stores the "reverse route" and the measurement data collected by the SP. It will travel along the "reverse route" which is computed by taking the corresponding SP's route, examining it from right (destination) to left (source), and removing any sequences of nodes which begin and end in the same node. For instance, the path $< a, b, c, d, a, f, g, h, c, l, m >$ will result in the reverse route $< m, l, c, b, a >$. Note that the reverse route is not necessarily the shortest reverse path, nor the one resulting in the best QoS. The route brought back by an ACK is used as the source route by subsequent DPs of the same QoS class having the same destination, until a newer and/or better route is brought back by another ACK. A *Mailbox (MB)* in each node is used to store QoS information. Each MB is organized as a Least-Recently-Used (LRU) stack, with entries listed by QoS class and destination, which are updated when an ACK is received.

The steps needed to establish a connection between some user U and a service S can then be listed as follows:

- U first searches for a directory; assuming he finds one, U formulates his request in the form of (SX, QY, PZ) meaning that he wants a service SX at QoS value QY

for a price of PZ. The directory either is unable to answer the request, or it provides one or more paths $\pi(U, SX, QY, PZ)$ which best approximate this request for several possible locations of the service.

- Assuming that the directory does provide the information, U sends out (typically via the INR) a sequence of smart packets SPs which have the desired QoS information, with several following each of the possible designated paths. The first SP for each of the paths will follow it to destination, with the purpose of verifying that the information provided by the directory is correct. Subsequent SPs on each route will be used to search for paths: they will invoke an optimisation algorithm at all or some of the INRs they traverse so as to seek out the best path with respect to the user's QoS and pricing requirements.
- INRs collect measurements and store them in mail boxes (MB). These can concern both short term measurements which proceed at a fast pace comparable to the traffic rates, and long term historical data. INRs will measure packet loss rates on outgoing links and on complete paths, delays to various destinations, possibly security levels along paths (when security is part of a QoS requirement), available power levels at certain mobile nodes, etc.. This constant monitoring can be carried out using the SPs and other user related traffic, or using specific sensing packets generated by the INRs.
- The network monitoring function can also be structured as a special set of users and services whose role is to monitor the network and provide advice to the users and to the directories.
- Each SP also collects measurements from the INRs it visits which are relevant to its users QoS and cost needs, about the path from the INRs which it visits.
- When a SP reaches a service SX, an acknowledgement ACK packet is sent back along the reverse path back to U; the ACK carries the relevant QoS information, as well as path information which was measured by the SP and by the ACK, back to the INRs and to the user U. The ACK may thus be carrying back a new path which was unknown to the directory.
- For a variety of reasons, both SPs and ACKs may get lost. SPs or ACKs which travel through the network over a number of hops (ERs or total number including routers within the clouds) exceeding a predetrmined fixed number, will be destroyed by the routers to avoid congesting the IN with "lost" packets.
- Note that the SPs and ACKs may be emitted by the directory itself, rather than by U. This would be an additional service offered by certain directories. One could also imagine that both users and directories have this capability so as to verify that the request is being satisfied.

5   Individual Versus Collective QoS Goals

The usual question that any normally constituted telecommunications engineer will ask with respect to the vision that we have sketched is what will happen when individual goals of users and services conflict with the collective goals of the system. We are allowing for users to set up the best paths they can find, from a selfish perspective, with services, and for services to actually do the same, in parallel with the behaviour of users. This has the potential for:

- Overloading the infrastructure, because services have an interest in maximising their positive response to user's needs, and they may even overdo it in terms of solliciting users; because of the possibility of billing, portions of the infrastructure itself may have an interest in getting overloaded.
- Creating traffic congestion and oscillations between hot spots, as users and services switch constantly to a seemingly better way to channel their traffic.
- Opening the door to malicious traffic whose sole purpose may be to deny service to legitimate users through the focused creation of overload in the services or the infrastructure (e.g. denial of service attacks).

The first of these points, which does not relate to malicious behaviour, can be handled through overall self regulation of the INRs, the users and services:

- When a new part of the infrastructure joins the IN, for instance a INR, it will be allocated an identity within the IN. We could have a virtual regulating agency (VRA) which sets up a dialogue with the INR to provide it with its identity, and which ascertains its type and nature from its technical characteristics. The VRA then enables the INRs operating systm with a set of parameters which in effect limit the number of resident processes and the amount of packet traffic that this particular INR can accept.
- Services and users which join the IN, also need to be identified by the VRA. Just as a shop rents a certain space in a building and on a particular street, the VRA can provide the service with a "footprint", depending on the rent it is willing to pay, and on the VRA's knowledge of currently available resources. This footprint can then determine the fraction and amount of processing power and bandwidth that it is allowed inside the IN and at any given INR.
- Note that the overall quality and seriousness of the VRA will make a particular IN more or less desirable to users and services.

The second point is related to dynamic behaviour. Each INR, in its role as a service support centre enabled by the VRA, will run the dynamic flow and workload control algorithms for each service and user that it hosts. However it will also run a monitoring algorithm which has IN-wide implications.

- For some user U assume that $RU(S)$ is the rank ordered set of best instantaneous choices for some decision (e.g. what is the best way to go to service S with minimum delay).
- At the same time, let $RN(U, S)$ be the rank ordered set of best instantaneous choices for the network (e.g. what is the best way to go to where service S is "sitting" so that overall traffic in the IN is balanced).
- The decision taken by the INR will be some weighted combination of these two rank orders. The weights can depend on the priority of the user, of the price it is willing to pay, and so on.
- Choices which are impossible or unacceptable to either of the two criteria (user or network) will simply be excluded. If there are no mutually possible choices, then the request will be rejected. When there are ties between choices, any one of the tied choices can be selected at random.

As an example, suppose that the ranking indicating the user's preference, in desecending order, among six possible choices is $\{1, 2, 3, 4, 5, 6\}$, while the network's preference ranking could be $\{5, 4, 2, 3, 1, 6\}$. If we use rank order as the decision criterion and weigh the INR and the user equally, then the decision will be to choose 2 whose total rank order is 5. If the network's role is viewed as being twice as important, we can divide the network's rank for some choice by 2 and add the resulting number to the rank that the user has assigned to that choice, which results in a tie between the three top choices $\{1, 2, 5\}$. If the network's role is three times more important, then we get a tie for the top choice between $\{1, 5\}$, and so on.

In the aproach that we have suggested for finding services, the user U formulates some request (SX, QY, PZ) for a service SX at quality level QY and for the price PZ. Both the quality of service value and the price constitute "goals" in the sense that the term is used in the CPN algorithm [39]. They may be treated as separate goals to be minimised, and combined in some manner as outlined above, or combined into some single common metric.

For instance, if QY is some non-negative number such as "loss" or "delay", we could combine the two considerations in a single metric such as $G = QY/PZ$ (quality for a given price), or as

$$G = PZ\ 1[QY < Qmax] + \frac{QY}{PZ}1[QY \geq Qmax] \qquad (1)$$

where $1[x]$ is the function which takes the value one if the predicate $x = true$ and takes the value zero if $x = false$. Thus (1) means, for instance, that as long as the delay is less than some maximum acceptable value $Qmax$, we are happy to minimise the price; however if the delay is larger than this maximum value, we want simply to minimise the delay per price unit that we pay for the service.

5.1   The Eternal Problem of Scalability

It is often said that the main impediment to the broad use of QoS mechanisms in the Internet is the issue of scalability. Indeed, if each Internet router were enabled to deal with the QoS needs of each connection, it would have to identify and track the packets of each individual connection that is transiting through it. The routing mechanism we propose for all requests through the IN is based on dynamic source routing[1]. In other words, the burden of determining the path to be used rests with the INR that hosts the service or user. In our proposed scheme, routers have two roles:

- The INR generates SPs for its own use that monitor the IN as a whole, and the user or service process resident at a INR generates the SPs and ACKs which are related to its connections to monitor their individual traffic.
- As a result of the information that it receives from SPs and ACKs, of the information similarly received by users and services that are resident at the INR, and of the compromise between global (IN) and local (user and service) considerations, the INR generates source routes for its resident users and services.

---

[1] Note that MPLS is a form of distributed virtual source routing where label switching at each node maps virtual addresses into physical link addresses.

– Each INR also provides QoS information to SPs and ACKs that are not locally generated but which are transiting through it, such as "what is the loss rate on this line", or "what time is it here now", or "what is the local level of security".

Thus we propose to avoid the scalability issue by making each INR responsible only for local users and services, much as a local telephone exchange handles its local users. Source routing removes the burden of routing decisions from all but the local INR, reducing overhead, and removing the need of "per flow" information handling except at INRs where the flows are resident. However, it comes at the price of being less rapidly responsive to changes that may occur in the network. This last point can be compensated by constant monitoring of the flow that is undertaken with the help of SPs and ACKs. Our scheme also requires that INRs be aware of the overall IN topology in terms of other INRs (but there is no need to know what is inside the "clouds"), although this can be mitigated if one accepts the possibility of staged source routing, i.e. with the source taking decisions up to a given intermediate INR, which then takes decisions as far as some other INR, and so on.

6   Protecting the IN Against Denial of Service Attacks

Denial of Service (DoS) attacks are known to the network research community since the early 1980s. Although initially DoS attacks were the act of hackers who wanted to demonstrate their ability and power over computer systems, they have now become an important weapon in the hands of cyber-criminals and for cyber-warfare. DoS attacks have reportedly been used against business competitors, for extortion purposes, for political reasons, and even as a form of "legitimate" protest. It is this variety of targets and types of attack that dictate the need for flexible defence systems which can react according to both the attacker's aims and the defender's needs.

A DoS attack is very often distributed (DDoS): the attacker takes control of a large number of networked computers and orders them to send a large number of packets to a specific target node, server or web site. Typical targets of an attack may include servers or web sites which accomplish some function in the public interest, or the servers of e-commerce web-sites which can suffer significant financial loss. Other targets may be news web-sites, corporate networks, banks, etc. Often, the attacker needs to conceal its identity and the nodes that are used in the attack. It can do this does by introducing fake IP addresses into the packets ("IP spoofing"), providing a false identity for the nodes that generate the attack. Indeed, in a seminal paper on some of the weaknesses of the TCP/IP protocol [1] it is said that (quote)"The weakness in this scheme (the IP protocol) is that the source host itself fills in the IP source host id, and there is no provision to discover the true origin of the packet".

As a result of an attack, the target's links and other routers in its neighbourhood are overwhelmed, and a number of legitimate users' traffic may be lost or backlogged. One of the first defensive measures proposed for DoS attacks is Ingress Filtering, which drops arriving packets if their IP addresses lie outside an acceptable range []. Another technique is IP traceback [8, 24, 33] uses probabilistic packet marking to allow the victim to identify the network path traversed by attack traffic without requiring interactive operational support from Internet Service Providers (ISPs). However false traceback messages can also be injected into the packet stream by attackers. IP traceback can also be used "post-mortem" [15] after an attack in order to understand how the attack has been conducted.

6.1 Protection and Defence Against DoS Attacks

Since attacks occur rapidly and unexpectedly, it is essential for the IN to incorporate an autonomic approach to DoS defence based on network self-observation and adaptive reaction. The effective protection of the IN from DoS attacks requires the combination of different elements:

- *Detection* of the existence of an attack. The detection can be either anomaly-based or signature-based, or a hybrid of those two. In anomaly-based detection, the system recognises a deviation from the standard behaviour of its clients, while in the latter it tries to identify the characteristics of known attack types.
- *Classification* of the incoming packets into valid (normal packets) and invalid (DoS packets). As in detection, one can choose between anomaly-based and signature-based classification techniques.
- *Response*. The protection system should either drop the attacking packets or redirect them into a trap for further evaluation.

As a consequence:

- A node which is undergoing a DoS attack (the victim node) must have the ability to detect or to be informed about the attack, based either on a local or distributed detection scheme. All nodes upstream, from the victim up to the source(s) of the attack, should also be informed of the ongoing threat and they should be incorporated as much as possible into the defence scheme. The detection scheme is always imperfect, so that both false alarms and detection failures are possible. Imperfections are possible both with regard to the detection of the attack as a whole, and the identification of the packets that belong to this attack. Thus some attacking packets will be missed and some non-attacking packets may be incorrectly dropped.
- The victim node and the informed nodes must react by *dropping packets* which are thought to be part of the attack, or by diverting them to a sink node whose role is to absord such attack packets.
- The attack itself, if not properly deviated or eliminated, will produce buffer overflows and saturation of network resources such as CPU capacity, due to the inability of the nodes or routers to handle the resulting overwhelming traffic.

6.2 Using CPN for Defence Against DoS Attacks

As we will briefly show in the sequel, a protocol such as CPN, because of its ability to react rapidly to changes in QoS, is particularly well adapted for DoS defence [50]. In CPN, each flow specifies its QoS requirements in the form of a QoS "goal". SPs associated with each flow constantly explore the network, and obtain routing decisions from network routers based on observed relevant QoS information. SPs store the identities of the nodes they visit, and collect local measurements such as times and loss rates. At each CPN node, the SP uses a local reinforcement learning algorithm based on measurements collected by previous SPs and ACKs, to elicit a decision from the node as to the next hop to travel to. When a SP reaches the destination node of the flow, an ACK packet is generated and returned to the source according to the opposite (destination to source) path traversed by the SP from which all repeated nodes are removed. When an ACK reaches the source, the forward route is now stored and used by subsequent payload or dumb packets (DPs), which are source-routed to the destination, until a new ACK packet brings back a new route.

The CPN-based DDoS defence technique exploits the ability of CPN to trace traffic going both down- and up-stream thanks to SPs and ACK packets. When a node detects an attack, it uses the ACKs to ask all intermediate nodes upstream to drop the packets of the attack flow. Each node is allowed to select the maximum bandwidth that it will accept from any flow that terminates at the node, and the maximum bandwidth that it allocates to a flow that traverses the node. These parameters may vary dynamically as a result of other conditions, and they can also be selected based on the identity and the QoS needs of the flows. When a node receives a SP or DP from a flow that it has not previously encountered (e.g. with a new source-destination pair, or a new QoS class), it sends a Flow-ACK packet back to the source along the reverse path, and informs the source of its bandwidth allocation. The node monitors the flows that traverse it, and drops packets of any flow that exceeds the allocation; it may also inform upstream nodes that packets of this flow should be dropped. Other possible actions include diverting the flow into a "honeypot", or to a special network.

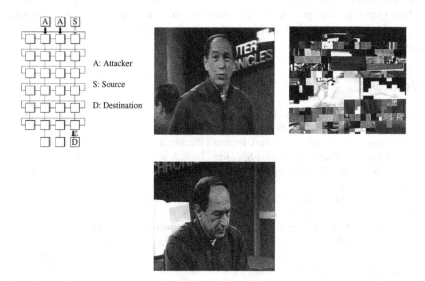

Fig. 2. Experimental evaluation of our defence scheme. The top-left figure shows the CPN testbed that is used in the experiment. Top-middle picture shows a frame of video when there is no attack, while the top-right image is the sams video frame as seen during an attack with no defense. At the bottom we see a video frame during an attack, but with the defence mechanism.

We illustrate the effect of this defence scheme with the following example. A fixed rate MPEG video stream is transferred from node 3 to node 30 in the CPN testbed shown in Figure 2 (top left). The video is initially uncorrupted by an attack, as shown in Figure 2 (top middle). Then, a DDoS attack is launched from nodes 1 and 2 to node 30. Without defence, the attack corrupts the video stream, making it unintelligible as shown in Figure 2 (top right). If node 30 does detect the attack because of the high traffic rate on incoming paths, it defends itself using the CPN trace-back mechanism, orders the packets to be dropped upstream, and drops the traffic coming into itself along those paths. The impact of the attack is reduced, which results in the clear video stream shown in Figure 2 (bottom right). The experiment shows CPN, or a similar adaotive network routing mechanism, coupled with the ability to detect an attack, can protect a sensitive real-time data stream.

7   Conclusions

We present an architecture for Intelligent Networks (INs) which offers a flexible and self-organising communication environment for users and services. The IN is composed of Intelligent Network Routers capable of supporting the user and service needs, and able to sense and adapt network paths and user to service connections dynamically as a function of network state and user and service quality of service needs. It uses smart packets for the search for services, as well as to offer QoS using on-line dynamic sensing and dynamically adaptive control. We suggest that these ideas can be supported by systems that incorporate some of the techniques offered by the CPN system. We also discuss the critical issue of denial of service attack and defence for such systems and show how QoS based adaptive techniques can help mitigate their effect. Other important issues, such as the energy efficient operation of wireless networks [41] have not been discussed in this paper.

References

[1] R.T. Morris. A Weakness in the 4.2BSD Unix TCP/IP Software. *Technical Report Computer Science #117*, AT&T Bell Labs, February 1985.
[2] P. Ferguson and D. Senie. Network Ingress Filtering: Defeating Denial of Service Attacks which Employ IP Source Address Spoofing. *Tech. Rep. RFC 2267*, January 1998.
[3] D. Williams and G. Apostolopoulos. QoS Routing Mechanisms and OSPF Extensions. RFC 2676, Aug. 1999.
[4] E. Bonabeau, M. Dorigo, G. Theraulaz. Swarm Intelligence: From Natural to Artificial Systems. New York, NY, Oxford University Press, 1999.
[5] E. Gelenbe, Z. Xu, E. Şeref. Cognitive packet networks. *Proc. 11th IEEE Int. Conf. on Tools with Artificial Intelligence (TAI99)*, 47-54, Chicago, Ill., 1999.
[6] S. Savage, D. Wetherall, A. Karlin, and T. Anderson. Practical Network Support for IP Traceback. *Proc. ACM SIGCOMM*, 295-306, Stockholm, Sweden, August 2000.
[7] C. Tschudin, H. Lundgren, H. Gulbrandsen. Active Routing for Ad Hoc Networks. IEEE Communications Magazine, April 2000.
[8] D. Song and A. Perrig. Advanced and Authenticated Marking Schemes for IP Traceback. *Proc. Infocom 2001*, ISBN: 0-7803-7016-3, vol. 2, pp. 878-886, Anchorage, Alaska, USA, 22-26 April 2001.

[9]   T. Berners-Lee, J. Hendler, O. Lassila. The Semantic Web. Scientific American, May 2001.

[10] G. P. Picco, A. L. Murphy, G. C. Roman. LIME: a Middleware for Logical and Physical Mobility. 22nd IEEE Intl. Conference Distributed Computing Systems, 2001.

[11] S. Ratsanamy,, P. Francis, M. Handley, R. Karp. A Scalable Content-Addressable Network. ACM SIGCOMM Conference, Aug. 2001.

[12] E. Gelenbe, E. Seref and Z. Xu. Simulation with learning agents. *Proceedings of the IEEE*, 89 (2), pp. 148-157, 2001.

[13] V. Paxson. An Analysis of Using Reflectors for Distributed Denial-of-Service Attacks. *ACM Computer Communications Review 31(3)*, July 2001.

[14] E. Gelenbe, R. Lent and Z. Xu. Design and performance of cognitive packet networks. *Performance Evaluation*, 46, pp. 155-176, 2001.

[15] G. Rice and J. Davis. A Genealogical Approach to Analyzing Post-Mortem Denial of Service Attacks. *Secure and Dependable System Forensics Workshop*, University of Idaho, September 23-25, 2002.

[16] BBC News. Mafiaboy hacker jailed. (September 13, 2001), http://news.bbc.co.uk/1/hi/sci/tech/1541252.stm.

[17] E. Gelenbe, R. Lent, and Z. Xu. Cognitive Packet Networks: QoS and Performance. *Proc. IEEE MASCOTS Conference*, ISBN 0-7695-0728-X, pp. 3-12, Fort Worth, TX, Oct. 2002.

[18] I. Horrocks, P. Patel-Schneider, F. van Harmelen. Reviewing the design of DAML+OIL: An ontology language for the semantic web" National Conference on Artificial Intelligence, Edmonton, Alberta, Canada, 2002.

[19] R. Mahajan, S. Bellovin, S. Floyd, J. Ioannidis, V. Paxson, and S. Shenker. Controlling High Bandwidth Aggregates in the Network. *ACM SIGCOMM Computer Communication Review*, ISSN: 0146-4833, Vol. 32, Issue 3, pp. 62–73, July 2002.

[20] E. Gelenbe, R. Lent, and Z. Xu. Cognitive Packet Networks: QoS and Performance. *Proc. IEEE MASCOTS Conference*, ISBN 0-7695-0728-X, pp. 3-12, Fort Worth, TX, October 2002.

[21] R. Albert, A. Barabasi. Statistical Mechanics of Complex Networks", Rev. Mod. Phys. 74 (47), 2002.

[22] C. Borcea, et al.. Cooperative Computing for Distributed Embedded Systems", 22nd International Conference on Distributed Computing Systems, Vienna (A), IEEE CS Press, 227-238, 2002.

[23] D. Estrin, D. Culler, K. Pister, G. Sukjatme. Connecting the Physical World with Pervasive Networks", IEEE Pervasive Computing, 1 (1): 59-69, Jan. 2002.

[24] A. Snoeren, C. Partridge, L.A. Sanchez, C.E. Jones, F. Tchakountio, B. Schwartz, S. Kent, and W.T. Strayer. Single-Packet IP Traceback. *IEEE/ACM Transactions on Networking*, ISSN: 1063-6692, Vol. 10, no. 6, pp. 721-734, December 2002.

[25] W.G. Morein, A. Stavrou, D.L Cook, A.D. Keromytis, V. Mishra, and D. Rubenstein. Using Graphic Turing Tests to Counter Automated DDoS Attacks against Web Servers. *Proc. 10th ACM Int'l Conference on Computer and Communications Security (CCS '03)*, ISBN: 1-58113-738-9, pp. 8-19, Washington DC, USA, October 27-30, 2003.

[26] L. Capra, W. Emmerich, C. Mascolo. CARISMA: Context-Aware Reflective mIddleware System for Mobile Applications", IEEE Transactions of Software Engineering Journal (TSE) 29(10):929-945, 2003.

[27] J. Kephart, D. Chess. The Vision of Autonomic Computing", IEEE Computer, 36 (1), 2003.

[28] F. Zambonelli, N. Jennings, M. Wooldridge. Developing Multiagent Systems: the Gaia Methodology", ACM Transactions on Software Engineering and Methodology, 12 (3):317-370, 2003.

[29] M. Papazoglou, M. Aiello, M. Pistore, J. Yang. XSRL: A Request Language for Web Services (www.webservices.org)", 2003

[30] M. Philipose, K. Fishkin, M. Perkowitz, D. Patterson, D. Fox, H. Kautz, D. Hahnel. Inferring Activities from Interactions with Objects", IEEE Pervasive Computing, 3(4):50-57, 2004.

[31] S. Jing, H. Wang, and K. Shin. Hop-Count Filtering An Effective Defense Against Spoofed Traffic. *Proc. ACM Conference on Computer and Communications Security*, pp. 30-41, ISBN 1-58113-738-9, Washington DC, October 2003.

[32] G. Mori and J. Malik. Recognizing objects in adversarial clutter - Breaking a visual CAPTCHA. *Proc. IEEE Computer Society Conference on Computer Vision and Pattern Recognition 2003 (CVPR '03)*, ISSN: 1063-6919, ISBN: 0-7695-1900-8, vol. 1, pp. 134-141, Madison, WI, USA, June 18-20, 2003.

[33] M. Sung and J. Xu. IP Traceback-Based Intelligent Packet Filtering: A Novel Technique for Defending against Internet DDoS Attacks. *IEEE Transactions on Parallel and Distributed Systems*, vol. 14, pp. 861-872, September 2003.

[34] R. Thomas, B. Mark, T. Johnson, and J. Croall. NetBouncer: client-legitimacy-based high-performance DDoS filtering. *Proc. DARPA Information Survivability Conference and Exposition*, vol. 1, pp. 14-25, April 22-24, 2003.

[35] A. Hussain, J. Heidermann, and C. Papadopoulos. A Framework for Classifying Denial of Service Attacks. *Proc. ACM SIGCOMM Conference on Applications, Technologies, Architectures, and Protocols for Computer Communication 2003*, ISBN: 1-58113-735-4, pp. 99-110, Karlsruhe, Germany, August 25-29, 2003.

[36] J. Mirkovic, P. Reiher, and M. Robinson. Forming Alliance for DDoS Defense. *Proc. 2003 workshop on New security paradigms*, 11-18, ISBN 1-58113-880-6, Ascona, Switzerland, August 2003.

[37] H. Liu, M. Parashar. Component-based Programming Model for Autonomic Applications. *Proc. First International Conference on Autonomic Computing*, New York, NY, USA, 2004.

[38] M. Mikic-Rakic, N. Medvidovic. Support for Disconnected Operation via Architectural Self-Reconfiguration. *Proc. First International Conference on Autonomic Computing*, (IEEE Computer Society), ISBN 0-7695-2114-2, New York, 2004.

[39] E. Gelenbe, M. Gellman, R. Lent, P. Liu, Pu Su. Autonomous smart routing for network QoS. *Proc. First International Conference on Autonomic Computing*, (IEEE Computer Society), ISBN 0-7695-2114-2, 232-239, New York, 2004.

[40] E. Gelenbe, R. Lent, A. Nunez. Self-aware networks and QoS. *Proceedings of the IEEE*, 92 (9), pp. 1478-1489, 2004.

[41] E. Gelenbe and R. Lent. Adhoc power aware Cognitive Packet Networks. *Ad Hoc Networks Journal*, Vol. 2 (3), pp. 205–216, 2004 (ISN: 1570-8705).

[42] E. Gelenbe. Cognitive Packet Network. *U.S. Patent No. 6,804,201 B1*, Oct. 12, 2004.

[43] J. Frey, G. Hughes, H. Mills, M. Schraefel, G. Smith, D. De Roure. Less is More: Lightweight Ontologies and User Interfaces for Smart Labs. *UK e-Science All Hands Meeting*, Nottingham, 2004.

[44] D.K.Y. Yau, J.C.S Lui, F. Liang, and Y. Yam. Defending Against Distributed Denial-of-Service Attacks With Max-Min Fair Server-Centric Router Throttles. *IEEE/ACM Transactions on Networking*, 13 (1): 29-42, February 2005.

[45] L. Tummolini, C. Castelfranchi, A. Ricci, M. Viroli, A. Omicini. Exhibitionists and Voyeurs do it better: A Shared Environment Approach for Flexible Coordination with Tacit Messages. *Environments for MultiAgent Systems*. LNAI 3374, Springer-Verlag, January 2005.

[46] F. Zambonelli, M.P. Gleizes, M. Mamei, R. Tolksdorf. Spray Computers: Explorations in Self Organization. *Journal of Pervasive and Mobile Computing*, 1 (1), May 2005.

[47] S. Kandula, D. Katabi, M. Jacob, and A. Berger. Botz-4-Sale: Surviving Organized DDoS Attacks that Mimic Flash Crowds. *Proc. 2nd USENIX Symposium on Networked Systems Design and Implementation (NSDI '05)*, Boston, MA, USA, May 2-4, 2005.

[48] J. Mirkovic and P. Reiher. D-WARD: A Source-End Defense against Flooding Denial-of-Service Attacks. *IEEE Transactions on Dependable and Secure Computing*, vol. 2, no. 3, pp. 216-232, July-September, 2005.

[49] P. Bouquet, L. Serafini, S. Zanobini. Peer-to-Peer Semantic Coordination. *Journal of Web Semantics*, 2 (1), 2005.

[50] E. Gelenbe, M. Gellman, and G. Loukas. An autonomic approach to denial of service defence. *Proc. of the IEEE Int. Symp. on a World of Wireless, Mobile and Multimedia Networks*, 537-541, June 2005.

MAC Protocol for Contacts from Survivors in Disaster Areas Using Multi-hop Wireless Transmissions

Poompat Saengudomlert, Kazi M. Ahmed, and R.M.A.P. Rajatheva

Asian Institute of Technology,
P.O. Box 4, Klong Luang, Pathumthani 12120, Thailand
{poompats, kahmed, rajath}@ait.ac.th

Abstract. We propose a simple multiple access control (MAC) protocol that allows survivors in disaster areas to establish contact to the base stations (BSs) provided by the rescue team. Our protocol relies on downstream broadcast single-hop wireless transmissions from the BSs to the survivors and upstream multi-hop wireless transmissions. For MAC, the protocol uses a combination of tree splitting and code division multiple access (CDMA). Compared to the single-hop approach for upstream transmissions, we show that the multi-hop approach can expend less transmission energy, especially when data aggregation is possible at relay points. In addition, the multi-hop approach requires less connection setup time. The energy and time efficiency makes the protocol attractive for communications for disaster management.

1 Introduction

Disasters such as the 9/11 attack and the tsunami incident have stimulated recent research interests on communication networks for disaster management. One attractive approach is to use the mobile units held by survivors, e.g. cellular phones or radio frequency identification (RFID) badges, as a mean to contact the rescue operation. Since existing communication infrastructures may be destroyed by the disaster, mobile units need to rely on one another to communicate in a multi-hop fashion to the available base stations (BSs), hence forming an ad hoc network.

Multiple access control (MAC) and energy efficient routing for ad hoc networks have been the subject of intense research (see [1], [2], [3], and [4] and references therein). Previous works in this area assume known identities of the survivors' mobile units. For example, survivors are subscribers of a common cellular service [5]. Alternatively, survivors have employee RFID badges [4]. In these scenarios, multiple access control (MAC) can be designed prior to the disaster based on time division multiple access (TDMA) [5], code division multiple access (CDMA) [6], and so on, to avoid transmission collisions.

In this work, we develop a protocol for survivors that need not belong to any particular service or organization to establish contact to a BS. We assume that the mobile units can be activated to operate on this common emergency protocol. The protocol can serve in the initialization phase of an emergency network. Once the

K. Cho and P. Jacquet (Eds.): AINTEC 2005, LNCS 3837, pp. 46–56, 2005.

existence of a survivor is known, future transmissions may follow a different MAC scheme such as using CDMA codes for individual mobile units.

Another unique aspect of our problem is the environment in which downstream transmissions from a BS to mobile units are single-hop while upstream transmissions are multi-hop. In addition to energy efficiency, we also consider time efficiency since transmission delays can be costly in emergency situations.

Section 2 describes the problem of MAC for emergency networks that we treat in this paper. Section 3 presents a baseline solution that employs single-hop transmissions for both upstream and downstream directions. Section 4 presents our proposed multi-hop MAC protocol. Section 5 compares the performances of the single-hop and multi-hop protocols in terms of energy and time efficiencies. Section 6 demonstrates the value of data aggregation at relay points. Finally, section 7 concludes the work.

2 Problem Description

Assume that after a natural disaster, an accident, or a terrorist attack, all communication infrastructures are destroyed. However, a mobile phone or an RFID badge held by each survivor is able to operate in the emergency mode after it is activated to do so. Such activations are not automatic so that we can distinguish survivors from victims.

A rescue team enters the field and sets up an emergency network containing a small number of BSs for wireless communications. Assume that the coverage areas of all BSs cover all the survivors, as shown in Fig. 1. In addition, assume that each survivor is in multiple coverage areas so that the survivor can estimate his or her location based on the received signal powers from different BSs [7]. In what follows, we shall refer to a survivor as a mobile unit.

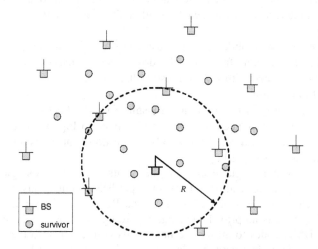

Fig. 1. Coverage areas of BSs cover all survivors

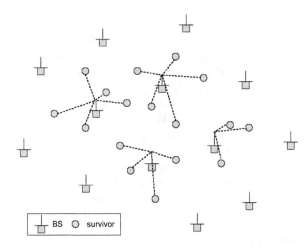

Fig. 2. Self-assignment of mobile units to BSs

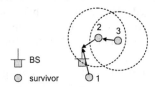

Fig. 3. Relay transmission for mobile unit 3

To avoid collisions among transmissions from BSs, we assume that the BSs use CDMA with known codes and reference bits. Each mobile unit can decode signals from the BSs by trying all (small number) CDMA codes and looking for reference-bits. Based on the receive powers, each mobile unit assigns itself to the closest BS, as shown in Fig. 2.

Transmissions from mobile units also use CDMA codes that are specific for individual BSs but are different from those for downlink transmissions. (Each BS has two CDMA codes associated with it.) As a result, there is little interference among different groups of mobile users.

To conserve energy in mobile units, transmit power is kept small. Consequently, uplink transmissions may need relaying, as illustrated in Fig. 3. In summary, uplink transmissions operate in a multi-hop mode, while downlink transmissions from the BSs are broadcast in a single-hop mode.

We shall focus on a single group of mobile units assigned to a single BS, keeping in mind that the protocol can operate on multiple BSs simultaneously. Our objective is to investigate how to connect all units to the BS in an energy and time efficient manner. Since there is no prior knowledge regarding survivors' identities, we cannot pre-assign CDMA codes to all mobile units. This is in contrast to the assumption in previous works on ad hoc networks in which node identities are known beforehand. We concentrate on the initialization period in which mobile units establish contact to report their existence as well as their locations to the BS. The BS in turn assigns

CDMA codes to the mobile units. At termination, the protocol will have assigned each mobile unit a CDMA code to be used for future uplink transmissions.

For simplicity, we assume that all BSs have the coverage radius R, while all mobile units have the coverage radius r with $r \leq R$. We assume that there are enough mobile units and that every unit has a multi-hop path to the BS. In addition, we assume that the BS has a good estimate of the survivor or mobile unit density (in person/m^2). For comparison, we shall use a single-hop MAC protocol with $r = R$ as the baseline protocol. Our challenge is to design a multi-hop MAC protocol with $r < R$.

We adopt the model in which time is divided into slots. A message packet can be transmitted in a single slot. In addition, through acknowledgement from the BS or a relay unit, mobile units can get immediate feedback on the transmission success before the next time slot. For protocol performances, we consider the expected maximum energy usage per mobile unit and the expected total connection setup time for all units. We assume that energy for message transmissions dominates the total energy usage of a mobile unit, which includes transmission, reception, and information processing.

3 Single-Hop MAC Protocol

If we set $r = R$, then each mobile unit can reach the BS in a single hop. In this case, one possibility is to use the tree splitting algorithm [8], [9]. Recall that, in the tree algorithm, if two message packets collide, then other packets will wait until the collision between these two packets is resolved. In the time slot after the collision, each of the two colliding packets is independently transmitted with probability 1/2. With probability 1/2, there will be only one transmission in this slot, leading to two consecutive successful transmissions. With probability 1/2, there will be zero or two transmissions in this slot. In this case, each packet is independently transmitted with probability 1/2 in the following slot, and so on until the collision is resolved.

For our purpose, we can consider that all mobile units have a message packet to transmit and are a part of a single collision resolution period (CRP) from the start. With the estimate of mobile unit density, the BS broadcasts a message notifying all mobile units to randomly split themselves into several subsets such that the expected number of units per subset is slightly greater than one. The tree splitting algorithm is then used on one subset at a time until all packets are transmitted. At the end of the CRP, all mobile units will successfully contact the BS. The throughput of this scheme was found to be approximately 0.4 [9].

Let ρ denote the throughput of the tree splitting algorithm. Let γ be the mobile unit density. Let P_0 be the minimum received power for reliable transmission. For transmission at distance l, we adopt the model where the minimum transmit power P_t is given by $P_t = \eta l^2 P_0$. Note that P_t is proportional to the square distance, i.e., l^2, with ηP_0 being the constant of proportionality. For simplicity, assume that each mobile unit transmits at power $\eta R^2 P_0$ in the single-hop protocol.

Let $\mathcal{E}$ be the expected maximum energy usage at a mobile unit, and T be the expected total connection setup time for all units. Since the throughput is ρ and the expected total number of mobile units in a radius-R coverage area is $\gamma\pi R^2$, $\mathcal{T}$ is given by [1]

$$\mathcal{T}_{\text{single-hop}} = \frac{\gamma\pi R^2}{\rho}. \tag{1}$$

For the tree splitting algorithm described above, most collisions result from two messages. Thus, the total number of transmissions (both successful and unsuccessful) is approximately $\gamma\pi R^2 \left(1 + 2(1/\rho - 1)\right) = \gamma\pi R^2 \left(2/\rho - 1\right)$. It follows that the expected number of transmissions per mobile unit is $2/\rho - 1$, yielding the expected energy usage per unit

$$\mathcal{E}_{\text{single-hop}} = \eta R^2 P_0 \left(\frac{2}{\rho} - 1\right). \tag{2}$$

Because of symmetry, the above expression is also the expected maximum energy usage at a mobile unit.

4 Multi-hop MAC Protocol

Roughly speaking, our proposed multi-hop MAC protocol proceeds in rounds. In the first round, all mobile units within one hop connect to the BS using tree splitting, similar to the single-hop protocol. In the second round, all mobile units within two hops connect (using tree splitting) to the one-hop units which in turn relay their information to the BS, and so on. Except in the first round, it is possible to have multiple successful transmissions simultaneously since $r < R$ and thus frequency reuse is possible. This parallelism allows us to reduce the expected total connection setup time $\mathcal{T}$. We shall shortly see that multi-hop transmissions can also reduce the expected maximum energy usage at a mobile unit.

We next present the multi-hop protocol in detail. In what follows, we refer to a mobile unit that can reach the BS in k hops as a k-hop unit.

Multi-hop MAC protocol

1. The BS broadcasts a message notifying all 1-hop units to contact, i.e., send its location information to, itself according to tree splitting. We regard all 1-hop nodes as being a part of a single CRP from the start. Whenever a 1-hop unit successfully contacts the BS, the BS responds with an assignment of CDMA code to that unit.

   Note that all downlink broadcast transmissions from the BS use the downlink CDMA code of the BS, while all uplink transmissions from the 1-hop units use the uplink CDMA code of the BS. At the end of the CRP, proceed to step 2.

---

2.   At the end of the previous round, say round $k - 1$, all $(k - 1)$-hop units have con-
     tacted the BS. From the information about their locations, the BS broadcasts a
     message to assign a subset S of $(k - 1)$-hop units whose coverage areas do not
     overlap to act as relay points. Each of these $(k - 1)$-hop units then broadcasts a
     message to all the units in its radius-r neighborhood. Among the units in this
     neighborhood, the ones that have not contacted the BS, i.e., k-hop units, then con-
     tact the $(k - 1)$-hop unit using tree splitting and regard this $(k - 1)$-hop unit as
     their parent, as illustrated in Fig. 4.

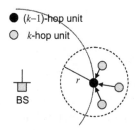

Fig. 4. A set of k-hop units and their $(k - 1)$-hop parent in round k

   At the end of the CRP, the $(k - 1)$-hop unit aggregates its children's information and transmits a message to its parent which in turn relays the message towards the BS. The BS, after receiving a message from all the $(k - 1)$-units in subset S, responds by broadcasting an assignment of CDMA code to each k-hop unit.
   Note that all downlink transmissions use the downlink CDMA code of the BS.
All uplink transmissions from k-hop units use the uplink CDMA code of the BS,
while further uplink transmissions use CDMA codes of the parents. Thus, there
is no collision among the transmissions by the parents.
   After subset S of $(k - 1)$-units, the BS assigns another subset of $(k - 1)$-hop
units to act as relay points, and so on until all the $(k - 1)$-hop units are assigned or
all the coverage areas of $(k - 1)$-hop units are already taken into account. Round k
is now completed. The protocol then proceeds to step 3.

3.   The BS checks whether there is any area inside its radius-R coverage area not yet
     covered by any of the previous rounds. If so, then proceed to the next round by
     going back to step 2. Else, the protocol terminates.

We next compute the values of the expected maximum energy usage $\mathcal{E}$ and the expected total connection setup time $\mathcal{T}$. First, we compute the total number of hops or
rounds H performed by the multi-hop protocol. From the viewpoint of a mobile unit
at distance l from the BS, we approximate the average distance towards the BS for
each transmission by considering that its parent is uniformly distributed in the semi-
circle centered at the unit with the diameter tangent to the radius-l circle centered at
the BS, as illustrated in Fig. 5. Note that this approximation becomes more and more
accurate as l increases.

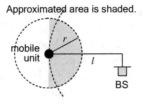

Approximated area is shaded.

Fig. 5. Approximation of the intersection of radius-*l* and radius-*r* circles by a semi-circle

From Fig. 5, using the polar coordinate, we can compute the average distance D_{avg} to be

$$D_{\text{avg}} = \int_{-\pi/2}^{\pi/2} \int_0^r \frac{s}{\pi r^2 / 2} s\, ds\, d\theta = \frac{2r}{3}. \tag{3}$$

Thus, the total number of hops to cover radius R is approximately $H = R/D_{\text{avg}} = 1.5R/r$. In the multi-hop protocol, a 1-hop unit uses the most energy compared to k-hop units, where $k > 1$. Part of the energy usage for a 1-hop unit is from tree splitting in the first round, and is equal to $\eta r^2 P_0 (2/\rho - 1)$ by the same argument used to establish (2). Another part is from acknowledging transmissions of its children in round 2 of the protocol, and is equal to $\eta r^2 P_0 \times \gamma \pi r^2 / \rho$. The last part is from relaying information from round 2 on. With data aggregation allowed, a 1-hop unit transmits once at the end of each round to the BS, yielding the energy for relaying equal to $(H-1)\eta r^2 P$. It follows that the total energy for a 1-hop unit is

$$\mathcal{E}_{\text{multi-hop}} = \eta r^2 P_0 \left(\frac{2}{\rho} - 1 \right) + \eta r^2 P_0 \frac{\gamma \pi r^2}{\rho} + (H-1)\eta r^2 P_0$$

$$= \eta r^2 P_0 \left(\frac{\gamma \pi r^2 + 2}{\rho} + \frac{3R}{2r} - 2 \right). \tag{4}$$

To compute T, it is useful to compute the average degree of frequency reuse in round k, $k > 1$. To do so, we consider that, after round $k - 1$, $(k - 1)$-hop units are on average located on the circle of radius $2r(k-1)/3$ based on D_{avg} in (3). With uniform locations for relay points, the degree of frequency reuse is approximately $2\pi \frac{2r(k-1)}{3} / 2r = \frac{2\pi(k-1)}{3}$, as illustrated in Fig. 6, where we approximate the length of the arc in each radius-*r* circle by $2r$. Note that this approximation becomes more and more accurate as k increases.

We now compute the time required in round k. To obtain the number of k-hop units, we compute the expected number of units in the radius-$2rk/3$ circle centered at the BS subtracted by the number of units in the radius-$2r(k-1)/3$ circle. With

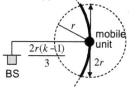

Arc to be approximated is bolded.

Fig. 6. Degree of frequency reuse in round k of the multi-hop MAC protocol

throughput ρ for tree splitting and degree of frequency reuse $2\pi(k-1)/3$, it follows that the expected time in round k for tree splitting is

$$T_{\text{splitting},k} = \frac{\gamma\pi\left(\dfrac{2rk}{3}\right)^2 - \gamma\pi\left(\dfrac{2r(k-1)}{3}\right)^2}{\rho\dfrac{2\pi(k-1)}{3}} = \frac{2\gamma r^2}{3\rho}\frac{2k-1}{k-1} \approx \frac{4\gamma r^2}{3\rho}.$$

For round k, the time for relaying information by the parents of k-hop units to the BS is $k-1$ transmissions. Therefore, the total time for connection setup after H rounds is

$$T_{\text{multi-hop}} = \frac{\gamma\pi r^2}{\rho} + \sum_{k=2}^{H}\left(\frac{4\gamma r^2}{3\rho}+k-1\right) = \frac{\gamma\pi r^2}{\rho} + \frac{4\gamma r^2}{3\rho}(H-1) + \frac{H(H-1)}{2}$$

$$\approx \frac{\gamma\pi r^2}{\rho} + \frac{4\gamma r^2}{3\rho}H + \frac{H^2}{2}.$$

Substituting $H = 1.5R/r$ yields

$$T_{\text{multi-hop}} = \frac{\gamma\pi r^2}{\rho} + \frac{2\gamma Rr}{\rho} + \frac{9R^2}{8r^2}. \tag{5}$$

5 Comparison Between Single-Hop and Multi-hop Protocols

Fig. 7 shows the ratio between $T_{\text{multi-hop}}$ and $T_{\text{single-hop}}$ based on the expressions in (1) and (5). The parameter values are $\gamma = .005$ person/m², $\eta P_0 = 1$ μJ/packet/m², $\rho = 0.4$ packet/transmission, and $r = 25$ and 50 m. The power parameter ηP_0 is based on the typical radio parameters obtained from [10]. We assume that each message packet contains 10,000 bits. In addition, the mobile unit density γ is chosen so that each r-radius coverage area contains approximately 10 persons for $r = 25$ and 40 persons for $r = 50$.

Notice that the ratio $T_{\text{multi-hop}}/T_{\text{single-hop}}$ decreases, i.e., further reduction in total time, as the value of R/r increases. From Fig. 7, we see that there is no improvement in the low end of R/r. However, the multi-hop protocol will operate on moderate values of R/r in practice. Overall, the results demonstrate a good potential in the reduction of T by the multi-hop protocol.

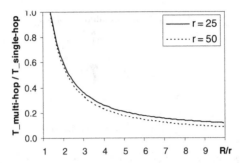

Fig. 7. The ratio $\mathcal{T}_{\text{multi-hop}}/\mathcal{T}_{\text{single-hop}}$ as a function of R/r

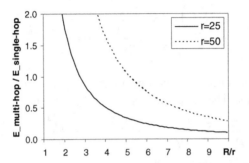

Fig. 8. The ratio $\mathcal{E}_{\text{multi-hop}}/\mathcal{E}_{\text{single-hop}}$ as a function of R/r

Fig. 8 shows the ratio between $\mathcal{E}_{\text{multi-hop}}$ and $\mathcal{E}_{\text{single-hop}}$ based on the expressions in (2) and (4). The parameter values are the same as above. From Fig. 8, we see that there is no improvement in the low end of R/r. However, for large enough R/r, i.e., sufficiently large number of hops, the multi-hop protocol can offer savings in energy consumption. Over all, the results demonstrate a potential in the reduction of $\mathcal{E}$ by the multi-hop protocol.

6   Value of Data Aggregation at Relay Points

This section discusses what happens to the expected maximum energy usage $\mathcal{E}$ when mobile units are *not* capable of aggregating information from their descendants before relaying it to their parents, i.e., each descendant requires one message transmission. Consider a 1-hop unit. Let N_l be the expected number of mobile units in the radius-l circle centered at the BS. The expected number of descendants that a 1-hop unit has is given by $\dfrac{N_R - N_{2r/3}}{N_{2r/3}} = \left(\dfrac{3R}{2r}\right)^2 - 1 \approx \dfrac{9R^2}{4r^2}$, where we use D_{avg} in (3) to estimate the radius for 1-hop units. This implies that the amount of energy required to relay descendants' information is $\dfrac{9R^2}{4r^2}\eta r^2 P_0$, yielding the total energy usage

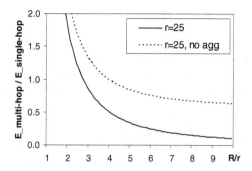

Fig. 9. The ratio $\mathcal{E}_{\text{multi-hop}}^{\text{no-agg}}/\mathcal{E}_{\text{single-hop}}$ as a function of R/r

$$\mathcal{E}_{\text{multi-hop}}^{\text{no-agg}} = \eta r^2 P_0 \left(\frac{2}{\rho} - 1 \right) + \eta r^2 P_0 \frac{\gamma \pi r^2}{\rho} + \eta r^2 P_0 \frac{9R^2}{4r^2}$$

$$= \eta r^2 P_0 \left(\frac{2 + \gamma \pi r^2}{\rho} + \frac{9R^2}{4r^2} - 1 \right). \tag{6}$$

Fig. 9 shows the ratio $\mathcal{E}_{\text{multi-hop}}^{\text{no-agg}}/\mathcal{E}_{\text{single-hop}}$ as a function of R/r for $r = 25$. For comparison, the ratio $\mathcal{E}_{\text{multi-hop}}/\mathcal{E}_{\text{single-hop}}$ is also given. The difference can be viewed as the value of data aggregation at relay points. Regarding the total connection setup time, we do not expect the total connection setup time to change significantly since a relay point can transmit each descendant's information even before the end of the CRP, unlike in the case with data aggregation.

7 Conclusion

We proposed a multi-hop MAC protocol that the survivors' mobile units or RFID badges can use to establish contact to the BS provided by the rescue team in an event of natural disaster, accident, or terrorist attack. We demonstrated the potential reduction in the expected maximum energy usage and the expected total connection setup time obtained by the multi-hop protocol compared to those of the single-hop protocol. Our results indicate that the multi-hop protocol can be energy and time efficient and is thus attractive for practical uses.

References

1. S. Kumar, V.S. Raghavan, and J. Deng, "Medium access control protocols for ad hoc wireless networks: a survey," *Ad Hoc Networks*, to appear. Available at: http://people.clarkson.edu/~skumar/MAC_Survey.pdf
2. Y.E. Sagduyu and A. Ephremides, "The problem of medium access control in wireless sensor networks," *IEEE Wireless Communications*, vol. 11, no. 6, pp. 44-53, Dec. 2004.

3. J.N. Al-Karaki and A.E. Kamal, "Routing techniques in wireless sensor networks: a survey," *IEEE Wireless Communications*, vol. 11, no. 6, pp. 6-28, Dec. 2004.
4. G. Zussman and A. Segall, "Energy Efficient Routing in Ad Hoc Disaster Recovery Networks," *Ad Hoc Networks*, vol. 1, no. 4, pp. 405-421, 2003.
5. T. Fujiwara, N. Iida, and T. Watanabe, "A hybrid wireless network enhanced with multi-hopping for emergency communications," *in Proc. of IEEE ICC 2004*, pp. 4177-4181, Jun. 2004.
6. A. Muqattash and M. Krunz, "CDMA-based MAC protocol for wireless ad hoc networks," *in Proc. of Mobihoc 2003*, pp. 153-164, Jun. 2003.
7. N. Bulusu, J. Heidemann, and D. Estrin, "GPS-less low-cost outdoor localization for very small devices," *IEEE Personal Communications*, vol. 7, no. 5, pp. 28-34, Oct. 2000.
8. J.I. Capetanakis, "Tree algorithms for packet broadcast channels," *IEEE Trans. on Information Theory*, vol. 25, no. 5, pp. 505-515, Sep. 1979.
9. D.P. Bertsekas and R.G. Gallager, *Data Networks*. Prentice-Hall, 1992.
10. W.R. Heinzelman, A. Chandrakasan, and H. Balakrishnan, "Energy-Efficient Communication Protocol for Wireless Microsensor Networks," *Proc. Hawaii International Conference on System Sciences*, Jan. 2000.

Performance Evaluation of Mobile IPv6 Wireless Test-Bed in Micro-mobility Environment with Hierarchical Design and Multicast Buffering

Yong Chu Eu[1], Sabira Khatun[1], Borhanuddin Mohd Ali[1], and Mohamed Othman[2]

[1]Department of Computer and Communication Systems Engineering,
Faculty of Engineering, University Putra Malaysia
[2]Department of Communication Technology and Network,
Faculty of Computer Science, University Putra Malaysia

Abstract. Mobile IPv6 allows mobile devices always addressable by its home address wherever it is located. In this paper the focus is given on a micro-mobility based test-bed development with improved scheme. The test-bed consists of both hardware and software including four personal computers and one laptop. The laptop is used as Mobile Node (MN), two personal computers are functioned as access points, one personal computer as Home Agent (HA) and the other one is used as Foreign Multicast Router (MR). This test-bed is used to analyze the Mobile IPv6 handover delay, packet delay and packet loss in real wireless environment. We measured two scenarios those are Mobile IPv6 handover and Enhanced Mobile IPv6 with multicast function and hierarchical design. Finally, it is concluded that the proposed handover scheme reduces the handover delay from 4.4s to 0.2s, packet delay from 0.7s to 0.109s and from 44 packet loss to zero packet loss during handover. This gives advantages of zero packet loss and lower handover delay which can significantly improve micro mobility performance 3rd and 4th Generation mobile telephone technology and for beyond.

1 Introduction

IP is a method or protocol by which data is sent from one computer to another on the Internet. Both ends of a TCP session (connection) need to keep the same IP address for the life of the session. IP needs to change the IP address when a network node moves to a new place in the network. In order to support IP mobility, IETF has proposed mobile IP based on IPv4. Later on as several drawbacks still persist in MIPv4; MIPv6 was proposed to solve these problems.

MIPv6 [1] provides comprehensive mobility management for the IPv6 protocol compared to MIPv4 [4]. MIPv6 manages all aspects of location updates and mobility management for active IPv6 hosts. It faces some problems due to its handover management. Mobile IPv6 provides an efficient handover in macro mobility management where the access points of the network are far from each other. The problem occurs when a mobile node moves from one access point to another access point in a small coverage area (micro-mobility). The handover occurs frequently and Mobile IPv6 is not suitable for such a micro mobility case. Mobile IPv6 generates significant

K. Cho and P. Jacquet (Eds.): AINTEC 2005, LNCS 3837, pp. 57–67, 2005.
© Springer-Verlag Berlin Heidelberg 2005

signaling traffic load in the core network, even for local movement followed by long interruptions. Studies and investigations to solve the drawbacks in Mobile IPv6 in micro-mobility should be considered here, especially in handover operation.

Several micro-mobility schemes have been proposed to overcome this shot coming namely Hierarchical Handover (HMIPv6) [5], Fast Handover (FMIPv6) [6], Cellular IP [7] and HAWAII [8]. HMIPv6 and FHIPv6 are extensions of MIPv6, which help to speed up the handover latency and provide uninterrupted service to roaming users. Cellular IP and HAWAII are micro mobility protocols relying on Mobile IP for the macro-mobility.

HMIPv6 reduces handoff latency by employing a hierarchical network structure in minimizing the location update signalling with external network. The mobility management inside the domain is handled by a Mobility Anchor Point (MAP), which acts as a local Home Agent. The delay is still high for HMIPv6 and the packet loss rate is high as well [2]. FMIPv6 introduces two new schemes that is anticipating the subnet handover by constructing a care of address on the new subnet and setting up a bi-directional tunnel between old and new subnet's Access Router in order to reduce packet drop. FMIPv6 may be very sensitive to any anomalies in the network, and it will only work correctly when all its assumptions hold [12] source-specific.

Cellular IP inherits cellular systems principles for mobility management, passive connectivity and handoff control but is designed based on the IP paradigm. HAWAII has a similar scheme for mobility management. Both of them use paging for reducing signal load and saving mobile device battery. The propagation of source-specific route within a single domain can significantly increase signalling complexity in Cellular IP and HAWAII scheme [13].

Although several schemes has been proposed but it seems some of them still in draft, and not the prefect solution so an enhanced scheme needs to be proposed that can combine the advantages of the above approaches and reduces the problems that have been faced by the above proposals.

The authors in [14] proposed a Multicast-based Re-establishment scheme, which reroutes connection in a crossover point near the base stations. Radio hints was used to identify the potential new base station in advance. Their work is quite generic since the IP Multicast still on the early stage at that time. They can be regarded as the pioneers of multicast based handover.

In [23] IETF Mobile IP is extended by implementing multicast. The Mobile IP Foreign Agents receive multicast addresses and packets by multicasting from the Home Agent to several Foreign Agents. The authors argue that the multicast extension improves the handover latency and packets loss for a Mobile IP handover, particularly for vertical handover. The current IETF Mobile IP specification [24,25] proposes remote subscription and bi-directional tunnelled multicast for handling multicast for mobile node. In the Mobile IP Foreign Agent Based Multicast proposal (Remote-Subscription Multicast), Mobile Node will subscribe its membership at the new foreign network with its new co-located care-of-address when it moves. The multicast router in the foreign network propagates this information for the MN. It is quite simple and no encapsulation is required. Remote Subscription does provide the most efficient delivery of multicast datagram, but this service may come at a high price for network involved and the multicast router must manage the multicast tree. The main problem here is that every foreign network must have multicast router.

In bi-directional tunnelled multicast [24,25], when a MN is roaming in foreign network, multicast packets will be encapsulated by the home agent and delivered to the MN by the same tunnelling mechanism as other unicast packets. MN only subscribes its multicast group membership to the multicast router in its home network. Its multicast group membership is transparent to any foreign network and the Foreign Agent wills forward the multicast packets in a similar way to unicast packets. Packets duplication and triangle routing is the main disadvantage of this proposal.

Mysore, J: et al in [15] explored IP Multicast as it is available today for host mobility, without any special change on the multicast. Their works include protocols such as ARP, ICMP, IGMP, TCP and UDP. They propose some design issues and implementation constraints that prevent the wide deployment of multicast for handover but some issues are still unresolved in the context of IP multicast. They concluded that using multicast could support host mobility.

The DATAMAN research group at Rutgers University proposes a multicast protocol for MN [17]. Their scheme was designed to support exactly-one multicast delivery, and assumes static multicast groups (membership of group does not change during the group's lifetime and the sender knowledge of the group membership), and thus does not extend easily to IP Multicast and the host group model.

Later, they [18] proposed extensions to IP multicast to support MNs. Their approach is based on the Columbia University Mobile IP protocol, and uses mobile support routers (which are similar to but not the same as the agents in IETF Mobile IP) provide multicast datagram delivery to mobile group members. They use DVMRP [19] in their implementations and believe that this can work with other multicast routing scheme.

The Columbia approach [20] was among the pioneering efforts to support mobility in the Internet. MN belongs to a virtual mobile network with a distinct network id. A collection of dedicated mobile support routers (MSRs) is used to provide packets forwarding and location management. MSR use tunnels to communicate with each other's. This approach use multicast to reduce packets loss, advance register performs resource reservation, and use the existing IP multicasting infrastructure to accomplish host mobility.

Helmy Ahmed [21] proposed another multicast based scheme in that a MN was assigned a multicast address, and the correspondent nodes send packets to that multicast node. As the mobile node moves to new location, it joins the multicast group through the new location and prune through the old location. Dynamic of the multicast tree provide for smooth handoff, efficient routing, and conservation of network bandwidth compared to the approaches that multicast to base stations around the location of mobile node.

In [22] a multicast routing protocol called Distributed Core Multicast with application to host mobility is proposed. DCM is designed for multicast with a high number of multicast groups and a low number of receivers. DCM avoids multicast group state information in backbone routers; it avoids triangular routing across expensive backbone links and scales well with the number of multicast groups. The authors argue that their protocol performs better than the existing sparse-mode multicast routing protocols. The approach of DCM and MOMBASA [16] are very similar. Nevertheless the focus of DCM is on the design of the multicast routing protocol, whereas MOMBASA stresses the mobility aspects. Moreover in contrast to MOMBASA,

DCM retains the classical IP multicast services model whereas in MOMBASA an extended service model is assumed.

The MOM proposal [24,26] came up with some new ideas to solve bi-directional routing problems. In the MOM protocol, when a Home Agent has more than one MNs residing at the same foreign network subscribed to the same multicast group, only one copy of multicast data is forwarded from HA to FA. The protocol solves the tunnel convergence problem by having the FA assigning a HA as the designated multicast service provider (DMSP) for a given multicast group to forward multicast packets to that FA.

The paper is organized as follows. Next section describes proposed scheme and its assumptions. Followed with test-bed design and specification, Measurement and finally conclusion.

2   Introduction to Hierarchical Design and Multicast Buffering in Mobile IPv6 Wireless Network

Our proposed scheme comes with enhanced Mobile IPv6 (RFC 3775) in micro level handover. This scheme integrates hierarchical concept [3,9] and multicast function [9,15,23,20]. We use hierarchical design to totally shield the micro mobility from macro mobility in order to reduce location update signal and signaling traffic within micro level network while multicast is used to send packets to MN through base stations that near to MN. That will reduce handover delay that causes packet loss when MN is roaming.

When MN roams to new foreign network, it is assigned a unique and global multicast address (care-of-address) [9,15,23,20]. MN will update CN and HA with Binding Update (BU) and will receive Binding Acknowledgement (BA) from them. A multicast group based on MN's global multicast address is being formed and Base Station (BS) that serves MN will invite BS near by to join this multicast group so that they can receive and buffer the same multicast packets that are ready for MN. Buffed packets will be deleted when lifetime is expired. When MN roams to new BS, it will send Handover Initiations (HI) with last packet sequence number that was received at previous BS, new BS will send Handover Acknowledgement (HACK) and forward the new multicast packet sequence to MN. MN continues receiving packet when roaming. So, no packet loss and low handover delay is shown. MN's global multicast address and group will remain the same as long as the MN within the same foreign network or different subnet or BS. Theoretically, no location updates signal needed between MN, HA and CN. Figure 1 shows proposed micro-mobility handover scheme.

2.1   Assumptions

1. Each of the network domains will be provided with a unique IPv6 Global Multicast Prefix Address.
2. All the routers must be multicast based, that means they have ability to multicast packets to MN.

Mobile Host (MN) connects to Access Point 1
(AP1) (Macro-Mobility) and forms a new global multicast care of address.

AP1 requests Access Point 2 (AP2) to join multicast group based on MN's global multicast care of
address.

MN sends Binding Update to HA and CH. CN
will send packet to MN's new care of address

Foreign Multicast Router will multicast packet
to both AP.

Both AP buffer the packets but AP1 forwards the
traffic to MN because MN associates with it.

When MN moves to AP2(Micro-Mobility), MN
sends notification together with last packet id to AP2.

AP2 will start forwarding new packet to MN
while AP1 buffers the packets for MN.

Fig. 1. Flow chart of proposed micro-mobility handover scheme

3. Here, the proposed scheme is based on Mobile IPv6 (RFC3775) for micro-
    mobility management while Mobile IPv6 for macro-mobility management
4. MN is able to make sure which new BS will provide a good reception and
    quality of signal to it through the beacon that was received before handover
    occurred.
5. The router must be able to authenticate signals from other routers, home
    agent and MN using IPSec to avoid forged messages.
6. A small program which has Buffering, Forwarding, Packet generator and
    Packet Receiver functions was developed so that the test-bed can run
    smoothly.

3 IPv6 Wireless Test-Bed Design and Specification

We setup a Wireless IPv6 test-bed to test and examine the performance of the proposed handover scheme compared to others related schemes. The test-bed is composed of hardware, software and network analyzes tools to monitor the handover operation. The hardware consists of 4 personal computers which act as multicast router, 2 access point, home agent and correspondent node. We use a notebook as MN. The overall test-bed architecture and design of the proposed scheme are shown in Figure 2. The used components including software and hardware are shown in Table1.

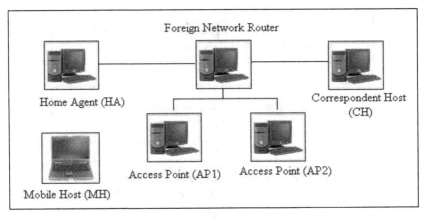

Fig. 2. Test-bed architecture and design

Table 1. Test-bed's components, software and hardware

Test-Bed Components	Software	Hardware
Wireless Home Agent Router-HA	Redhat Linux kernel 2.4.26[27], with Linux Ethernet Bridge 0.8.9[28], Host Access Point (HOSTAP) 0.2.6[29], Mobile IPv6 in Linux (MIPL) 1.1[30], radvd[31]	Personal Computer with Network Card and Wireless Network Card
Multicast Router-Foreign Network	Fedora 3.0 with kernel 2.6.9, Linux-Multicast-Forwarding 0.1c[32], USAGI patch[33]	Personal Computer with 4 Network Card
2 Access Point act as Base Station (AP)	Redhat Linux 9.0 with Kernel 2.4.26 with Bridge 0.8.9, HOSTAP 0.2.6, radvd	2 Personal Computer with 1 Network Card and 1 Wireless Network Card
Correspondent Node (CN)	Redhat Linux 2.4.26, MIPL 1.1	Personal Computer with Network Card
Mobile Node (MN)	Redhat Linux 2.4.26, MIPL 1.1, HOSTAP 0.2.6	Notebook with 1 Wireless Network Card

4 Performance Measurements

Here, the three performance evaluation entities are packet loss, packer delay and handover delay. Packet loss means the total number of packets sent minus the number of packets received by receiver. Handover delay [23] is the difference in time between the arrival of the first new packet from the new access point and the arrival of the last packet from the old access point. Packet delay means the time difference between the packet sent and the packet received. Since we gathered the results from real environment, so wired and wireless media delay were taken into account in our performance evaluation entities.

Prior to the test-bed performance measurement, time synchronization of CN (packet generator) and MN (packet receiver) is done with Network Time Protocol (NTP) [33] to make sure the accuracy if the collected data. First, we analyzed packet loss, packet delay and handover delay in Mobile IPv6 test-bed (conventional Mobile IPv6) as shown in Figure 3. Unicast packets are generated with a small packet generator program [34] (each packet has sequence number and time stamp) from CN to MN. 100-byte-long TCP packets are sent periodically in every 100ms. Essid of MN is changed to force MN to do handover to AP1 then to AP2. Packet receiver program on MN captures all packets destined to MN. Figure 4 shows the average packet delay is 0.7s, the handover delay for the first handover (macro-mobility) is 4.4s (44 packet loss) and 4.2s (42 packet loss) during the second handover (micro-mobility). Apart

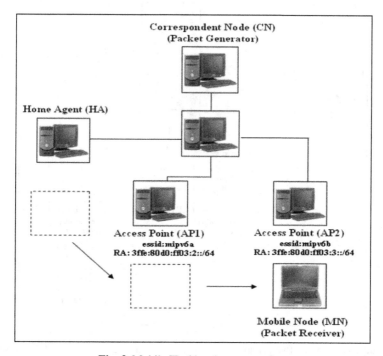

Fig. 3. Mobile IPv6 handover scenario

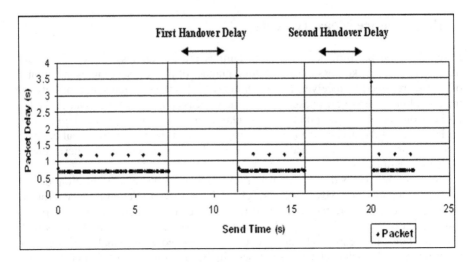

Fig. 4. Mobile IPv6 handover delay

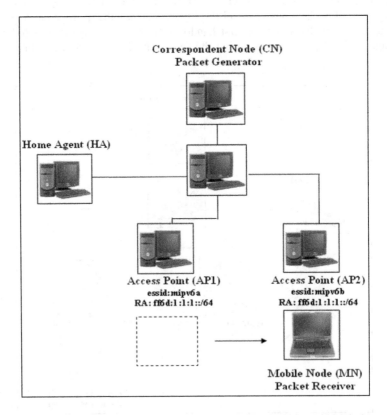

Fig. 5. Enhanced mobile IPv6 handover with multicast function and hierarchical design scenario

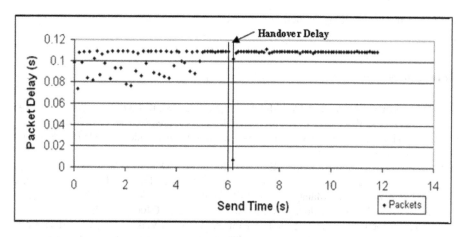

Fig. 6. Enhanced mobile IPv6 handover with multicast function and hierarchical design handover delay

from that, after occurrence of each handover, the packet delay is very high: 3.6s after the first handover and 3.4s after the second handover.

Secondly (as shown in Fig 5), the packet loss, packet delay and handover delay in Mobile IPv6 test-bed with enhanced handover scheme and multicast function is measured. Again, multicast packets were generated with a small packet generator program (each packet has sequence number and time stamp) from Multicast Router to MN through both Access Points. In this case, 100-byte-long UDP packets are sent periodically within 100ms. A buffer program in both access points stored all packets belonged to MN and forwarded them in time when MN associated with it. At the beginning, MN received multicast packet at AP1, later, essid of MN is changed to force it roam from AP1 to AP2. As MN roamed to AP2, it sent the last packet sequence number to AP2, AP2 forwards the buffered packets starting from the last packet received at AP2. Packet receiver program at MN captured all received packets. At this point, a graph on packet loss, packet delay and handover delay was plotted as shown in Figure 6 From the graph, we can see that the average packet delay is 0.109s; the handover delay is 0.20s with not packet loss.

By comparing the results in Figure 4 and Figure 6, it can be conclude that the proposed handover scheme reduces the handover delay from 4.4s to 0.2s, packet delay from 0.7s to 0.109s and from 44 packet loss to zero packet loss during handover. Hence, the efficiency of the proposed scheme is apparent.

5 Conclusion

This research proposes an enhanced micro mobility handover scheme and the development of a test-bed for evaluating the performance and effectiveness of the proposed scheme over the existing one. Based on the obtained results we conclude that our solution provides an effective improvement of handover performance in micro-mobility environment for Mobile IPv6. That gives the advantages of zero packet loss and lower handover delay, which can be used to improve micro mobility on 3G, 4G and beyond.

References

[1] D.Johnson, C.Perkins, J. Arkko, Mobility Support in IPv6, RFC3775, Jun 2004.
[2] Uyless Black, Voice over IP (Upper Saddle River, N.J. Prentice Hall PTR, 2000).
[3] Indra Vivaldi, mproved Handover Routing Scheme In Hierarchical Mobile IPv6 Network, Master Thesis, May 2003.
[4] C.Perkins, IP Mobility Support, IETF RFC 2002, Okt 1996.
[5] Hesham Soliman, Claude Castelluccia, Karim El-Malki, Ludovic Bellier, Hierarchical Mobile IPv6 Mobility Management (HMIPv6), IETF RFC 4140, August 2005
[6] Rajeev Koodli, Fast Handovers for Mobile IPv6, IETF RFC4068, July 2005
[7] A.T.Campbell, J.Gomez, C-Y.Wan, S.Kim, Z.Turanyi, A.Valko, Cellular IP, IETF Internet Draft <draft-ietf-mobileip-cellularip-00.txt>, December 1999, Work in Progress
[8] R.Ramjee, T.La Porta, S.Thuel, K.Varadham, L.Salgarelli, IP micro-mobility support using HAWAII, IETF Internet Draft <draft-ietf-mobileip-hawaii-00.txt>, June 1999, Work in Progress
[9] Cheng Lin Tan, Stephen Pink, Kin Mun Lye, A Fast Handoff Scheme for Wireless Network, 2nd ACM international workshop on Wireless mobile multimedia, Seattle, Washington, United States, 83 – 90, 1999
[10] S.Thomson and T.Narten, IPv6 Stateless Address Autoconfiguration, IETF RFC 2462, Dec 1998.
[11] J.Bound et al. Dynamic Host Configuration Protocol for IPv6 (DHCPv6), IETF RFC 3315, July 2003
[12] Janne Lundberg, An Analysis of the Fast Handovers for Mobile IPv6 Protocol, Mobile networks based on IP protocols and unlicensed radio spectrum Seminar on Internetworking, Espoo, May 27, 2003
[13] Subir Das, Archan Misra, Prathima Agrawal, TeleMIP:Telecommunication-Enhanced Mobile IP Architecture for Fast Intradomain Mobility, Personal Communications, IEEE, Volume: 7, Issue: 4, Aug. 2000, Pages:50 - 58
[14] Keeton, M: et al, Providing Connection-oriented Network Services to MNs, Proc. Of the USENIX Symposium, Oct 93
[15] Mysore, J: et al, A New Multicasting-based Architecture for Internet Host Mobility. Proceedings of ACM Mobicom'97, Oct 97
[16] A.Festag, A.Wolisz, MOMBASA:Mobility Support-A Multicast-based Approach, Proceedings of European Wireless 2000 together with ECRR 2000, September 2000, Dres den, Germany
[17] A.Achatya and B.Badrinath, Delivering multicast messages in networks with mobile hosts, in Proc. 13th.Conf.Distributed Computing Systems, Pittsburgh, PA(May 1993), pp.292-299.
[18] A.Achatya A.Bakre and B.Badrinath, IP multicast extensions for mobile internetworking, in Proc.IEEE INFOCOM'96, San Francisco, CA (March 1996)
[19] D. Waitzman, C Patridge and S.Deering, eds., Distance vector multicast routing protocol, RFC 1075, BBN STC and Stanford University (Oct 1988)
[20] J. Ioannudis, D.Duchamp. and G.Q.Maguire Jr. IP-based Protocols for Mobile Internetworking, Proceedings of ACM SIGCOMM, Sept 1991

[21] Ahmed Helmy Multicast Based Architecture for IP Mobility: Simulation Analysis and comparison with Basic Mobile IP Networked Group Communication, Proceedings of NGC 2000 on Networked group communication

[22] Blazevic,L.; et al.: Distributed Core Multicast (DCM) : A Routing Protocol for IP with Application to Host Mobility. ACM SIGCOMM Computer Communication Review, Vol.29, No.5, Sept.99

[23] Stemm, M.; et al.: Vertical Handoffs in Wireless Overlay Network. ACM Mobile Networking (MONET), Special Issue on Mobile Networking in the Internet, Winter 1998

[24] Sahar Al-Talib, Multicast-Based Mobile IPv6 Join/Leave Mechanism Software Master Thesis, University Putra Malaysia, NOV 2002

[25] Vineet Chikarmane, Carey L. Williamson, Richard B. Bunt and Wayne L.Mackrell Multicast Support for Mobile Host using Mobile IP: Design Issues and Proposed Architecture, Baltzer Science Publishers 1998

[26] T.G.Harrison, C.L.Williamsom, W.L. Mackrell and R.B.Bunt, Mobile Multicast (MOM) protocol: Multicast support for mobile hosts, in: Proc.of ACM/IEEE mobiCom (Sept 1997)

[27] Redhat Linux or Fedora Core 3.0, http://www.redhat.com

[28] Linux Ethernet Bridge http://bridge.sourceforge.net

[29] Host AP driver for Intersil Prism2/2.5/3 http://hostap.epitest.fi

[30] Mobile IPv6 for Linux, http://www.mobile-ipv6.org/

[31] Linux IPv6 Router Advertisement Daemon http://v6web.litech.org/radvd

[32] Linux IPv6 Multicat Forwarding http://clarinet.u-strasbg.fr/~hoerdt/ linux_ipv6_ mforwarding/

[33] USAGI Project, http://www.linux-ipv6.org/

[34] The Network Time Protocol, http://www.ntp.org/

[35] Zoltan Lajos Kis, Zsolt Kovacshazi, Peter Kersch, Csaba Simon, Adaptation of IPv6 Multicast Protocols to heterogeneous mobile networks, EURICE 2004

Prioritisation of Data Partitioned MPEG-4 Video over GPRS/EGPRS Mobile Networks

Mehdi Jafari[1] and Shohreh Kasaei[2]

[1] University of Kerman, Iran
[2] Sharif University, Iran

Abstract. With the advance of multimedia systems and wireless mobile communications, there has been a growing need to support multimedia services (such as mobile teleconferencing, mobile TV, telemedicine, and distance learning) using mobile multimedia technologies. Despite the research done in the field of mobile multimedia, delivery of real-time interactive video over noisy wireless channels is still a challenge for researchers. This paper presents a method for prioritising data partitioned MPEG-4 video in a way suitable for transmission over a mobile network. The effectiveness of the technique is demonstrated by examining its performance when the transport of the prioritized video streams can be accomplished using packet switching technology over the enhanced general packet radio service access network infrastructure.

Keywords: GPRS, EGPRS, MPEG-4, Prioritisation, Error resilience.

1 Introduction

The *third generation* (3G) wireless mobile networks will be able to support a greatly enhanced range of services compared to those available on the current second generation mobile networks. In addition to conventional voice communication services (provided by the second generation *global mobile system* (GSM)[1] networks), the 3G mobile networks will support a high data rate transmission that will enable the support of a wide range of real-time mobile multimedia services (including combinations of video, speech/audio and data/text streams) with the *quality of service* (QoS) control. With 3G, mobile users will have the ability to remotely connect to the Internet while retaining access to all its facilities (such as e-mail and Web browsing sessions). Mobile terminals will be enabled to access remote websites and multimedia-rich databases with entering the Web browsers of these terminals.

All mentioned services require a real-time transmission of video data over fixed and mobile networks with varying bandwidth and error rate characteristics. Due to the huge bandwidth requirements of raw video signals, they must be compressed before transmission in order to optimize the required bandwidth to provide a multimedia service.

Image and video coding technology has witnessed an evolution, from the first generation canonical pixel-based coders to the second-generation segmentation-, fractal-, and model-based coders to the most recent third-generation content-based

K. Cho and P. Jacquet (Eds.): AINTEC 2005, LNCS 3837, pp. 68 – 82, 2005.

coders. Both the ITU and *international organization for standardization* (ISO) [2] have released standards for still image and video coding algorithms that employ waveform-based compression techniques to trade-off between the compression efficiency and the quality of reconstructed signals. After the release of the first still image coding standard (namely, JPEG in 1991), ITU recommended the standardization of its first video compression algorithm (namely, ITU H.261) for low bit rate communications over the ISDN, in 1993. Intensive work has since been carried out to develop improved versions of this ITU standard, and this has culminated in a number of video coding standards (namely, MPEG-1 (1991) for audiovisual data storage on CD-ROM, MPEG-2 (or ITU-T H.262, 1995) for HDTV applications, and ITU H.263 (1998) for very low bit rate communications over the PSTN networks). The first content-based object-oriented audiovisual compression algorithm (namely, MPEG-4 (1999), for multimedia communications over mobile networks) was developed. In this paper, the MPEG-4[2, 3] data partitioning property is used for video error resilience over error–prone mobile communication channels.

Since the MPEG-4 coded video data is highly sensitive to information loss and channel bit errors, the decoded video quality is bound to suffer dramatically at high channel bit error ratios. Perhaps the most significant problem is the effect of high error channel conditions experienced by mobile users. This quality degradation is unacceptable when no error control mechanism is employed to protect coded video data against the hostility of error-prone environments. In certain cases, these streams are required to travel across a number of asymmetric networks until they get to their final destinations.

Consequently, the coded video bit streams have to be transmitted in the form of packets whose structure and size depend on the underlying transport protocols. During transmission, these packets and the enclosed video payloads are exposed to channel errors and excessive delays; and hence to information loss. Lost packets impair the reconstructed picture quality if the video decoder does not take any action to remedy the resulting information loss.

The effects of a bit error on the decoded video quality can be categorized into three different classes as follows.

1. A single bit error on one video parameter does not   have any influence on segments of video data other than the damaged parameter itself. In other words, the error is limited in this case to a single *macro block* (MB) that does not take part in any further prediction processes. (One example of this category is encountered when an error hits a fixed-length intra DC coefficient of a certain MB which is not used in the coder motion prediction process.) Since the affected MB is not used in any subsequent prediction, the damage will be localized and confined only to the affected MB. This kind of error is the least destructive to the QoS.

2. The second type of error is more problematic because it inflicts an accumulative damage in both time and space due to the prediction process. (One example of this category is encountered when the prediction residual of motion vectors is sent; and hence the bit errors in the motion code words propagate until the end of frame.) Moreover, the error propagates to subsequent inter coded frames due to the temporal dependency induced by the motion compensation process. This effect can be

mitigated if the actual *motion vectors* (MVs) are encoded instead of the prediction residual.

3. The worst effect of bit errors occurs when the synchronization is lost and the decoder is no longer able to figure out that the received information belongs to which part of the frame. This category of error is caused by the bit rate variability characteristics. When the decoder detects an error in a variable length code word, it skips all the forthcoming bits (regardless of their correctness) when searching for the first error-free synch word to recover the state of synchronization. This paper is focused on the two last error effects and presents some error resilience techniques based on *unequal error protection* (UEP) for MPEG-4 coding over the GPRS/EGPRS mobile networks.

Since real-time video transmissions are sensitive to time delays, the issue of re-transmitting the erroneous video data is impractical. Therefore, other forms of error control strategies must be employed to mitigate the effects of errors inflicted on coded video streams during transmission. Some of these error control schemes employ data recovery techniques that enable decoders to conceal the effects of errors by predicting the lost (or corrupted) video data from previously reconstructed error-free information. These techniques are decoder-based and incur no changes on the employed transport technologies. Moreover, they do not merge any redundancy on compressed video streams and thus are referred to as zero redundancy error concealment techniques. Concealment at the decoder is based on exploiting the spatial and temporal data redundancy to obtain an estimate of the lost texture data. The efficiency of error concealment depends on the amount of redundancy in texture and compressed bit streams that are not removed by the source encoder. To be more specific, error resilience for compressed video can be achieved through the addition of suitable transport and error concealment methods, as outlined in the system block diagram shown in Figure 1.

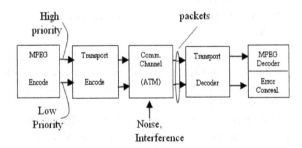

Fig. 1. Block diagram of a visual communication system

In this paper, in addition to prioritization error control, a simple error concealment technique sets both the MVs and the texture blocks of the concealed inter frames to zero, while it copies the same MB from the previous frame to the current error-concealed intra frames.

Another error control scheme used for MPEG-4 allows multi-layered video coding. The compressed bit stream of each *video object plane* (VOP) in the video sequence consists of a number of layers; namely the base layer and a number of enhancement layers. The base layer contains information that is essential for texture reconstruction, while the enhancement layers contribute to improve the perceptual quality at the expense of additional overhead bits. The compression ratio of enhancement layers is a compromise between the coding efficiency and video quality. If the channel can handle high bit rates, then more enhancement layers can be accommodated to improve the output quality. Conversely, in situations such as congestion of network links, only the base layer is transmitted to avoid traffic explosion and to guarantee the maximum possible video quality. In addition to its scalability benefits, the layered video coding has inherent error-resilience benefits; particularly when the base layer can be transmitted with higher priority and the enhancement layers with lower priority. The layered video coding is usually accompanied by the use of unequal error protection to enable the high-priority base layer to achieve a guaranteed QoS and the enhancement layers to produce quality refinement. This approach is known as layered coding with transport prioritization, and is used extensively to facilitate error resilience in video transport systems.

A similar method to improve the quality of video transport over networks is the prioritization of different parts of the video bit stream by sending data as different separate streams. This enables the video encoder to demand the network to send the data using channels with different priorities, allocating more important and error-sensitive data to more reliable and secure channels. For object-based coding used in MPEG-4[4], prioritization of objects can be used to improve the system performance. Although the approach to implement the UEP in MPEG-4 described in [4] gives a good improvement in system performance, the implementation of it at the application layer would leave all network and transport layer headers unprotected. Also, the UEP mechanism applied to the prioritized video information does not provide any protection against the packet loss. To overcome the above mentioned shortcomings, this paper presents an efficient use of video prioritization schemes in mobile networks. These schemes makes use of the data partitioning technique employed by MPEG-4 in each corresponding video packet to sent the partitions in two separate video streams and use different GPRS/EGPRS channel protection schemes. This allows the encoder to allocate a higher priority to the more error-sensitive data and transmit it over higher quality and more reliable channels. This paper presents a data partitioning approach for prioritization, using MPEG-4 over a GPRS/EGPRS channel model. Section 2 presents MPEG-4 data packetisation. Section 3 discusses the error sensitivity of different MPEG-4 parameters. Section 4 describes the method of testing over multislot GPRS/EGPRS channels. Finally, Section 5 concludes the paper.

2 MPEG-4 Packetization

MPEG-4 is an open standard of ISO [3]. The most important application of MPEG-4 will be in multimedia environment. The media that can use the coding tools of MPEG-4 includes computer networks, wireless communication networks, and the Internet. Perhaps the most fundamental shift in the MPEG-4 standard has been towards object-based or content-based coding, where a video scene can be handled as a set of foreground and background objects rather than just as a series of rectangular frames. This type of coding opens up a wide range of possibilities (such as independent coding of different objects in a scene, reuse of scene components, compositing, and a high degree of interactivities). Unlike block-based video coders (such as MPEG-1,-2, H.261, and H.264), MPEG-4 detects entities in the video frame that the user can access and manipulate, hence providing the user with content-based functionalities for the processing and compression of any video scene. MPEG-4 defines a syntactic description language to describe the exact binary syntax of an audio-visual object bitstream, as well as that of the scene description information. A typical video syntax hierarchy is shown in figure 2.

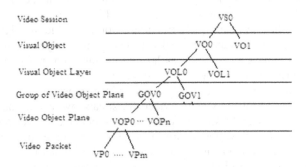

Fig. 2. MPEG-4 video syntax hierarchy

The *video session* (VS) is the highest syntactic structure of the coded video bitstream. A video sssion is a collection of one or more *visual objects* (VOs). The VO header information contains the start code followed by profile and level identification and a VO identification to indicate the type of object, which may be a still texture object, a mesh object, or a face object. A VO can consist of one or more *visual object layers* (VOLs). In the VOL, the VO can be coded with spatial or temporal scalability. *Group of Video* (GOV) is like the *group of picture* GOP [4] in MPEG-1 and -2, and *visual object plane* (VOP) is a video frame.

As shown in Figure 3, a VOP (or video frame) consists of a VOP header and several *video packets* (VPs). The VOP and VP headers contain the synchronization code and compression parameters. Each video frames starts with a start code. There are a number of start codes defined by MPEG-4 to make the decoding process clear and efficient [5]. Start codes are unique combination of bits that never occur in the video data. Each start code consists of a start code prefix followed by a start code value. The start code prefix is a string of twenty-three bits with the value zero followed by a single bit with the value one. The start code value is an eight bit integer, which identifies the type of start code. The VOP time parameter represents the number of the seconds elapsed since synchronization point marked by time stamp of the previously decoded *intra-* (I-) or *predictive-* (P-) VOP, in the decoding order. After the time parameter, the VOP quantization is added [6],[8]. Apart from the start code, each video frame contains resynchronization markers at the boundaries of video packets. Also, data partitioning in MPEG-4 divides the VP into two parts. The header, motion, and shape data are coded in the first partition, while the less important texture information is placed in the second partition [8]. A VP header consists of *variable length coded* (VLC) macro block number, quantization scale parameter, and an optical *header extension code* (HEC) as shown in Figure 3(b). Each VP is partitioned into two portions separated by a DC-marker (in case of I-VOPs) or a motion marker (in case of P-VOPs). The MPEG-4 in addition to coding the texture and motion information traditionally encountered in block-based video coders, codes the shape of each VOP, as illustrated in Figure 3(b), so that the composition of objects can be done at the end decoder.

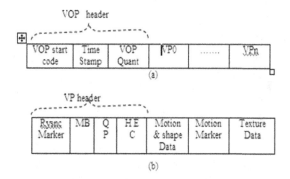

Fig. 3. MPEG-4 Structure, (a) Video object plane (VOP) and (b) Video packet (VP)

Different partitions of MPEG-4 packets have variable sensitivity with respect to errors. The worst effect of bit errors occurs when the synchronization is lost. Also motion vectors and shapes are more sensitive to errors than texture data. Next section describes the sensitivity of different partitions and the proposed prioritized technique for video error resilience.

3   Error Sensitivity of MPEG-4 Parameters

Errors have a detrimental effect on the decoded video quality due to spatial and temporal inter-dependencies of video data. Similar to the block-based video coders, the effects of errors on object-oriented compressed video streams (such as MPEG-4) depended on the type of the corrupted video parameter and the sensitivity of this parameter to errors. However, object-based video coded streams contain synchronisation word, motion vectors, shape data, and texture; within which the synchronisation lost has the worst effect of bit errors occurrence, then the MV and shape are more sensitive with respect to errors. The errors are described below.

Synchronisation word error: This category of error is caused by the bit rate variability characteristic. When the decoder detects an error in a VLC word, it skips all the forthcoming bits, regardless of their correctness, when searching for the first error-free synchronization word to recover the state of synchronization. Therefore, the corruption of a single bit is transformed into a burst of channel errors. The most basic error resilience option places the coded video data into regular sized packets, with resynchronization words between each packet. As such, any single packet can be decoded independently of all of the other packets within that frame. Insertion of resynchronization words at regular bit intervals eliminates error propagation in coding (such as in H.263), where insertion of such words is performed at regular texture block intervals (such as in MPEG-4). Figure 4 demonstrates the impact of synchronisation of an H.263 decoder on reconstructed video quality (in mean square sense) for *walking person* sequences.

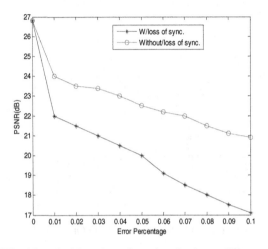

Fig. 4. PSNR with and without loss of synchronization at different error rates

	MV2	MV3
MV1	MV	

Fig. 5. Neighboring MVs to predict current MV

MV: Similar options exist to the intra-, predictive-, and *bidirectional-* (B) frames in MPEG-1 and MPEG-2. The I-VOP is encoded without any motion compensation process. The P-VOP is predicted using MVs from a past I- or P-VOP. The B-VOP is predicted using MVs from a past and future I- or P-frame. When the inter mode is chosen, the *differentially coded motion vectors* (MVDs) are to be transmitted. The horizontal and vertical components of the MV are coded differentially by using three neighboring MV candidate predictors as shown in Figure 5. The formula is given as:

$$
\begin{aligned}
MVDx &= MVx - Px \\
MVDy &= MVy - Py \\
Px &= Median\ (MV\ 1x, MV\ 2x, MV\ 3x) \\
Py &= Median\ (MV\ 1y, MV\ 2y, MV\ 3y)
\end{aligned}
\tag{1}
$$

The object-based motion prediction process in block-based video coding algorithms makes the compressed bit stream more sensitive to errors and information loss. For instance, a bit error in differential MV of a MB or one of its candidate MV predictors can lead to incorrect reconstruction of the MB. The reason is that the decoder becomes unable to compensate for the motion of the currently processed MB with respect to its best-match matrix in the reference frame. An erroneous MV leads to incorrect reconstruction of its corresponding MB and other MBs, whose MV depend on the erroneous MV as a candidate predictor. This MV dependency is the main reason for video quality degradation in both spatial and temporal domains. This effect can be mitigated if the actual MV are encoded instead of the prediction residual. In Figure 6, for *walking person* sequence encoded at 30 kbit/s, the quality of reconstructed frames using MVD is compared with that of actual MV value.

Shape coding: Shape coding (used in MPEG-4) is required to specify the boundaries of each non-rectangular object in a scene. Shape information may be in binary or gray scale format. Gray scale shape information has values in the range 0 to 255 (that are compressed using block-based DCT and motion compensation in MPEG-4, therefore are high sensitive in error-prone environment.). Although gray scale information is more complex and requires more bits to be encoded, it introduces the possibility of overlapping semi-transparent VOPs.

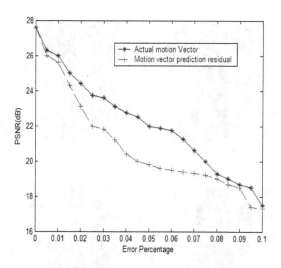

Fig. 6. PSNR values at different error rates with and without motion vector prediction

Texture: Frame intensity values and motion-compensated residual values within a VOP are coded as texture. The basic tools for coding the texture are similar to the MPEG-1 and MPEG-2. Thus, they are transformed using the DCT, and then scalar quantized and entropy coded by a VLC. Usually, the majority data of each packet is made up of texture information; the loss of which causes much less distortion at decoder than if motion or header data is lost. It is obvious that MVs and shapes are more sensitive to errors than texture data. MPEG-4 data partitioning succeeds in ensuring that much of the packet data is not very sensitive to error. It also places the most sensitive data at the beginning of the packet. Here, *walking person* sequence is used to analyze the error sensitivity of data in the first and second partitions of an MPEG-4 video packets. It is a CIF formatted 30 fps sequence in which the person moves quickly and the camera follows him. At the decoder, a simple error concealment technique sets both MVs and texture blocks of the concealed, I-frames to zero, while it copies the same MB from the previous frame to the current error-concealed I-frames. Figure 7 shows that while texture errors can be concealed with reasonable efficacy, the concealment of motion and shape data results in reconstructed frames that contain a high degree of distortion. The subjective results shown in Figure 7 confirm the above mentioned error sensitivities. These errors are demonstrated by the PSNR values of Figure 8. Corruption of texture produces little effect in terms of visible distortion until the bit stream is subjected to high error rates. On the other hand, shape data proves to be highly sensitive (as corruption of shape in the sequence leads to perceptually unacceptable quality).

In this work we use data partitioning prioritisation using MPEG-4 by which the error resilience is improved by prioritisation of different parts of the video bitstream by sending the data as two separate streams. Section 4 describes the proposed prioritization method using MPEG-4 over a GPRS/EGPRS channel model.

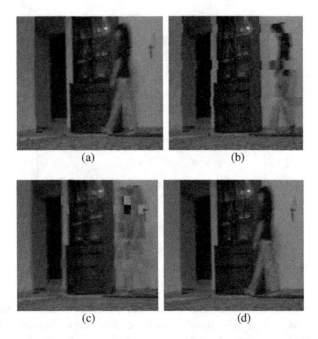

Fig. 7. (a) Error free sequence, (b) Shape data, (c) Motion data, (d) texture data , all corrupted at BER=5×10^{-4}

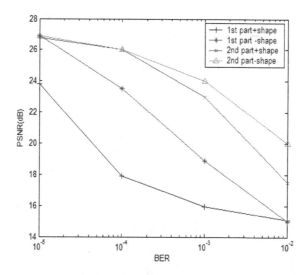

Fig. 8. Sensitivity to errors of MPEG-4 video parameters generated by *walking person* sequence, with corruption of the first and second partitions with and without shape information

4 MPEG-4 Prioritisation over GPRS/EGPRS Channel Model

The 2G mobile cellular networks, namely GSM, do not provide sufficient capabilities for routing of packet data. In order to support packet data transmission and allow the operator to offer efficient radio access to external IP-based networks (such as the internet and corporate intranets), the *general packet radio service* (GPRS) has been developed by the *European telecommunication standards institute* (ETIS) and added to the GSM. The GPRS permits packet mode data transmission and reception based on IP technology. For data packet transmission in the GPRS network, at the time the session is set up, the mobile terminal is identified by an IP address assigned to it either permanently or dynamically. In addition to the packet structure, the QoS of video communications over the future mobile networks depends on a number of other parameters, namely the available throughput and the employed channel coding schemes. For example, the GPRS data is transmitted over the *packet data traffic channel* (PDTCH) after being error-protected using one of four possible channel protection schemes, namely CS-1, CS-2, CS-3, and CS-4. Table 1 shows the data rates per timeslot for each of the four channel protection schemes. The total throughput, for all combinations of timeslots (TS) and channel coding schemes allowed by GPRS, is depicted in table 2. The first three coding schemes (CS) use convolutional codes and block check sequences of different strengths to produce different protection rates. The CS-2 and CS-3 use punctured versions of the CS-1 code, thereby allowing for a greater user payload at the expense of reduced performance in error-prone environments. However, the CS-4 only provides error detection functionality and is therefore not suitable for video transmission purposes. The four channel coding schemes supported by the GPRS can be used to offer different levels of protection for separate video streams.

Table 1. PRS data rates per timeslot for each of the four channel protection schemes

Scheme	Code rate	Radio block payload (bits)	Data rate/slot (kbit/s)
CS1	½	181	8.0
CS2	2/3	268	12.35
CS3	3/4	312	14.25
CS4	1	428	20.35

Table 2. Video source throughput In KBIT/S for all GPRS timeslot/CS combinations

Scheme	1 TS	2 TS	3 TS	4 TS	5 TS	6 TS	7 TS	8 TS
CS1	6.8	13.6	20.4	27.2	34	40.8	47.6	54.4
CS2	10.5	21	31.5	42	52.2	63	73.5	84
CS3	12.2	24.4	36.6	48.8	61	73.2	85.4	97.6
CS4	17.2	34.4	51.6	68.8	86	103.2	120.4	137.6

The GPRS service introduced in the GSM system is an intermediate step towards the third-generation UMTS network. EGPRS (Enhanced GPRS) is an enhanced version of GPRS that allows for a considerable increase in throughput availability to a single user given enough traffic availability from active sources and benign interface conditions. This implies that EGPRS can provide video services with higher data rates than is possible with GPRS. Like GPRS, EGPRS supports its own nine joint modulation-coding schemes which are referred to as MCS-1 to MCS-9. One vital difference between the coding schemes used in EGPRS and those employed by the GPRS PDTCHs is that the ratio block headers are encoded separately from the data payload. The data rates allowed per timeslot and presented to the LLC layer for each of the MCS schemes employed by EGPRS are depicted in table 3. Using the multislotting capabilities of the radio interface, the video source can have multiples of these data rates, as shown in table 4. This reflects the large spread the values of

Table 3. GPRS data rates allowed per timeslot for each of the nine channelprotection

Scheme	Code rate	Radio block payload (bits)	Data rate/slot (kbit/s)	Header code rate
MCS-1	0.53	176	8.8	0.53
MCS-2	0.66	224	11.2	0.53
MCS-3	0.8	296	14.8	0.53
MCS-4	1.0	352	17.6	0.53
MCS-5	0.37	448	22.4	1/3
MCS-6	0.49	592	29.6	1/3
MCS-7	0.76	2*448	44.8	0.36
MCS-8	0.92	2*544	54.4	0.36
MCS-9	1.0	2*592	59.2	0.36

available throughput for video services over EGPRS. The choice of a suitable CS-TS combination for video services over mobile networks depends highly on the activity of the video source and error characteristics of the radio network.

Table 4. Video source throughput in kbit/s fo all egprs ts/mcs combinations

Scheme	1 TS	2 TS	3 TS	4 TS	5 TS	6 TS	7 TS	8 TS
MCS-1	7.5	15	22.5	30	37.5	45	52.5	60
MCS-2	9.6	19.2	28.8	38.4	48	57.6	67.2	76.8
MCS-3	12.6	25.2	37.8	50.4	63	75.6	88.2	100.8
MCS-4	15	30	45	60	75	90	105	120
MCS-5	19	38	57	76	95	114	133	152
MCS-6	25.2	50.4	75.6	100.8	126	151.2	176.4	201.6
MCS-7	38	76	114	152	190	228	266	304
MCS-8	46.2	92.4	138.6	184.8	231	277.2	323.4	369.6
MCS-9	50.31	100.6	150.9	201.2	251.5	301.8	352.1	402.4

In this work, the stream prioritization scheme described above was tested on simulated GPRS (EGPRS) mobile access channel, so that each stream will be transported over a different mobile bearer channel as offered by the underlying network. The simulated scenario is shown in Figure 9, in which the video frame is MPEG-4 encoded and separated into two segments, with high- and low-priority. Then, the video data are produced by the encoder and the RTP/UDP/IP protocols are encapsulated to them. They are then passed through different schemes of wireless GPRS (EGPRS) links and sends to the end system. The first and second partitions are streamed using the CS-1 (MCS-2) and CS-3 (MCS-5), respectively. At the end system the two different radio bearer channels are decoded and combined to reconstruct the data.

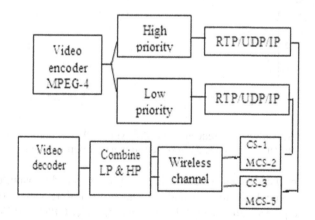

Fig. 9. Tested scenario

The comparison is made among partitioning the proposed method using CS-1(MCS-2), CS-3(MCS-5), and single streaming over CS-2(MCS-3). For real-time

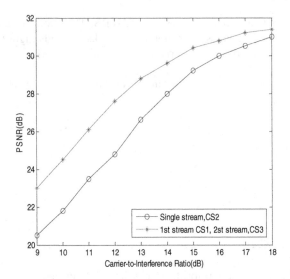

Fig. 10. PSNR for *walking person* sequence encoded with MPEG-4 translated over a GPRS channel using both the single stream and prioritized stream transport mechanisms

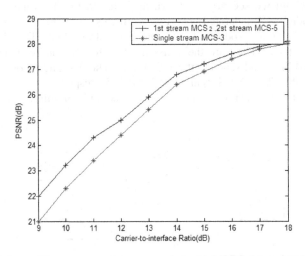

Fig. 11. PSNR for *walking person* sequence encoded with MPEG-4 translated over a EGPRS channel using both the single stream and prioritized stream transport mechanisms

operation, video packets are encapsulated into RTP packets and transmitted using the UDP transport protocol. Figure 10 and 11 show the PSNR values obtained for *walking person* sequence coded with MPEG-4 and sent over GPRS and EGPRS using the prioritized video transport technique described above, respectively. Here, the two data-partitioned output streams are generated using *mobile multimedia systems* (MoMuSys) and MPEG-4 verification model software. They are subsequently encapsulated into RTP packets for real-time transmission. The CS-1 (MCS-2) is used to protect the high-priority video stream and CS-3 (MCS-5) is used to protect the low-priority stream for the prioritized video transport of partitioned video data. However, only the CS-2 (MCS-3) is used to protect the single stream output with no prioritization. The PSNR values (see Figure 10 and 11) show that the prioritization of video streams for UEP protection and transport over two GPRS/EGPRS radio bearer channels offers a better objective performance than the single stream case with EEP and no transport prioritization. Figure 12 and 13 show the objective quality resulted from frame 71, for single and prioritized stream transmissions over GPRS/EGPRS.

(a)   (b)

Fig. 12. Frame 71 of *walking person*. (a) Single stream transmission. (b) Prioritized stream transmission over GPRS.

(a)   (b)

Fig. 13. Frame 71 of *walking person*. (a) Single stream transmission. (b) Prioritized stream transmission over EGPRS.

The PSNR values shown in Figure 10 and 11, demonstrates the effectiveness of the proposed error resilience tools in improving the quality of a video service over the GPRS/EGPRS.

5   Conclusion

In this paper, we proposed a prioritized streaming technique for video translation over simulated GPRS/EGPRS mobile networks. Subjective and objective results have been analyzed. The results have shown that the proposed method provides considerable improvement in QoS when transmitting image sequences over GPRS/EGPRS mobile channels.

References

[1] ETSI/SMG, GSM 03.64 1998 "Overall Description of the GPRS Radio Interface Stage 2", V.5.2.0.

[2] MPEG-4 Committee Draft,"Information technology Coding of Audio-Visual Objects: Visual ISO/IEC 144496-2," ISO/IEC JTC/SC29/WG11 N22o2, Tokyo, Mach 1998.

[3] T.Sikora, and Chiariglione,"The MPEG-4 Video Standardand its Potential for Future Multimedia Applications," Proceeding of IEEE, ISCAS conference, Hongkong, June 1997.

[4] A. Puri and A. Eleftherials. MPEG-4: An object-based multimedia coding standard supporting mobile applications. ACM Mobile Network and Applications, Vol. 3,June 1998, pp. 5-32.

[5] W. Rabmer, M. Budagavi, and R. Talluri, proposed extensions to DMIF for supporting unequal error protection of MPEG-4 video over H.324 mobile networkes. DOC M4135. MPEG Atlantic, City meeting, October 1998.

[6] H.Schulzrinne,S.Casner.R.Frederick, and V. acobson. RTP:A Transport Protocol for Real-Time Applications , RFC1889, Audio-Video Transport Working Group.January 1996.

[7] "Coding of Moving Pictures and Audio", ISO/IEC 14496- 2:1999/Arnd.1 :2000(E), 31-01-2000.

[8] Z. Ahmed, S. Worrall, A. H. Sadka, A. Kondoz," A Novel Packetisation Scheme for MPEG-4 over 3- G Wireless System", VIE. 2005, Glasgow, UK, 4-6 April 2005, PP. 309-314.

Load Balancing QoS Multicast Routing Protocol in Mobile Ad Hoc Networks

Mohammed Saghir, Tat Chee Wan, and Rahmat Budiarto

Network Research Group,
School of Computer Science,
University of Science Malaysia,
11800 Penang, Malaysia
mohammed@nrg.cs.usm.my, {tcwan, rahmat}@cs.usm.my

Abstract. Recently, multimedia and group-oriented computing become increasingly popular for the users of ad hoc networks. The study of QoS issues in MANETs is vital for supporting multimedia and real-time applications. MANETs can provide multimedia users with mobility they demand, if efficient QoS multicasting strategies were developed. In this paper, we study QoS requirements, illustrate advantages and limitations of existing QoS routing protocol and propose a QoS Multicast Routing protocol (QMR) with a flexible hybrid scheme for QoS multicast routing. The hybrid scheme contains some mechanisms that provide fix-reservation and shared-reservation bandwidth to guarantee QoS multicast routing. The proposed protocol uses forward nodes to apply QoS multicast routing from source(s) to a group of destinations and support load balancing. Analysis results show the ability of QMR to exploit residual bandwidth efficiently without effect on the reserved bandwidth and provide a balance between performance gains and design complexity.

1 Introduction

In Mobile Ad hoc NETworks (MANETs), all communications are done over wireless media without the help of wired base stations. Distant nodes communicate over multiple hops and nodes must cooperate with each other to provide routing. The challenges in ad hoc networks are attributed to mobility of intermediate nodes, absence of routing infrastructure, low bandwidth, and computational capacity of the nodes.

The Quality of service (QoS) routing in MANETs is difficult because the network topology may change constantly. Another challenge with supporting QoS for real-time applications is associated with the design of a decentralized medium access control (MAC) protocol. Because of these challenges, best effort distributed MAC controllers are widely used in existing wireless ad hoc networks [5]. Because the dynamic topology and limited bandwidth, the design of QoS multicast routing protocols is more complicated than that in traditional networks; so traffic load must be distributed to eliminate hot spots and provide *load balancing* in the network [6].

QoS in MANETs is highly dependent upon routing and MAC layers [13]; delivering packet from source to destination with service quality in MANETs is

K. Cho and P. Jacquet (Eds.): AINTEC 2005, LNCS 3837, pp. 83–97, 2005.
© Springer-Verlag Berlin Heidelberg 2005

intrinsically linked to the performance of the dynamic routing protocol [17]. Multicast routing is more efficient in MANETs because it is inherently ready for multicast due to their broadcast nature which avoids duplicate transmission. Packets are only multiplexed when it is necessary to reach two or more destinations on disjoint paths. This advantage is saving bandwidth and network resources [7].

Several protocols have been developed to perform ad hoc multicast routing, i.e. CAMP, FGMP, ODMRP, and SOM. However, these multicast protocols did not address the QoS aspect of ad hoc communication. There are several studies for unicast routing protocols with QoS in MANETs in the literature [21][1][20] but QoS support for a multicast protocols should be designed differently from unicast protocols. The main difference between QoS unicast routing and QoS multicast routing is related to the resource reservation between a source and its destination(s). QoS multicast routing should provide QoS paths to all destinations, so QoS multicast should cope with large number of destinations and be able to utilize them. A lantern-tree [23] addresses QoS multicast routing, this protocol uses a lantern-tree as a topology for multicast group and CDMA/TDMA model at MAC layer; however lantern-tree takes a long time at startup to find all paths and to share time slots. Although multiple paths have some advantages, i.e. path diversity may provide higher aggregate bandwidth through spatial reuse of the wireless spectrum, the use of a higher number of links creates more contention at MAC layer, and the complexity for maintaining multiple routes is higher [3]. Supporting multiple paths in MANETS is challenging: a set of maximally disjoint paths with uncorrelated links must be created, despite the high complexity of the search algorithm and delay and complexity of packet re-sequencing [18].

Our work in this paper focuses on one critical issue in future MANETS: QoS multicast routing support. We propose a new QoS multicast routing protocol named QMR – it is a mesh-based protocol which is established on-demand to connect group members and provide QoS paths to multicast groups. We define *Forward Nodes* (FNs) as a subset of the network topology that provides at least one path from each source to each destination in the multicast group. The proposed protocol integrates bandwidth reservation functions into a multicast routing protocol and provides bandwidth reservation for multicast group with the assumption that bandwidth information is available from the underlying MAC layer.

This paper is structured as follows: Section 2 describes our QoS multicast routing protocol, while performance evaluation and simulation results are presented in Section 3. Finally, Section 4 concludes this paper and makes mention of future work.

2   QoS Multicast Routing Protocol (QMR)

It has been proven that on demand multicast routing with mesh topology is superior to other strategies such as tree topologies in MANET [25]. Mesh topology provides alternative paths to deal with dynamic topology changes in MANET, so we propose an on demand multicast routing protocol that uses *forward nodes* to apply multicast routing with QoS from source(s) to a group of destinations. In the QMR protocol, we propose a flexible *hybrid* scheme that combines some features from the high quality

QoS of IntServ and service differentiation of DiffServ. Firstly, FN provides IntServ and reserves bandwidth for every source that accepts its *QoS route request* (QREQ); after that, FN provides DiffServ when it receives data packets from other sources (no bandwidth reservation will be done for such sources) if it has extra bandwidth. At this point, usable bandwidth is partitioned into *fix reserved* for sources with QREQ entries, and *shared* for all other sources. Salient features of QMR include efficient use of residual bandwidth, and reduced control overhead by dropping QREQs that were not accepted due to resource constraints, and reduced redundant data transmissions that occur as a result of using the forward group (FG) scheme found in ODMRP [19]. The *hybrid* scheme switches between the use of redundant paths (using FGs) if the network has sufficient bandwidth for all or some sources and the use of unipaths found in other multicast routing protocols if sufficient bandwidth is not available. Integrating the resource reservation protocol into an on-demand routing protocol would be an efficient means to search for a path with the required bandwidth [10]. By considering QoS constraints while performing on demand routing, the QMR protocol is therefore more efficient.

2.1 Session Initiation and Destruction

A node that has data to send starts a session by broadcasting a session initiation as a QREQ with max hop (MH) greater than zero. Intermediate nodes rebroadcast QREQs if they have enough bandwidth and decrement MH. These processes will continue until QREQs arrive at the destinations or MH equals zero. Destination nodes receive QREQs and send route replies (RREPs) to the source. In several applications such as news service and video broadcasting, the destinations are free to join and leave the group dynamically, and source does not need to be completely aware of destination group membership [4]. In QMR, source and destination nodes use soft state for leaving; source and destination nodes can leave session by stopping the periodic transmission of QREQs and RREPs respectively.

2.2 Data Structures

The QMR protocol utilizes the following data structures:

> Message duplication: identifies each received QREQ or data packet (Fig. 1).
> Bandwidth reservation table: identifies each reserved bandwidth, type of reservation (allocate or reserve) and time of reservation (Fig. 2).

Source ID	Packet Sequence Number	Multicast ID

Fig. 1. Multicast message duplication

Source ID	B.W Requirement	Type of Reserve
Time of Reserve		Multicast ID

Fig. 2. Bandwidth reservation

Source ID	Sequence Number	Hop Count	QoS Requirements	Multicast ID

Fig. 3. Format of QREQ packet

Source ID	Sequence Number	Next Hop	Multicast ID

Fig. 4. Format of RREP packet

Source ID	Sequence Number	Data

Fig. 5. Format of data packet

Source ID	Next Hop	Destination ID	Multicast ID

Fig. 6. Format of acknowledgment packet

2.3 Operation

Similar to the operation of on-demand routing protocols, QMR comprises of a request phase and a reply phase. The request phase invokes a route discovery process to find QoS routes to all destinations. The following sections describe the request phase, admission control, reply phase, data forwarding, route recovery, and Bandwidth reservation.

2.3.1 QoS Route Request Phase

This section discusses the QoS route request phase of our protocol. It starts when a source node wishes to join the group; it broadcasts a QREQ to search and discover paths to all destinations in the multicast group. Fig. 3 shows the format of QREQ packet. When an intermediate node receives QREQs from the source, it records the source ID and the sequence number in its duplication table. Intermediate nodes rebroadcast QREQs if three conditions are satisfied: the request is not a duplicate, intermediate node has enough available bandwidth and the max hop is greater than zero. The routing table is updated with the node ID in the QREQ for use as a reverse path to the source. To prevent looping and receiving of multiple copies of the same QREQ, the identification of each received packet must be compared with those stored in its duplication table. Fig. 1 describes the contents of multicast message duplication table.

2.3.2 Admission Control

It has been proven in [24] that using multiple constraints to provide QoS routing is a NP-complete problem. In MANETs, node mobility causes frequent and constant changes in topology, which places a great challenge on the QoS routing protocols. Because of this, we focus on bandwidth as a main component in fulfilling QoS requirements. Admission control will use available bandwidth calculation to determine if QREQs can be accepted. Using distributed admission control at every intermediate node prevents false admission control that occur when multiple sources simultaneously initiate QREQs since QREQs are dropped if there is insufficient bandwidth to support those requests. After a route is discovered, admission control policy should guarantee the minimum bandwidth requested for each source [15].

In QMR, admission control prevents race conditions [8] that appear when multiple reservations happen simultaneously at an intermediate node. Available bandwidth has three states in QMR: free, allocate, reserved. Algorithm 1 explains how QMR prevents race conditions. In our approach, when source updates *forward nodes*, the paths will also be updated. Intermediate nodes will then re-estimate the available bandwidth, so all changes caused by node movement will be taken into consideration.

```
When a node receives a QREQ packet, it checks its
available bandwidth
If (available bandwidth is enough) then
    Node accepts QREQ and rebroadcast it
Else
      If (the type of reserve bandwidth is allocate) then
           If (wait time for RREP is finished) then
                Node free bandwidth;
         Else    drop QREQ packet.
      Else    drop QREQ packet.
```

Algorithm 1. Sub algorithm at intermediate node to prevent race conditions

Most multicast applications have the characteristic that the number of sources is less than the number of destinations. In this situation, source advertising is more efficient than destination advertising [13]. Based on this, we apply source advertising in our protocol. New destinations in QMR do not affect old destinations because FN can forward data packet to any number of destinations. QMR uses distributed admission control at every intermediate node. When the intermediate node receives a QREQ packet, it must calculate its available bandwidth and rebroadcast QREQ packets if it has enough available bandwidth, otherwise it drops the packet to avoid flooding the network with unnecessary packets.

In addition, the intermediate node temporarily updates its available bandwidth with the current QoS conditions before it rebroadcasts the QREQ packet. With this rule, nodes do not accept more traffic than the available bandwidth. Fig. 7 shows QREQ phase with admission control and describes how intermediate nodes behaves when it receives QREQs. The problem with the admission control solutions in most previous studies is that they are one-time procedures performed before the flows start; it does not take into account changes in capacity of the wireless channel after accepting requests. Capacity of wireless channels may change dramatically and available bandwidth estimated before accepting requests will be affected due to dynamic mobility of nodes and due to fading and outside interference [22].

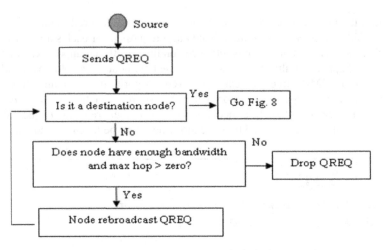

Fig. 7. QoS route request phase and admission control

2.3.3 Reply Phase and Forward Nodes Selection

Initially, a multicast destination initiates the reply phase by choosing the next hop to the source and sending a reply. The destination node will receive QREQs from several nodes. It selects the node ID in the first received QREQ as the next hop to the source and sends route reply (RREP). Fig. 4 shows the format of RREP; reply packets contain one or more replies for one or more sources. When an intermediate node receives a RREP, it checks if the next hop ID in RREP matches its own ID. If it does, the node realizes that it is on the reverse path to the source and it is a part of the *forward nodes*, so it sets the FN flag. The next hop node ID field is filled by

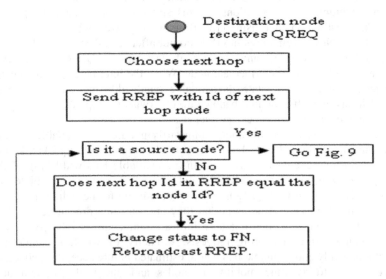

Fig. 8. Reply phase and FN establishment

extracting information from its routing table. In this way, each intermediate node propagates RREPs until they reach the multicast source. An acknowledgment packet is used to guarantee reply arrival. Fig. 6 shows the format of an acknowledgment packet. This whole process constructs *forward nodes* that forward data packets from source to destinations. The source sends data packets if it receives at least one reply from its destinations. Fig. 8 shows route reply phase from destination to source and describes how intermediate nodes behaves when it receives a route reply.

2.3.4 Data Forwarding

Here we consider how data is transmitted through *forward nodes*. When an intermediate node receives data packet, it forwards the packet if it is a FN for the packet source or there is enough residual bandwidth. With FNs, forwarding data packets for other sources does not affect bandwidth that was reserved for other sources, whereas in FG used in ODMRP, forwarding data packets for some sources affects forwarding data packet for other sources. Fig. 5 shows the format of a data packet and Fig. 9 shows the forward data packet phase.

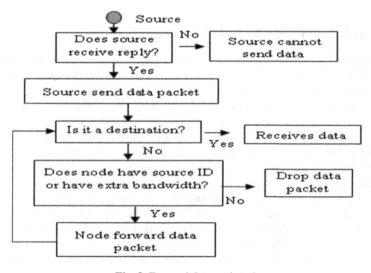

Fig. 9. Forward data packet phase

2.3.5 Route Recovery

As we mention in section 2.3.2 that we use source advertising instead of destination advertising so in our proposed routing protocol, each source periodically sends QREQ that apply route recovery by updating *forward nodes*, similar to the scheme used in ODMRP [19]. The value of the interval for sending QREQs is very important. Several tests were carried out to determine this value, and five seconds was chosen as the interval for sending QREQs since it provides the best tradeoff between throughput and overhead.

2.3.6   Bandwidth Reservation

In a non-QoS scheme, the forward node does not compute its bandwidth before accepting a request packet, so some forwarding nodes become heavily overloaded; as the number of sources grows, requests are accepted without considering the available bandwidth.

In QMR when an intermediate node receives the first QREQ and has enough available bandwidth, it allocates bandwidth for this request and set the time of allocation. When the intermediate node receives a reply (RREP) for the request, it changes the status of the reserved bandwidth from *allocate* to *reserved*. The reserved bandwidth will stay in reserved status until the *forward node* is reset. If an intermediate node receives a second QREQ, it follows Algorithm 1 in Section 2.3.2 for bandwidth reservation. Note that we do not use a timeout timer to change the status of reserved bandwidth from *allocate* to *free* after the waiting time for the corresponding RREP is finished. Instead, the allocated bandwidth is freed when it receives a new QREQ and it needs the allocated bandwidth and the waiting time has finished. This scheme gives RREPs a higher chance of success. In QMR, new QREQs may elect new FNs if the old one changes its location or there is no available bandwidth.

2.4   Using Forward Nodes for Load Balancing

In the proposed protocol, the idea of *forward nodes* is used to support multicast routing with QoS and provide *load balancing*. Considering the QoS support, the bandwidth on a single path might not be enough if there are many sources sending to the multicast group simultaneously. Nodes in single path may be overloaded and many packets may be discarded. We propose a practical solution where the data packets from different sources may go through different paths to the same group. When an intermediate node drops QREQs, it may arrive at destinations through other intermediate nodes that have enough available bandwidth. In addition, congestion prevention and *load balancing* occur as a result of updating *forward nodes*. The number of *forward nodes* will go up or down depending on: number of sources, number of destinations and the state of the network. When network bandwidth is strictly limited, the number of *forward nodes* will be large, whereas when the network bandwidth is sufficient, the number of *forward nodes* will be small. The number of destination nodes will be limited if there is no *forward node* to forward to them.

3   Performance Evaluation

We simulated the proposed protocol in GLOMOSIM [26] and conducted experiments to evaluate the effectiveness of the proposed protocol. The overall goal of our simulation study is to analyze the behavior of our protocol under a range of various mobility scenarios, available bandwidth, and data packet sizes. Our simulations have been run using a MANET with 50-100 nodes moving over a rectangular 1000 m × 1000 m space for over 600 seconds of simulation time. The multicast traffic sources in our simulation are constant bit rate (CBR) traffic. Each traffic source originates 512-byte data packets, using a rate of 2 packets/s. Nodes in our simulation move according to the Random Waypoint mobility model. The range of mobility speed is 0-20 m/s and pause time is equal to 30s. The performance evaluation is an average of the results of 310 different simulations using random seed values from 0-9. In order to observe the behavior of the routing protocol in a simple environment, we considered a scenario with 3 multicast source and 15 multicast destinations in all experiments (assuming that all destinations interesting for receiving from all sources and sources use same bandwidth requirements). Each node has the same transmission range of 250m and raw data rate of 2Mb/s. Note that for a 2Mb/s channel, 0.5 Mb/s is the maximum data rate that can be supported [9], so we assume that the maximum bandwidth available is 0.5Mb/s. The minimum bandwidth requirements are 0.1, 0.2, 0.4 Mb/s which means that any node can accepts requests from 5, 2, or 1 sources respectively. IEEE 802.11 Distributed Control Function (DCF) is used for the MAC layer. In many previous studies that proposed for unicast and multicast routing with QoS, a CDMA/TDMA channel model was used as the MAC layer [21][1][19][23]. It is difficult to realize such centralized MAC schemes in a dynamic wireless environment; whereas IEEE 802.11 is widely used [11]. In addition, CDMA/TDMA is difficult to be implemented in a real network due to issues of code and time synchronization between nodes [16]. Re-assignment of slots after topology changes makes TDMA scheme is very inefficient [2].

The efficiency of our proposed QoS multicast routing protocol is evaluated through the following performance metrics:

> *Success rate (SR):* The Ratio between the number of replies (successful QoS multicast routes) received at the source and the total number of QoS multicast requests which were initiated from source to all destinations.

> *Packet Delivery Ratio:* The Ratio between the number of data packets received at the destinations and the number of data packets that should have been received. This metric indicates the reliability of the proposed protocol.

> *Control Overhead:* Number of transmitted routing control packet (QREQ, RREP, ACK) per data packet delivered. Control packets are counted at each hop.

> *Average Latency:* The average end-to-end delivery latency is computed by subtracting the packet generation time at the source from the packet arrival time at the destinations. This metric indicates the performance of the proposed protocol.

3.1 Performance of Success Rate (SR) vs. Mobility

The simulation results of SR vs. mobility are given in Fig.10. The average SR is obtained by multiplying the number of QREQs sends from source by the number of destinations divided by the number of replies received at the source.

3.1.1 Effect of the Number of Mobile Hosts
Each value in Fig. 10 (a) is obtained by assuming that the minimum bandwidth requirement is 0.2Mb/s, number of mobile hosts is 50, 75, 100 and mobility is

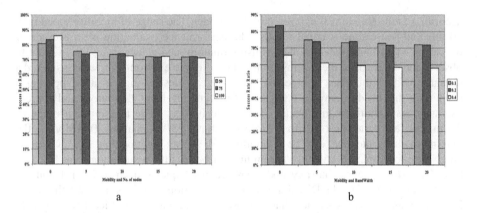

Fig. 10. Performance of SR vs. of (a) the No. of mobile hosts, and (b) the size of the required bandwidth for a given flow

0-20m/s. Fig. 10 (a) shows the success rate of searching for QoS multicast routes vs. various number of mobile hosts. With no mobility, SRs are 80.8%, 83.6%, 86.1% with number of nodes respectively; with 20m/s mobility of mobile host, SRs are 71.6%, 71.9%, 71.03% for 50, 75, 100 nodes. When the number of mobile hosts increase, the SR ratio increases for mobility 0m/s. This is due to the construction of a denser mesh, constituting a larger number of *forward nodes*, which provides more chances for QREQ to be forwarded instead of dropped. The SR ratios remain relatively constant for different mobility speed; although it is lowest than the 0m/s. Periodic sending of requests frees unusable reserved bandwidth and re-estimates the available bandwidth, which gives sources another chance by accepting new requests.

3.1.2 Effect of Size of Bandwidth Requirement
Each value in Fig. 10 (b) is obtained by assuming that the bandwidth requirements are 0.1, 0.2, 0.4 Mb/s, mobility is 0-20m/s and number of mobile host is 75. Fig. 10 (b) shows the SR of searching for QoS multicast route vs. the bandwidth requirement. With no mobility, SRs are 82.6%, 83.6%, 65.7% for 0.1, 0.2, 0.4 Mb/s, with 20m/s mobility, SRs are 72.05%, 71.9%, 58% for 0.1, 0.2, 0.4 Mb/s respectively, for 0.1 and 0.2 Mb/s, the SR remains relatively constant over different mobility speed, however, for 0.4Mb/s, the SR decreases more rapidly with increased mobility speed.

3.2 Performance of Packet Delivery Ratio (PDR) vs. Mobility and Data Size

The performance of PDR vs. mobility is given in Fig. 11. Three kinds of effects are observed.

3.2.1 Effect of the Number of Mobile Hosts
Each value in Fig. 11 (a) is obtained by assuming that the bandwidth requirement is 0.2Mb/s, number of mobile hosts are 50,75, 100 and mobility is 0-20m/s. Fig. 11(a)demonstrates the PDR vs. various number of mobile host. When mobility of host is 0m/s (static), PDRs are 93.4%, 95.1%, 96% with respect to the number of host,

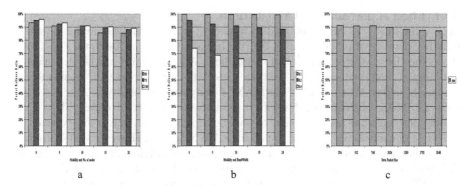

a   b   c

Fig. 11. Performance of PDR vs. effect of (a) No. of mobile hosts, (b) the size of the required bandwidth for a given flow, and (c) the size of data packet

when the mobility of host is 20m/s, PDRs are 85.3%, 88.2%, 89.3% with respect to number of host. When the number of mobile host increases, the PDR increases because the chances increase for data packet to be forwarded instead of being dropped. PDR is still high for high mobility although SR is decrease to about 70% in (Fig. 10 (a)); this is because some *forward nodes* have enough residual bandwidth to forward data packets for other sources. This feature reflects that QMR is able to exploit the residual bandwidth efficiently without effect on reserved bandwidth.

3.2.2 Effect of Size of Bandwidth Requirement

Each value in Fig. 11 (b) is obtained by assuming that the bandwidth requirements are 0.1, 0.2, 0.4 Mb/s, mobility is 0-20m/s and number of mobile hosts is 75. Fig. 11 (b) demonstrates the PDR vs. bandwidth requirement. When mobility of host is 0m/s (static), PDRs are 99.8%, 95%, 73.9% with respect to the minimum bandwidth requirement. When mobility of host is 20m/s, PDRs are 99%, 88.3%, 64% with respect to the bandwidth requirement. The results show that PDR decreases when bandwidth requirements increases dramatically, this is because intermediate nodes do not have enough available bandwidth to accept QREQs and do not have enough residual bandwidth to forward data packets for other sources. Some applications prefer low data rate coding with minimum bandwidth and high packet delivery ratio for transmitting video other than high data rate coding with maximum bandwidth, to avoid congestion in the network and dropping of large number of packets [12].

3.2.3 Effect of the Size of Data Packet

Each value in Fig. 11 (c) is obtained by assuming that the bandwidth requirement is 0.2Mb/s, mobility is 10m/s and No. of mobile hosts is 75. Fig. 11 (c) demonstrates the PDR vs. the size of data packet, the results of PDR is 91.2% with data packet size 256B, 90% with data packet size 1024B and 87% with data packet size 2048B.

3.3 Performance of Control Overhead (OH) vs. Mobility

The control overhead (OH) vs. mobility results is given in Fig. 12. To study the performance of OH, two kinds of effects are analyzed.

3.3.1 Effect of the Number of Mobile Hosts

Each value in Fig 12 (a) is obtained by assuming that the bandwidth requirement is 0.2Mb/s, number of mobile hosts is 50, 75, 100 and mobility is 0-20m/s. Fig. 12 (a) shows the OH vs. various number of mobile host. When topology of MANET is static (mobility is 0m/s), control OHs are 0.454packet, 0.453p, 0.451p with respect to the number of nodes. When topology of MANET is dynamic (mobility is 20m/s) control OHs are 0.536p, 0.462p, 0.495p with respect to the number of nodes. Many QoS routing protocols use multiple paths to provide a high success rate at the expected of having high control overheads as we discussed in Section 1. However QMR protocol is efficient in providing QoS multicast routing protocol with low control overhead. At this time we cannot do a quantitative comparison to other protocols e.g. Lantern-tree-based [23] because we do not have sufficient details to implement and simulate the protocol in GLOMOSIM. Qualitatively, the overhead in Lantern-tree-based protocol increases dramatically when mobility increase, whereas for QMR overhead remains relatively constant for different mobility speed.

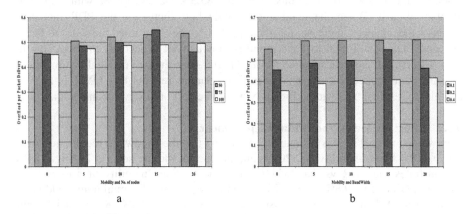

a   b

Fig. 12. Performance of OH vs. effect of (a) No. hosts, (b) size of bandwidth requirement

3.3.2 Effect of Size of Bandwidth Requirement

Each value in Fig. 12 (b) is obtained by assuming that the bandwidth requirements are 0.1, 0.2, 0.4 Mb/s, mobility is 0-20m/s and number of mobile hosts is 75. Fig. 12 (b) shows the OH vs. the bandwidth requirement. With no mobility, the control OHs is 0.552packet, 0.453p, 0.356p with respect to the bandwidth requirements. With 20m/s mobility, the control OHs is 0.595p, 0.462p, 0.417p with respect to the bandwidth requirements.

3.4 Performance of Average Latency (AL) vs. Mobility and Data Size

The performance of AL vs. mobility is given in Fig. 13. Three kinds of effects are discussed.

3.4.1 Effect of the Number of Mobile Hosts

Each value in Fig. 13(a) is obtained by assuming that the bandwidth requirement is 0.2Mb/s, number of mobile hosts is 50, 75, 100 and mobility is 0-20m/s. Fig. 13(a)

shows the AL vs. number of mobile host. When mobility of host is 0m/s, the ALs are 0.059s, 0.050s, 0.044s with respect to the number of nodes. When mobility of host is 20m/s, AL are 0.078s, 0.076s, 0.077s with number of nodes respectively. When mobility increases, the topology changes frequently and packet arrive at destinations through large number of *forward nodes*.

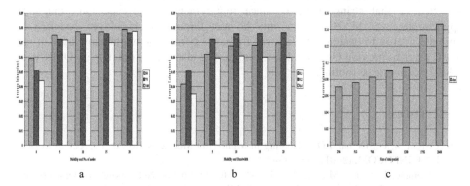

a   b   c

Fig. 13. Performance of AL vs. effect of (a) No. of mobile host, (b) the size of bandwidth requirement, and (c) the size of data packet

3.4.2 Effect of Size of Bandwidth Requirement
Each value in Fig. 13 (b) is obtained by assuming that the bandwidth requirements are 0.1, 0.2, 0.4 Mb/s, mobility is 0-20m/s and number of mobile hosts is 75. Fig. 13 (b) shows the AL vs. the bandwidth requirement. When mobility is 0m/s, the ALs are 0.041s, 0.050s, 0.034s with respect to the bandwidth requirements. When mobility is 20m/s, the ALs are 0.069s, 0.076s, 0.059s with respect to the bandwidth requirements.

3.4.3 Effect of Size of Data Packet
Each value in Fig. 13 (c) is obtained by assuming that the bandwidth requirement is 0.2Mb/s, mobility is 10m/s and number of mobile hosts is 75. Fig. 13 (c) shows AL vs. data packet size. The AL is 0.070s for data packet size 256B, 0.090s for data packet size 1024B and 0.14s for data packet size 2048B.

4 Conclusion and Future Work

This paper presents the QoS multicast routing protocol (QMR) for a wireless ad hoc network. QMR protocol uses *forward nodes* as a subset of mesh. Our protocol uses redundant paths when the network bandwidth is sufficient or unipath when the network bandwidth is strictly limited. Periodic sending of requests frees unusable reserved bandwidth and re-estimates the available bandwidth, which gives the source another chance by accepting new requests. The *hybrid* scheme in QMR reflects that QMR protocol exploits the residual bandwidth efficiently without effect on the reserved bandwidth. QMR protocol is efficient in providing QoS multicast routing capability with low control overhead. It should be noted that control overhead increases very slowly significantly for increases in mobility of hosts which means that reconstructing failed paths or searching for new paths does not affect control overhead. In general, the simulation results reflect a good success rate ratio and high packet delivery ratio associated with low control overhead, this come as a result of preventing overload and providing *load balancing* for *forward nodes*. QoS multicast routing needs an efficient way to estimate available bandwidth and network feedback to applications to adapt its traffic rate when bandwidth requirements cannot be met. We believe that stringent performance requirements to provide QoS multicast applications over MANETs can only be met through a cross-layer design, Our ultimate goal is to design a cross-layer framework from the application layer to the MAC layer to provide QoS multicast routing. We also intend to compare QMR with other QoS multicast routing.

References

1. C.-R. Lin: On-Demand QoS Routing in Multi-Hop Mobile Networks. Proceedings of the IEEE INFOCOM, (2001) 1735–1744
2. Chlamtac, I., Conti, M., and Liu, J. J.-N: Mobile Ad hoc Networking: Imperatives and Challenges. Vol. 1(1) Ad hoc Networks Publication, (2003) 13-64
3. E. Setton, T. Yoo, X. Zhu, A. Goldsmith, and B. Girod: Cross-layer Design of Ad hoc Networks for Real-Time video Streaming. To Appear IEEE Wireless Communications Magazine, (2005)
4. E.Pagani,G.P.Rossi: A Framework for the Admission Control of QoS Multicast Traffic in Mobile Ad hoc Networks. Proc. Fourth ACM International Workshop on Wireless Mobile Multimedia Roma (2001) 3-12
5. G. S. Ahn, A. T. Campbell, A. Veres and L.H. Sun: SWAN: Service Differentiation in Stateless Wireless Ad hoc Networks. In Proc. IEEE INFOCOM, (2002)
6. Guojun, Wang. Jiannong, Cao1: A Novel QoS Multicast Model in Mobile Ad hoc Networks. Proceedings of the 19th IEEE International Parallel and Distributed Processing Symposium (2005)
7. Hasana, M. and Hoda, L.: Multicast Routing in Mobile Ad Hoc Networks. Telecommunication Systems, Kluwer Academic Publishers (2004) 65–88
8. I. Jawhar and J. Wu: A Race-Free Bandwidth Reservation Protocol for QoS Routing in Mobile Ad hoc Networks. Proceedings of the 37th Annual Hawaii International Conference on System Sciences, IEEE Computer Society (2004)
9. J. Li, C. Blake, D. D. Couto, H. Lee, and R. Morris: Capacity of Ad hoc Wireless Networks. In Proc. 7th ACM Int. Conf. Mobile Comput. Netw (2001) 61–69
10. J. Ng, C.P. Low, and H.S. Teo: On-demand QoS Multicast Routing and Reservation Protocol for MANETs. Proc. 15th IEEE International Symposium on Personal, Indoor and Mobile Radio Communications Spain (2004)
11. K. Xu, K. Tang, R. Bagrodia, M. Gerla, M. Bereschinsky: Adaptive Bandwidth Management and QoS Provisioning in Large Scale Ad hoc Networks. Proceedings of MILCOM, Boston, MA (2003)
12. L. Chen and W. Heinzelman: QoS-aware Routing Based on Bandwidth Estimation for Mobile Ad hoc Networks. IEEE Journal on Selected Areas of Communication, Special Issue on Wireless Ad hoc Networks, Vol. 23. (2005)
13. L. Mohammed: Ad hoc wireless networks. CRC. (2003)
14. N. Nikaein, H. Labiod, and C. Bonnet: Error Recovery Scheme for Multicast Transmission over Wireless Networks, IEEE ICC, New Orleans, USA (2000)

15. Q. Xue and A. Ganz: Ad hoc QoS On-demand Routing (AQOR) in Mobile Ad hoc Networks. Journal of Parallel Distributed Computing vol. 63 (2003) 154-165

16. R. Gupta et al: Interference-Aware QoS Routing for Ad-Hoc Networks. ICC. (2005)

17. S.B.Lee et al: INSIGNIA: an IP-Based Quality of Service Framework for Mobile Ad hoc networks. Journal of Parallel and Distributed Computing, Special issue on Wireless and Mobile Computing and Communications. Vol. 60 (2000) 374-406.

18. Shiwen, Mao: Video Transport over Ad hoc Networks. Multistream Coding with Multipath Transport, IEEE Journal on Selected Areas in COM. Vol. 21. (2003)

19. Sung-Ju, Lee, William Su, Mario Gerla: On-Demand Multicast Routing Protocol (ODMRP) for Ad hoc Networks. <draft-ietf-manet-odmrp-02.txt>

20. Y.-S., Chen and Y.Yu: Spiral-Multi-Path QoS Routing Protocol in a Mobile Ad hoc Network. IEICE Transactions on Communications (2004) 104–116

21. Y.-S. Chen, Y.-C. Tseng, J.-P. Sheu and P.-H. Kuo: On-demand, Link-State, Multi-Path QoS Routing in A wireless Mobile Ad hoc Network. In Proc. European Wireless (2002)

22. Yaling Yang Kravets, R: Distributed QoS Guarantees for Real-Time Traffic in Ad hoc Networks Sensor. Ad hoc Communications and Networks, IEEE SECON (2004)118-127

23. Yuh-Shyan Chen and Yun-Wen Ko: A Lantern-Tree Based QoS on Demand Multicast Protocol for A wireless Ad hoc Networks. IEICE Transaction on Communications Vol.E87-B. (2004) 717-726

24. Z. Wang and J. Crowcroft: QoS Routing for Supporting Resource Reservation. IEEE Journal on Selected Areas in Communications (1996) 1228–1234.

25. S.-J. Lee, W. Su, J. Hsu, M. Gerla, and R. Bagrodia. A Performance Comparison Study of Ad Hoc Wireless Multicast Protocols. In Proceedings of IEEE INFOCOM 2000, 565–574

26. http://pcl.cs.ucla.edu/projects/glomosim/obtaining_glomosim.html

A Framework for the Comparison
of AODV and OLSR Protocols

Dmitri Lebedev

LIX, École Polytechnique, France
lebedev@lix.polytechnique.fr

Abstract. The goal of this paper is to elaborate a framework for a comparison of two ad hoc network protocols: AODV and OLSR. We propose a model to study the reaction of two protocols on faulty links in the network. The principle idea is to test two protocols with the same useful load and the proportion of faulty links and observe the amount of traffic the two protocols will create.

The goal of this paper is to elaborate a framework for a comparison of two ad hoc network protocols: AODV [1] and OLSR [2]. We propose a model to study the reaction of two protocols on faulty links in the network. AODV is using flooding to repair its paths which generates heavy load on the network structure. The paths found by the flooding are longer in the number of hops, therefore some bandwidth is lost. On the other hand shorter links are more robust to the interference. OLSR, in its turn, has its paths optimised and traffic consumes less bandwidth, but on the reverse side its links may be less tolerant to noises.

The difficulty of such comparison resides in different nature of these two protocols: one is reactive and another one is proactive. AODV is a classical example of a reactive routing protocol: the paths in AODV are found "on demand". It means that if there is a node A that needs to communicate with a node B and A does not know the route to B, then A starts path search process. This process provides a route to B, if it exists. Therefore, the node A has no route to B before it decides to send it a message.

OLSR is an example of a proactive protocol where the paths are ready before a node decides to send a message. In general case, the nodes know the routes to all other nodes participating in the network. One way to achieve such capabilities is to regularly send topology control (TC) messages, that informs all the nodes about the route available route.

The basic path search mechanism of AODV is flooding. Flooding can induce a high traffic charge on the network and also the paths obtained are not optimal in the number of hops. OLSR is tending to optimise its paths. It appears that such optimisation has two sides: on one hand shorter paths optimise the traffic load as the messages need less relaying. But on the other hand longer paths are more easily blocked by noise and therefore they are less reliable.

K. Cho and P. Jacquet (Eds.): AINTEC 2005, LNCS 3837, pp. 98–111, 2005.

The idea of this study is to test two protocols with the same useful load and the proportion of faulty links and observe the amount of traffic the two protocols will create. In the following sections, we will address these problems and will build analytical models where these phenomena can be seen. In section 1, we introduce the model of the physical layer of the network, which describes the signal propagation and the node distribution. In section 2, we introduce our simplified model of OLSR and AODV protocols. The section 3 contains the results of the flooding analysis. Finally in section 4, we compare the performance of the protocols.

1   Physical Layer Model: Links and Node Distribution

We base our study on the physical model of the wireless networks where a link availability depends not only on the distances between the nodes but also on the interference with their active neighbours. We consider a discrete (slotted) model of time.

1.1   Link Availability

The signal attenuation function, which describes the power of the signal received from node A by node B, is modeled

$$W_{A,B} = Qr^{-\alpha} \tag{1}$$

where r is the distance between A and B, Q is the power of the signal of the sender and α is a constant depending on the environment the network is working in. We assume that all the nodes transmit with a constant and equal power and therefore we take $Q = 1$. In vacuum α constant is equal to 2; in outdoor environment with limited obstacles it is taken between 2 and 4. Here, we put $\alpha = 2.5$, which is a typical value in similar studies.

As the receiving node B can be surrounded by other nodes that are transmitting their messages in the same time, there is a condition on decoding the signal coming from A based on the Signal-to-Interference-Ratio(SIR) :

$$SIR(A, B) = \frac{W_{A,B}}{\sum_{k \neq A} W_{k,B}} \tag{2}$$

This is a simplified formula as the more complex expressions can obtained, for example, by including the background noise into the consideration. It will be then added to the denominator of (2). The protocol multiplexing scheme is another example, it is sometimes included as a coefficient to the sum. The condition that the signal from A can be successfully received by B is

$$SIR(A, B) > K \tag{3}$$

where K is constant threshold often to be taken equal to 10 (1dB).

1.2   Node Distribution and Neighbourhoods

A big challenge for ad hoc network models is the mobility of the network. In our case we avoid this difficulty, as it is done in a number of other works, by considering such slowly moving (fixed) network that its snapshot does not change for some reasonable time. The nodes are distributed accordingly to a Poisson point process of intensity ν, therefore all points are independent and the number of points in some bounded region of area s is distributed accordingly to the Poisson distribution with mean $s\nu$. We assume that the total surface of the network is a bounded region of area S, therefore the average number of nodes in the network is $N = S\nu$.

Networks are modeled using graph representation corresponding to their structures. Although in general case the wireless network the links can not be represented as a fixed set, because of the constraints of the interference: the neighbouring links can not be active at the same time. This aspect is captured by the condition (3) on SIR. Nevertheless, we can loose this constraint on the link modeling choosing the following condition: there is a link between two node A and B if $Pr[SIR(A, B) > K] > 1/3$. It creates a graph model of the network with the vertex set being the node set and edges being the links defined through this condition. This modeling was presented in [3], where it is considered in greater details. It is similar to a unit disk model but with the radius of covering region depending on the current traffic level.

Consider the network with average λ messages per unit area of surface per time slot. This value λ is called *traffic density*. As the traffic density can be seen as an intensity of a new Poisson point process, obtained from the initial one by thinning, the average distance at which an active node successfully received can be calculated (see [3]) and it is expressed as

$$r(\lambda) = \frac{r_0}{\sqrt{\lambda}}, \qquad (4)$$

where r_0 is a constant. Assuming $\alpha = 2.5$ ([4]), r_0 is approximately equal to 0.12. The expression (4) can lead to many interesting results including, for example, an expression of the average number of neighbours M:

$$M(\lambda) = \nu \frac{\pi r_0^2}{\lambda} \qquad (5)$$

2   Overall Traffic Equations

2.1   OLSR Traffic Model

OLSR is a proactive protocol and as a consequence there is a lot of periodic activity to establish the paths across the network. The "Hello" messages mechanism informs the nodes about their 2-hop neighbourhood. Then every node fixes a subset (Multi Point Relay nodes, MPR) of its neighbours allowed to re-transmit its messages. This subset should be able to reach with its transmissions all the nodes at a two hop distance and it is also optimised in size. The node choosing its MPRs is called *MPR selector*.

After this step, every node advertise itself across the network. This is performed by a flooding the MPR neighbours of the node, then MPRs of MPRs, etc. Therefore, compared to the classical AODV flooding the number of the nodes and messages involved is much less. A good optimisation is achieved using the following protocol rule: a node forwards a message only received *for the first time* from its selector. We include into our simplified model the following packets of OLSR (see [5], [6] for similar models)

- "Hello" messages. Their frequency rate is denoted by χ. In the case, when it is necessary to avoid a confusion with the AODV notations we will use subscripts $_{AODV}$ and $_{OLSR}$.
- Topology Control (TC) messages. They compose the OLSR flooding and contain the information about the issuing node and its MPR nodes. We denote the rate by τ.
- Normal information packets. We denote the number of communicating pairs by Λ, the average distance between sender and receiver by $L = C_1\sqrt{S}$, where S is the area of the network and the average hop length we denote by h. The rate of the communication is ω.

We can express the traffic density in the whole network over some long period of time by

$$\lambda_{OLSR} = \frac{\chi N}{S} + \frac{N_{MPR}N\tau}{S} + \frac{L\Lambda\omega}{hS}, \tag{6}$$

where N_{MPR} is the number of nodes participating in MPR flooding. The study [6] of the MPR nodes in the network showed that for M neighbours there are about $5M^{1/3}$ of MPR nodes. Therefore we can estimate N_{MPR} in the following way: it is the probability of every node to become an MPR multiplied by the number of nodes,i.e. $\frac{5M^{1/3}N}{M}$. As we have seen it in (5), the number of neighbours in its turn is determined by the node density and the traffic density.

We suppose that the equation (6) describes the behaviour of the network in equilibrium. When a perturbation occurs the network readjusts itself, if necessary, it recreates new routes. Under such assumptions we can conclude that the hop length of the routes in OLSR corresponds to a current level of the traffic density and therefore can be taken as $r_0/\sqrt{\lambda_{OLSR}}$.

Re-writing (6) (dropping the subscript of λ_{OLSR}) we obtain

$$\lambda = \chi\nu + 5\left(\nu\frac{\pi r_0^2}{\lambda}\right)^{-2/3} N\nu\tau + \frac{\sqrt{\lambda}C_1\Lambda\omega}{r_0\sqrt{S}} \tag{7}$$

$$\lambda = \chi\nu + 5\left(\pi r_0^2\right)^{-2/3}\lambda^{2/3}N\nu^{5/3}\tau + \sqrt{\lambda}\frac{C_1\Lambda\omega}{r_0\sqrt{S}} \tag{8}$$

Although this equation is not solvable in radicals for λ, we can obtain numerical approximation, that will be used in the comparison of two protocols in section 4.

2.2   AODV Traffic Model

AODV is a complex protocol with several types of control packets circulating between the nodes. In our analysis, we will concentrate on the following subset of them:

- "Hello" messages. They were introduced into AODV to maintain the local connectivity. These messages are transmitted periodically. We denote this frequency rate by χ.
- Route Request (RREQ) messages. These messages are participating in the discovery of new routes and in repairing the old ones. Therefore they are "on demand" packets which compose the flooding. We do not include in consideration the local repair mechanism of AODV and new path creation. The latter can be easily included into the additive traffic density expression. We suppose that over some long period of time the frequency of flooding waves is ϕ, which depends on the rate of the errors which occur due to the wireless traffic congestion or interference problems.
- Normal traffic packets. They depend on the communication pattern of each network. We assume that in our case we have Λ paths regularly communicating, with the frequency ω.

These messages compose the main part of the traffic. The equation of the traffic density over some long period of time in the AODV network can be expressed in following way.

$$\lambda_{AODV} = \frac{N\chi}{S} + \frac{\phi N}{S} + \frac{L\Lambda\omega}{hS}, \text{ or} \tag{9}$$

$$\lambda_{AODV} = \nu\chi + \phi\nu + \frac{C_1\Lambda\omega}{h\sqrt{S}} \tag{10}$$

where N is the number of nodes, S is the area ($N/S = \nu$), L is the average euclidean distance between sender-destination pair of Λ paths (it is of order $O(\sqrt{S})$, i.e. $L = C_1\sqrt{S}$, where the constant C_1 is characterise the region), h is an average hop length. In this expression there are two parameters that depends on other ones: flooding rate ϕ and hop length h.

The flooding rate ϕ should be proportional to the number of useful packets $L\Lambda/h\omega$. These packets or links can be interrupted by two events: by a random perturbation (thermal noise) or by a flooding that interrupted the link. The latter can be considered as self-inflicted damage. Let us denote the thermal component of ϕ by θ and the self-inflicted one by ϵ.

$$\phi = \theta + \epsilon \tag{11}$$

θ is a parameter of our model, whereas ϵ has to be determined using a flooding model we present in the next section.

The hop length h also depends on the flooding characteristics, node density and the current traffic density λ. This is because flooding is exactly the way AODV finds the paths.

3   Path Robustness

In this section, we study a model of the flooding in a wireless network. The flooding is a network broadcast mechanism where every node transmits a message to all its neighbours, who repeat it to their neighbours and so on. Every node transmits this message just once. In AODV, flooding is used to find a path to a node whose location is unknown. The flooding is loop free and it forms a spanning tree over the network nodes. We will consider flooding from two points of view: one considers its load on the network whereas the other reflects the properties of the discovered paths.

3.1   The Flooding Model

One of the main parameters of a flooding is the node density ν. The more the network is dense , the higher is the load per unit area. In terms of traffic density flooding is the heaviest charge for the network. Consider a situation when all nodes in some neighbourhood almost simultaneously start re-transmitting a message. If there is no controlling MAC protocol, the traffic density raise up to highest level - ν, which means that all nodes transmit at the same time. The flooding in this case would stop after a first step or two. Due to their own interference the retransmitting nodes would not have any neighbours capable of receiving them.

There are multiple solutions for sharing the media by the waiting nodes. In case of the 802.11 MAC layer, it is the Distributed Coordination Function (DCF [7]): prior to initiating a transmission, a node chooses a random delay from a fixed interval. Then, at the end of its waiting period, if the node finds the network still busy it chooses another random waiting time from the double of the initial interval. This is called an *exponential backoff algorithm*. Some protocols can use their own random delay (*jitter*) mechanisms.

What characteristics of the flooding are we interested in? From the point of view of the network load it is the evolution of the traffic density, when a wave of flooding goes over some bounded region S'. In particular, we consider the period of time necessary for the traffic to decrease to pre-flooding level. One of the major assumptions in this work is that on average the flooding wave, after several initial steps, moves over the network with a constant speed. We denote this speed by v. Intuitively, a flooding spreads, in average, in all directions at the same pace. This assumption seems reasonable as a small arc of a circle can be approximated by a straight line. We have performed several simulations to see what happens to a network when the flooding arrives.

3.2   Simulation Conditions

In order to simplify our simulations we consider a model where all nodes try to transmit the message with a constant probability $p = 0.025$. This algorithm plays the role of the coordination function between the nodes.When a node succeeds in transmitting a message it does not repeat it any more. The transmit power of all nodes is 1 and the signal attenuation is $r^{-\alpha}$, where $\alpha = 2.5$. The network is contained in a square of size $200m \times 200m$. The total number of nodes is 1000 ($\nu = 0.025$). The simulation starts with 20 nodes on the left edge ready to transmit a request message.

We also impose on the network a homogeneous background noise of power $0.0001dB$. As we can see from (3), this level of noise allows the direct communication at distance $d \approx 15.8m$ or about 20 neighbours. In other terms we can find the traffic density that would allow such radius of communication from (4): $\lambda_{noise} = r_0^2/d^2 \approx 0.0000576$. This corresponds to 2.3 messages in average in time slot over all network of 1000 nodes. This provides a reasonable value of the threshold λ_{noise} that we can set as the parameter for the flooding. We will be interested in the time until the traffic density of the flooding falls under this level. The results of the simulations are presented in Fig. 1 and 2.

Let us discuss Fig. 1 in more details. By our assumption of linearity of the flooding progression this curve should be composed of two straight lines: one corresponds to the event when the flooding is gaining new nodes and another one to the event when there are no new nodes left. This is not the case in this simulation due to the border effects. Nevertheless, approximating the straight part of this curve on the interval [20,80] we can calculate that the flooding is moving with the speed about 8.08 nodes per time slot. Let us now analyse the diagram on Fig. 2

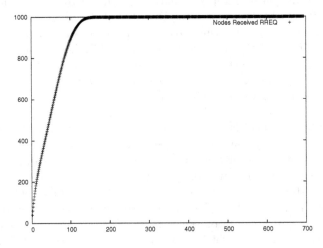

Fig. 1. The number of nodes having received the message as a function of time

On Fig. 2, we present the traffic density during the simulation. The horizontal line on the graph corresponds to the background noise level. On the graph we can observe how flooding crosses a bounded region. The second point, where the curve intersects the background noise level, is the moment when the flooding traffic density has decreased sufficiently. For this simulation it takes 164 steps.

This value, of course, depends on the probability p. When this probability is higher, the peak of the traffic density is also higher, but it takes less time for the flooding to pass. Nevertheless, there is a bound after which some nodes of the network can never receive the message. As we have seen it in the example earlier, when $p = 1$, the flooding will stop almost instantly as there would be too much noise created by the neighbouring nodes for anyone to hear them. Therefore, the process of increasing p up to the bound at which flooding is yet possible can be seen as the way to optimise the network.

3.3   Analytical Explanation of the Simulation

The curve on the Fig. 2 can be explained as follows. First of all, it decomposes into two parts: ascending and descending. Ascending part corresponds to the progress of the wave through the network, whereas the descending one reflects the event that there is no new nodes receiving the message and the number of nodes wishing to retransmit decreases at rate p. Also it is worth noting here that under the conditions of this model the number of nodes awaiting to transmit the message is proportional to the number of transmitting nodes (with coefficient p).

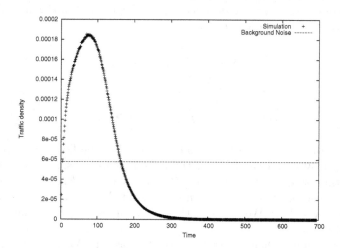

Fig. 2. Traffic density evolution (number of packets per unit square) with time

During the ascending phase, if we denote by x_n the number of nodes awaiting to transmit the message, there is following recursive expression for x_{n+1}: $x_{n+1} = x_n(1-p)+X$, where X is the average number of new nodes gained by the flooding at an arbitrary step (here $X = 8.08$). We take $x_1 = X$ because of the linearity assumption. Therefore

$$x_{n+1} = \frac{1 - (1-p)^{n+1}}{p} X \qquad (12)$$

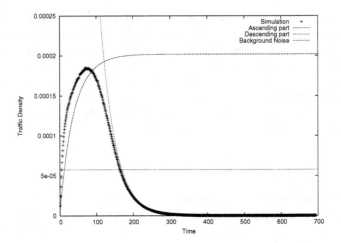

Fig. 3. Traffic density estimated by the analytical curves

Let us denote by n_0 the moment when all N nodes received the message. As some of them have already re-transmitted it, we have $x_{n_0} < N$. At the moment, when the sequence x_n starts to decrease: $x_{n+1} = x_n(1-p)$ (for $n > n_0$). Therefore the sequence $\{x_n\}$ can be written as

$$x_0 = X \tag{13}$$

$$x_{n+1} = \frac{1 - (1-p)^{n+1}}{p} X, \quad 0 < n \le n_0 \tag{14}$$

$$x_{n+1} = x_{n_0}(1-p)^{n-n_0}, \quad n > n_0 \tag{15}$$

The value of n_0 can be derived from the speed of the flooding wave that we have denoted by v and the form of the region. If we consider, as we did it in the simulation, a square region of size $R \times R$, then $n_0 = R/v$. The speed of the flooding can also be expressed in terms of the number of new vertices receiving the message: $X = Rv\nu$. We compare these two curves of x_n with the simulation results on Fig. 3.

To find the point when the traffic level decreases to λ_{noise} we need to solve the following equation:

$$\lambda_{noise} = x_{n_0}(1-p)^{n-n_0} \frac{p}{S}$$

We put the coefficient p/S here to scale the number of nodes, wishing to transmit, to the traffic density (number of transmitting nodes per unit area). We obtain

$$n = n_0 + \frac{\log(\lambda_{noise} S) - \log(x_{n_0} p)}{\log(1-p)} \tag{16}$$

The area of the region is $S = R^2$, therefore using the speed of the flooding wave we can calculate the value of x_{n_0}:

$$x_{n_0} = \frac{1 - (1 - p)^{\sqrt{S}/v+1}}{p} \sqrt{S} v \nu \qquad (17)$$

Substituting (17) into (16), we obtain

$$n = n_0 + \frac{\log(\lambda_{noise} S) - \log\left(\left(1 - (1 - p)^{\sqrt{S}/v+1}\right) \sqrt{S} v \nu\right)}{\log(1 - p)} \qquad (18)$$

Using (18) we can determine the average interval of time necessary for flooding to pass. It will be used to estimate the probability for the flooding to fail an active link. When this period increases the probability is higher. We do not exact distribution for x_{n_0}, only the average value. This value we will compare with a parameter of AODV that defines the broken links (NEXT_HOP_WAIT in RFC 3561).

3.4 Hop Length Estimation

The simulations, run with the same parameters as in section 3.2, show that the average value of the hop length is between the hop length corresponding to the background noise and the average hop length of the flooding peak. For purposes of this work we traced a simple curve approximating the hop length. h_{AODV} depends on the node density ν, background noise λ_{noise} and probability p, which is one of the characteristics of the model. As in this model ν and p are fixed, the dependence we need to evaluate is h_{AODV} a a function of λ_{noise}. We approximate the simulation results on Fig. 4 with the following expression:

$$h_{AODV}(\lambda_{noise}) = \frac{8.2}{\lambda_{noise}^{0.1}} - 11.7 \qquad (19)$$

4 Protocol Comparison

In this section we compare the traffic density of two protocols with the same useful load Λ. For OLSR we will use directly the equation (8). Though the AODV equation (10) needs some further calculations.

We assume that the network is in equilibrium. Therefore, there are Λ working connections with the average traffic density λ_{AODV}. We suppose that the network uses for the flooding packets the same *probability* coordination and we can apply the model of the section 3 to AODV. We fix probability $p = 0.025$ and λ_{noise} is the average traffic density across the network, i.e. $\lambda_{noise} = \lambda_{AODV}$. In this case we can use the estimation for the hop length from (19), as the values of the traffic density traced in the approximated simulations are the same as λ_{AODV}.

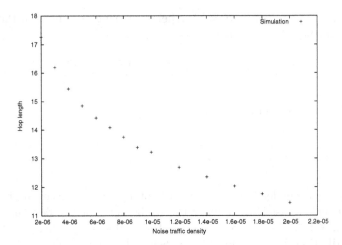

Fig. 4. Hop length dependence on the background noise

The error rate ϵ is the second parameter of AODV and we have not found its explicit form. This is the rate of the failures caused by floodings. It can be estimated as it corresponds to the probability (rate) of the active link meeting a wave of the flooding.

$$\epsilon = \frac{C_1 \Lambda \omega \sqrt{S} \phi}{h} P_e \qquad (20)$$

where P_e is the probability of the event that flooding will interrupt the transmission at arbitrary node. The interpretation of (20) is the following: the number of errors caused by floodings is proportional to the number of links (the number of useful connections multiplied by the average path length) $\Lambda \omega C_1 \sqrt{S}/h$ and to the flooding rate ϕ. Substituting in the expression (11) the value of ϵ by (20) and solving the resulting equation we obtain

$$\phi = \frac{\theta}{1 - \frac{C_1 \Lambda \omega \sqrt{S}}{h} P_e} \qquad (21)$$

At this point we need another assumption to continue further our investigation. We assume that P_e is a probability function which depends on the duration of the peak of the flooding wave in some bounded region (taken from (18)) and it depends on T^*, the period which AODV waits before declaring a packet lost. P_e is ascending in flooding peak duration and descending in T^*. We choose P_e to correspond these constraints as following:

$$P_e = \exp\left(C_2\left(n_0 + \frac{\log(\lambda_{noise}S) - \log\left(\left(1 - (1-p)^{\sqrt{S}/v+1}\right)\sqrt{S}v\nu\right)}{\log(1-p)}\right)/T^* + C_3\right) \qquad (22)$$

where C_2 and C_3 are normalising constants. This function can influence the form of the final curve. But as long as the thermal noise θ is small enough, this influence is limited.

Value	AODV	OLSR	Description
	0.00001	0.00001	Time slot duration
N	1000	1000	Number of nodes
S	200	200	Surface
C_1	0.5	0.5	$C_1 = S/\sqrt{L}$
r_0	0.12	0.12	
ω	1/10000	1/10000	Useful load rate
χ	1/2000	1/2000	Hello rate
τ	NA	1/20000	OLSR TC refresh rate
T^*	0.005		AODV Link failure detection time
v	1.616		Speed of the flooding wave
p	0.025		Probability of the node transmission

Fig. 5. The different parameters of two protocols used in comparison

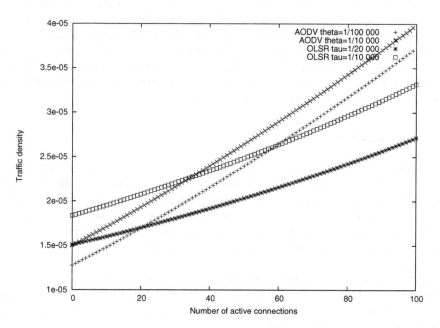

Fig. 6. Traffic density of AODV ($\theta = 1/10000$ and $\theta = 1/100000$) and OLSR ($\tau = 1/20000$ and $\tau = 1/10000$)

On Fig. 6 we see the final results that follow from OLSR (8) and AODV traffic density equations. We restrain from writing the AODV function in whole as it is composed of multiple parts: equation (10), (18),(19), (22), (21) and the parts corresponding to P_e (22) and to h_{noise} (19) are quite frivolous. Nevertheless, these curves allows to estimate the comparative behaviour of two protocols.

The x-axe is the percentage of the nodes involved in some active connection (number of the connections is then $xN/100$). Most of the parameters are shown in the table in Fig. 5. The curves of AODV are calculated through a numerical computation with the values of $\theta = 1/100000$ and $\theta = 1/10000$. Whereas, OLSR has two curves with $\tau = 1/20000$ and $1/10000$. These two parameters also make the comparison difficult. The values of θ and τ are related in the sense that when there is θ error rate in the network, OLSR should raise the rate τ sufficiently high, to repair the broken links with a minimal delay.

On Fig. 6 the slope of AODV is greater than those of OLSR. The reason for this is that the length of the paths of AODV are longer than those of OLSR. This is the L/h part in the initial equations for both protocols (6) and (9). When θ increases AODV curve makes almost a parallel shift up. OLSR has a similar reaction on τ.

We can derive from the graphs that at some values of θ and τ the protocols have a load at which they perform equally. Before this point AODV performs better in terms of the overhead. After this point it is OLSR that has the less overhead. It is not too surprising as we can see it on two extreme examples. In one case there is a zero useful traffic, therefore AODV does nothing, which is much better than OLSR establishing the structure of the network. In the other case the useful load is high, that makes the paths in AODV longer and the error rate higher, therefore OLSR optimisation of the resources allows a better overhead. The approach to estimate the traffic density of OLSR and AODV we present here allows to consider the case in greater details.

5   Conclusion

We presented a framework for the comparison of two ad hoc network protocols, AODV and OLSR. This is not a complete analytical model as there are two important parameters that are approximated: AODV hop length and probability of flooding to break a link. Although we propose some inside on the mechanism of the interaction between flooding and the existing links and this approach allows us to compare the curves of the traffic level of these two protocols.

We have chosen to model the flooding in the network where the conflict resolution is based on a parameter p, where the nodes decide to transmit independently with probability p. For the future work, of course, it will be interesting to integrate 802.11 DCF into this framework.

References

1. Perkins, C.E., Royer., E.M.: Ad hoc on-demand distance vector routing. Proceedings of the 2nd IEEE Workshop on Mobile Computing Systems and Applications (1999) 90–100
2. Clausen, T., Jacquet, P., Laouiti, A., Muhlethaler, P., Qayyum, A., Viennot, L.: Optimized link state routing protocol. In: IEEE INMIC, Pakistan (2001)
3. Jacquet, P.: Elements de theorie analytique de l'information, modelisation et evaluation de performances. Technical Report 3505, INRIA (1998)
4. Adjih, C., Baccelli, E., Jacquet, P.: Link state routing in wireless ad-hoc networks. Technical Report 4874, (INRIA)

5. Jacquet, P., Viennot, L.: Overhead in mobile ad-hoc network protocols. Technical Report 3965, (INRIA)
6. Jacquet, P., Laouiti, A., Minet, P., Viennot, L.: Performance of multipoint relaying in ad hoc mobile routing protocols. In: NETWORKING. (2002) 387–398
7. P802.11. IEEE standard for Wireless LAN Medium Access Control (MAC) and Physical Layer (PHY) Specifications (1997)

Distributed Node Location in Clustered Multi-hop Wireless Networks

Nathalie Mitton and Eric Fleury

INRIA/ARES - INSA de Lyon - 69621 VILLEURBANNE Cedex,
Tel: 33-(0)472-436-415
firstname.lastname@insa-lyon.fr

Abstract. Wireless routing protocols in MANET are all flat routing protocols and are thus not suitable for large scale or very dense networks because of bandwidth and processing overheads they generate. A common solution to this scalability problem is to gather terminals into clusters and then to apply a hierarchical routing, which means, in most of the literature, using a proactive routing protocol inside the clusters and a reactive one between the clusters. We previously introduced a cluster organization to allow a hierarchical routing and scalability, which have shown very good properties. Nevertheless, it provides a constant number of clusters when the intensity of nodes increases. Therefore we apply a reactive routing protocol inside the clusters and a proactive routing protocol between the clusters. In this way, each cluster has $O(1)$ routes to maintain toward other ones. When applying such a routing policy, a node u also needs to locate its correspondent v in order to pro-actively route toward the cluster owning v. In this paper, we describe our localization scheme based on Distributed Hashed Tables and Interval Routing which takes advantage of the underlying clustering structure. It only requires $O(1)$ memory space size on each node.

Keywords: wireless networks, localization, DHT, interval routing.

1 Introduction

Wireless multi-hop networks such *ad hoc* or sensor networks are mobile networks of mobile wireless nodes, requiring no fixed infrastructure. Every mobile can independently move, disappear or appear at any time. A routing protocol is thus required to establish routes between terminals which are not in transmission range one from each other. Due to the dynamics of such wireless networks (terminal mobility and/or instability of the wireless medium), the routing protocols for fixed networks do not efficiently fit. *Ad hoc* routing protocols proposed in the MANET working group are all flat routing protocols: there is no hierarchy, all the terminals have the same role and are potential routers. If flat protocols are quite effective on small and medium size networks, they are not suitable for large scale or very dense networks because of bandwidth and processing overheads they generate [22]. A common solution to this scalability problem is to introduce a hierarchical routing. Hierarchical routing often relies on a specific partition of the network, called *clustering*: the terminals are gathered into clusters according to some criteria, each cluster is identified by a special node called *cluster-head*. In most of the literature, a hierarchical routing means using a proactive[1] routing protocol inside the

---

[1] Nodes permanently keep a view of the topology. All routes are available as soon as needed.

K. Cho and P. Jacquet (Eds.): AINTEC 2005, LNCS 3837, pp. 112–127, 2005.

clusters and a reactive[2] one between the clusters [6, 9, 19, 20]. In this way, nodes store full information concerning nodes in their cluster and only partial information about other nodes. We previously introduced a clustering algorithm [13]. Its builds clusters by locally constructing trees. Every node of a same tree belong to the same cluster, the tree root is the cluster-head. This algorithm has already been well studied by simulation and theoretical analysis. It has shown to outperform some other existing clustering schemes regarding structure and behavior over node mobility and link failure. It may also be used to perform efficient broadcasting operations [15]. Nevertheless, it provides a constant number of clusters when the node intensity increases. Thus, there still are $O(n)$ nodes per cluster and using a proactive routing scheme in each cluster as in a classical hierarchical routing, would imply that each node still stores $O(n)$ routes, which is not more scalable than flat routing. Therefore, we propose to use the reverse approach, *i.e.*, applying a reactive routing protocol inside the clusters and a proactive routing protocol between the clusters. Indeed, as the number of clusters is constant, each cluster has only $O(1)$ routes to maintain toward other ones. As far as we know, only the SAFARI project [21] has proposed such an approach, even if most of the clustering schemes [1, 19] present an increasing number of nodes per cluster with an increasing node intensity and still claim to apply a proactive routing scheme inside the clusters. When applying such a routing policy, a node u first needs to locate the node v with which it wants to communicate, *i.e.*, it needs to know in which cluster node v is. Once it gets this information, u is able either to pro-actively route toward this cluster if it is not the same than its, or to request a route toward node v inside its cluster otherwise. So, a localization function which returns for any node u, the name of the cluster to which it belongs is needed. In this paper, we introduce our localization scheme which takes advantage of the tree/cluster structure. It is based on Distributed Hashed Tables (DHT) and Interval Routing. DHT are known to be scalable and allow to each node u to register its cluster identity over the network on several rendezvous points which will be contacted by every node looking for node u. Interval routing is already known to be a highly memory efficient routing for communication in distributed systems. In addition, this scheme also takes advantage of the broadcasting feature of the wireless communications to reduce even more the memory size. Yet, each node only requires a memory size in $O(1)$.

The remaining of this paper is organized as follows. The description of the initial clustering algorithm as well as interesting features for the localization algorithm are given in Section 2. Then, we describe in Section 3 how we propose to take advantage of the underlying structure of our organization to perform our localization algorithm. Then, we describe in Section 4 our proposition. Lastly, in Section 5, we discuss some improvements and future works.

2   Our Cluster Organization

In this section, we summarize our previous clustering work on which we apply our localization scheme. Only basic features which are relevant for localization and routing are mentioned here. For more details, please refer to [13, 14, 16].

---

[2] Routes are searched on-demand. Only active routes are maintained.

Let's first introduce some notations. We classically model a wireless multi-hop network by a random geometric graph $G = (V, E)$ where V is the set of mobile nodes ($|V| = n$) and $e = (u, v) \in E$ represents a bidirectional wireless link between a pair of nodes u and v. If $dist(u, v)$ is the Euclidean distance between nodes u and v, then $\exists(u, v) \in E$ iff $dist(u, v) \le R$. R is thus the transmission range. If $d(u, v)$ is the distance in the graph between nodes u and v (minimum number of hops needed to reach v from u), we note $\Gamma_k(u)$ the set of nodes v such that $d(u, v) = k$. Note that node u does not belong to $\Gamma_k(u) \forall k$. $\delta(u) = |\Gamma_1(u)|$ is called the *degree* of u. We note $\mathcal{C}(u)$ the cluster owning u. Let $\mathcal{P}(u)$ denote the parent node of node u in a tree and $Ch(u)$ the set of children of u, *i.e.*, the set of nodes v such that $\mathcal{P}(v) = u$. Note that u is a leaf iff $Ch(u) = \emptyset$. Moreover, we note $sT(u)$ the subtree rooted in node u. We say that $v \in sT(u)$ if $v = u$ or if u is the parent of node v ($\mathcal{P}(v) = u$) or if the parent of node v is in the subtree rooted in u ($\mathcal{P}(v) \in sT(u)$): $\{v \in sT(u) \cap \Gamma_1(u)\} \Leftrightarrow \{v \in Ch(u)\}$ or $\{v \in sT(u) \cap \bar{\Gamma}_1(u) \setminus \{u\}\} \Leftrightarrow \{\mathcal{P}(v) \in sT(u)\}$.

2.1   The Clustering Heuristic

Our initial goal was to propose a way to use multi-hop wireless networks over large scales. We proposed a clustering algorithm motivated by the fact that in a multi-hop wireless environment, the less information exchanged or/and stored, the better. First, we wanted a cluster organization with no-overlapping clusters with a flexible radius (Many clustering schemes [5, 11] have a radius of 1, in [1, 7] the radius is set a priori.), able to adapt to the different topologies. Second, we wanted the nodes to be able to compute the heuristic from local information, only using their 2-neighborhood knowledge. (In [1], if the cluster radius is set to d, the nodes need to gather information up to d hops away before taking any decision.) Finally, we desired an organization robust and stable over node mobility, *i.e.*, which do not need to be recomputed at each single change in the topology. For it, we introduced a new metric called *density* [13]. The notion of density of a node u (noted $\rho(u)$) characterizes the "relative" importance of u in the network and within its neighborhood. This link density smooths local changes down in $\Gamma_1(u)$ by considering the ratio between the number of links and the number of nodes in $\Gamma_1(u)$.

Definition 1 (density).
The density of a node $u \in V$ is: $\rho(u) = \frac{|\{e = (v, w) \in E \mid w \in \{u\} \cup \Gamma_1(u) \text{ and } v \in \Gamma_1(u)\}|}{\delta(u)}$

Because of page restrictions, we only give here a sketch of the cluster/tree formation, but the algorithm and an example can be found in [13]. On a regular basis, each node locally computes its density value and regularly locally broadcasts it to its 1-neighbors (*e.g.*, using `Hello` packets). Each node is thus able to compare its density value to its 1-neighbors' and decides by itself whether it joins one of them (the one with the highest density value) or it wins and elects itself as cluster-head. The node Id are used to break ties. In this way, two neighbors can not be both cluster-heads. We actually draw a tree $T' = (V, E')$ which is a subgraph of G, such that $E' \subset E$. T' is actually a directed acyclic graph (DAG). A DAG is a directed graph that contains no cycles, *i.e.* a directed tree. The node which density value is the highest within its neighborhood becomes the root of the tree and thus the cluster-head of the cluster. If node u has joined node w, we say that w is node u's parent (noted $\mathcal{P}(u) = w$) in the clustering tree and that node u is a child of node w (noted $u \in \mathcal{C}h(w)$). A node's parent can also have joined another node and so on. A cluster then extends itself until it reaches another cluster. If none of the nodes has joined a node u ($\mathcal{C}h(u) = \emptyset$), u becomes a leaf. All the nodes belonging to a same tree belong to the same cluster. We thus build the clusters by building a spanning forest of the network in a distributed and local way. As proved in [16], at the end of three message exchange rounds, each node is aware of its parent in the tree, at the end of four message rounds, it knows the parent of each of its neighbors and thus is able to determine whether one of them has elected it as parent and thus learns its condition in the tree (root, leaf, regular node). A node is a leaf if no other node has chosen it as its parent; a node is a cluster-head if it has chosen itself as parent and all its 1-neighbors have joined it; a node is a regular node otherwise. It has also been proved that in an expected constant and bounded time, every node is also aware of its cluster-head identity and of the cluster-head identity of its neighbors. It thus knows whether it is a border node. A node is a frontier node if at least one of its neighbors does not belong to the same cluster than itself.

2.2   Some Characteristics of Our Clustering Algorithm

The cluster formation algorithm stabilizes when every node knows its *correct* cluster-head value. In [16], it has been proved by theory and simulation to self-stabilize within a low, constant and bounded time. It also has been proved that a cluster-head is aware of an information sent by a frontier node in a constant and bounded time since the tree depth is bounded. The number of clusters built by this heuristic has been studied analytically and by simulation. It has shown to be upper bounded by a constant asymptote when the number of nodes in the network increases. Compared to other clustering schemes as DDR [19] or Max-Min d cluster [1], our cluster organization has revealed to be more stable over node mobility and arrivals and to offer a better behavior over non-uniform topologies (see [13]). Moreover, our algorithm presents a smaller complexity in time and messages as it only needs information regarding the 2-hop neighborhood of a node while Max-Min needs information to d hops away and DDR nodes need to store information about all the nodes belonging to the same cluster than themselves.

Other interesting features for routing and locating obtained by simulations are gathered in Table 1. These characteristics illustrate some of our motivations for our

Table 1. Some cluster and clustering trees characteristics

	500 nodes	600 nodes	700 nodes	800 nodes	900 nodes	1000nodes
# clusters/trees	11.76	11.51	11.45	11.32	11.02	10.80
Cluster diameter	4.99	5.52	5.5	5.65	6.34	6.1
Cluster-head eccentricity	3.01	3.09	3.37	3.17	3.19	3.23
Node eccentricity	3.70	3.75	3.84	3.84	3.84	3.84
Tree depth	3.27	3.34	3.33	3.34	3.43	3.51
Degree in the tree of non-leaves	3.82	3.99	4.19	4.36	4.51	4.62
% leaves	73,48%	74,96%	76,14%	76,81%	77,71%	78,23%

proposition as explained later in Section 3. The eccentricity of a node is the greater distance in number of hops between itself and any other node in its cluster. We can see in Table 1 that the tree depth is low and close to the optimal (cluster-head eccentricity). That means that the routes in the trees from the cluster-head to any other node within its cluster are close to the shortest paths in the network. Clustering trees present some interesting properties as a great proportion of leaves and a small amount of non-leaf children per node. This feature and the *"Degree in the tree of non-leaf nodes"* entry in Table 1 show that, in average, an internal node does not have a lot of children.

3   Basic Ideas

In fixed networks, routing information is embedded into the topological-dependent node address. For instance, an IP address both identify a node and locate it since the network prefix is included in the IP node address. In wireless networks, nodes may arbitrarily move, appear or disappear at any time. So, the permanent node identifier can not include dynamic location information and thus, it has to be independent from the topology, which implies an indirect routing between nodes. A routing operation is referred as indirect when it is performed in two steps: *(i)* first **locate** the target and then *(ii)* **communicate** with the target. This allows the network to dissociate the location of a node from the location itself. With this approach, the routing information can be totally distributed, which is important for achieving scalability in large scale networks. Figure 1 illustrates such a routing process.

We propose to use such an indirect routing scheme for routing in large scale multihop wireless networks. We are motivated by the fact that we want a very scalable solution, thus our proposition aims to store as less information on nodes as possible. We also want to avoid situations where the distance between the source/requester (node u in Figure 1) and the rendezvous node (node w) is much greater than the distance between the source (node u) and the destination (node v). Indeed, if the request has to cross twice the whole network before two nodes are able to directly communicate, we waste bandwidth and latency. Distributed Hash Tables (DHT) are a basis for indirect routing. They provide a general mapping between any information and a location. They use a virtual addressing space $\mathcal{V}$. Partitions of this virtual space are assigned to nodes in the network. The idea is to use a *hash* function to first distribute node location information among rendezvous points. This same *hash* function is known by every node and may then be used by a source to identify the rendezvous point which stores the position of the target node. Each information is *hashed* into a key ($hash(v) = key_v \in \mathcal{V}$) of this virtual addressing space $\mathcal{V}$ and is then stored on the node(s) responsible for the partition of the virtual space this key belongs to.

DHT have been applied at two different levels: at the application level and at the network level. Applying DHT at the application layer is widespread in peer-to-peer networks. The information hashed in such file-sharing systems is the identity of a file. The node responsible for the $key = hash(file)$ stores the identifier of the nodes which detain that file. DHT nodes form an overlay network on which the lookup queries are routed. The main difference among the many proposals is the geometry of this

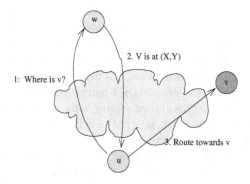

Fig. 1. Indirect Routing: Node u needs to ask w where v is

overlay [24, 8, 12]. At the network layer, DHT are applied to distribute node location information throughout the topology and are used to identify a node which is responsible for storing a required node location. This is the way we intend to use DHT. When a node u needs to send an information to a node v, it first has to know where v is. To get this information, it first asks a node w in charge of the key $k = hash(v)$ and thus knowing where v is. In DHT-independent routing schemes, *i.e.*, the virtual address is not used for the routing operation. The nodes generally know their geographic coordinates, either absolute (by using a GPS for example) or relative, which is the location information they associate to the key. By performing $hash$(target), a node u gets the geographical coordinates of a rendezvous area $\mathcal{A}$. u then applies a geographic routing protocol to join a node v laying in $\mathcal{A}$ and which is aware of the geographical coordinates of the target node. From it, node u is able to reach the destination by performing a geographical routing again. This is the case for instance in [2, 17, 18], in the Terminodes project [3] or in the Grid project [10]. As we do not want our nodes to depend on a positioning system, we can not apply that DHT utilization. In DHT-dependent routing schemes, the virtual space of the DHT is used not only for locating but at the same time for routing toward the destination. The virtual address is dependent of the location. In this way, the coherency of the routing protocols relies on the coherent sharing of the virtual addressing space among all nodes in the network. The routing is performed over the virtual structure. In such scenarii, a node u performing $hash(w)$ gets the virtual address of the rendezvous point. From it, u routes in the virtual space to v which gives it the virtual address of w. u thus is able to reach w by performing a routing again in the virtual space. The routing scheme used is generally a greedy routing: "Forward to the neighbor in the virtual space whose virtual address is the closest to the virtual address of the destination". This is for instance the case of Tribe [26] or L+ [4] on which is based SAFARI [21]. The main challenge here is to disseminate the partitions of the virtual space in such a manner that the paths in the virtual space are not much longer than the physical routes.

Thus, in DHT-based systems, we can consider two phases of routing: *(i)* a routing toward the rendezvous node which stores the needed information (Arrow 1 on Figure 1) and *(ii)* a routing toward the final destination node which location had been obtained by the lookup operation (Arrow 3 on Figure 1). In all proposals cited above, both routing phases are performed in the same way, either in the physical network (for DHT-independent routing proposals), or in the virtual one (for DHT-dependent routing proposals). In the approach we propose, the two routing steps are completed in two different manners. The first routing step is performed by using the virtual address of the rendezvous point (DHT-dependent) whereas the routing toward the final destination is performed over the physical network (DHT-independent). Indeed, as we propose to distribute a virtual space over each cluster, routing to the destination in the virtual space is not possible as the destination node may not be in the same cluster as the sender node and thus be in a different virtual space.

In our proposal, we propose nodes register their cluster Id as location information. As seen in Section 2, we have a tree structure. We propose to partition the virtual space $\mathcal{V}$ each tree and that each node registers on each cluster in order to add redundancy. In this way, when a node v looks for a node u, it just has to search the information in its cluster. As the node eccentricity is low (Section 2), the latency is reduced. Moreover, partitioning the virtual space in each cluster rather than once in the whole network avoids situations where the distance between the source and the rendezvous point is much greater than the distance between the source and the destination, as in this way, the source and the rendezvous point always are in the same cluster whereas the destination may be anywhere in the network.

To distribute the partitions of the virtual space of the DHT in such a way that, given a virtual address, a node u is able to find the node responsible for without any additional information, we use a tree Interval Labeling Scheme to then allow an Interval Routing on the virtual space. Interval Routing is was introduced in wired networks by Santoro and Khatib in [23] to reduce the size of the routing tables. It is based on representing the routing table stored at each node in a compact manner, by grouping the set of destination addresses that use the same output port into intervals of consecutive addresses. The main advantage of this scheme is the low memory requirements to store the routing on each node u: $O(\delta(u))$. The routing is computed in a distributed way with the following algorithm: at each intermediate node x, the routing process ends if the destination y corresponds to x, otherwise, it is forwarded with the message through an edge labeled by a set I such that $y \in I$. The Interval Labeling Scheme (ILS) is the manner to assign the intervals to the edges of each node, in order to perform an efficient interval routing with routes as short as possible. Yet, the authors of [25] showed that undirected trees can support an Interval Routing Scheme with shortest paths (in the tree) and with only one interval per output port when performing a Depth-First-Search ILS. In wired networks, nodes have to store an interval for each of its output edge. Therefore the size of the routing table is in $O(\delta(u))$. But in wireless environments, a transmission by each node can reach all nodes within radius distance from it. Edges actually are hyper-edges (see Figure 2). Thus the problematic is a bit different as querying unicast transmission actually are broadcast transmission. Indeed, as from a node u, there is only one hyper-edge, nodes can store only one interval for it and thus for all their neighbors. In our proposal, nodes only store the interval for which their subtree is responsible (and the intervals of each of their neighbors). This gives a table routing size in $O(1)$. When a query is sent, as all neighbors receive it in any way, only the one(s) concerned by it answer(s).

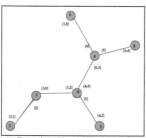
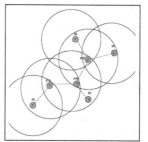

(a) Routing in wired environments. Nodes store one interval per output edge.

(b) Routing in wireless environments. Nodes store only one interval (one per hyper-edge).

Fig. 2. Edges in wired networks (a) Vs hyper-edges in wireless networks (b)

Summary and complexity analysis. To sum up, we propose to apply an indirect routing scheme over our clustered network by using DHT which associate each node identifier to a virtual address of a space $\mathcal{V}$. The set of virtual addresses $\mathcal{V}$ is partitioned d times over the nodes of each cluster. As the number of cluster is constant and clusters are homogeneous, each node finally stores $O(1)$ location information.

When a node u wants to communicate with a node v, it first uses the DHT to find out the virtual address of v: $hash(v) = key_v \in \mathcal{V}$. Then, by using an Interval Routing over the virtual space $\mathcal{V}$ of its own tree, it reaches at least one node in its cluster responsible for storing the location of v, *i.e.*, $\mathcal{C}(v)$. As the intervals of the neighbors are not stored on the node, this one only stores its own interval and the size of its routing table is in $O(1)$. As the number of clusters is constant when the intensity of nodes increases, each cluster has $O(1)$ routes to maintain toward other ones for the proactive routing phase.

4   Our Proposition

In this section, we describe how we wish to use DHT and Interval Routing over our cluster organization. However, because of page restriction, several details have been eluded but can be found in [14].

Virtual space partitioning. In this section, we present the way we distribute the partitions of $\mathcal{V}$, which leads to a Depth-First Search (DFS) ILS, *i.e.*, the optimal ILS for a tree . Let $I(u)$ be the partition of $\mathcal{V}$ node u is assigned. $i(u)$ is the first element of $I(u)$: $I(u) = [i(u), ...[|i(u) \neq hash(u)$. $i(u)$ is used as the virtual identifier of node u in the virtual space $\mathcal{V}$. Let $I_{tree}(s\mathcal{T}(u)) = \bigcup_{v \in s\mathcal{T}(u)} I(v)$ be the interval/partition of $\mathcal{V}$ of which the subtree of node u is in charge. $|I|$ is used to refer to the size of interval I. Partitions of $\mathcal{V}$ are distributed in such a way that, for every node $u \in V$:

– The intervals of the nodes in $s\mathcal{T}(u)$ form a contiguous interval.
– The size of the interval a subtree is in charge of is proportional to its size: $|I_{tree}(s\mathcal{T}(u))| \propto |s\mathcal{T}(u)|$. We thus have, for every node $v \in s\mathcal{T}(u)$, $|I_{tree}(s\mathcal{T}(u))| \geq |I_{tree}(s\mathcal{T}(v))|$.

- $\mathcal{V}$ is completely shared among the nodes of the cluster: $\mathcal{V} = \bigcup_{v \in \mathcal{C}(u)} I(v)$.
- Regions are mutual exclusive: $\forall v \in \mathcal{C}(u), \forall w \in \mathcal{C}(u), v \neq w \ I(v) \cap I(w) = \emptyset$.

We propose a parallel interval distribution over the different branches of each tree. This distribution can be qualified of quasi-local according to the taxonomy established in [27] as each node u needs information up to $d_{tree}(u, \mathcal{H}(u))$ only, where d_{tree} $(u, \mathcal{H}(u))$ is the number of hops in the tree between u and its cluster-head $\mathcal{H}(u)$ in the tree. Our algorithm runs in two steps: a step up the tree (from the nodes to the cluster-head) and a step down the tree (from the cluster-head to the nodes). The complexity in time of our distribution algorithm for a cluster/tree is $2 \times (Tree\_depth)$. As the tree depth is bounded by a constant, the complexity in time is $O(1)$. Each step has a time complexity of $O(Treedepth)$. A node u which has been assigned an interval $I(u)$ is responsible for storing location information of all nodes v such that $hash(v) \in I(u)$. Note that, as $\mathcal{V}$ is much smaller than the domain of the node identifiers, they are several nodes v such that $hash(v_i) = hash(v_j)$. As each internal node only has few children to which distribute the partitions of the virtual space (Section 2), this ILS does not include a lot of computing on nodes.

Step 1. As seen in Section 2, every node u might be aware in an expected bounded time, of the parent of each its neighbors. It thus is able to determine whether one of them has elected it as parent and thus learns its condition in the tree (root, leaf, regular node). Thus, in a low and constant time, each internal node in the tree is aware of the number of its children. If every node sends its parent the size of its subtree $(|sT(u)| = |u \bigcup_{v \in Ch(u)} sT(v)| = 1 + \sum_{v \in Ch(u)} |sT(v)|)$ up to the cluster-head, each node is expected to know the size of the subtree of each of its neighbors in a low time and so the cluster-head.

Step 2. Once the cluster-head is aware of the size of the subtree of each of its neighbors/children, it shares $\mathcal{V}$ between itself and its children. Each node v is assigned a partition of $\mathcal{V}$: $I_{tree}(sT(v))$ proportional to the size of its subtree. Each internal node then re-distributes the partition its parent assigned it, between itself and its own children, and so on, till reaching the terminal leaves. Once an internal node u has assigned partitions of the virtual space among its children, it only stores the interval $I_{tree}(sT(v))$ for which its subtree is responsible (and not intervals each of its children is in charge of). Then, it stores location information for nodes which key is in $I(u)$ only (and not for all keys in $I_{tree}(sT(u))$).

Departures and arrivals. When a node arrives in a tree, it is responsible for none interval for a while. When a node leaves, the information for which it was responsible is lost (but is still expected to be found in other clusters). Each internal node u is aware of the departures and arrivals of its children. When it sees too many changes among its children, it locally re-distributes $I_{tree}(sT(u))$ among itself and its children. When intervals are re-assigned, in order to maintain the previous information stored by the nodes and not to loose it, every node keeps the latter information it was responsible for, in addition to the new one, for a period time $\Delta(t)$, $\Delta(t)$ being the period at which nodes register their location.

Routing in the virtual space. In this section, we detail how the Interval Routing is performed in a tree. The routing is performed till reaching the node responsible for this key. In our model, each node u has a unique identifier $Id(u)$. As in every DHT scheme, we assume that every node knows a specific function $hash$ which associates each node identifier to a value in the logical space $\mathcal{V}$: $hash : \mathbb{R} \rightarrow \mathcal{V}$, $Id(u) \rightarrow hash(u)$. As $\mathcal{V}$ is much smaller than the domain of the node identifiers $\mathbb{R}$, several nodes may have the same value returned by the $hash$ function. We use the following tuple as a key for a node x: $\{hash(x), id(x)\}$. In the following, we may use only x instead of $Id(x)$. A node u uses that kind of routing when it needs to reach a node responsible for a given key, which can happen for three reasons:

- u **wants to register a position:** u may need to register its position. In this case, u is looking for the node responsible for its own virtual address: $hash(u)$. u then sends a Registration Request (RR) $\langle RR, key = \{hash(u), u\}, \mathcal{C}(u), flag \rangle$.
- u **needs to locate** x: in this case, u is looking for the node responsible for the virtual address of x: $hash(x)$. u sends a Location request (LR) $\langle LR, key = \{hash(x), x\}, i(u), flag \rangle$. $i(u)$, which is the identifier of u in the virtual space, will then be used to reply to node u.
- u **needs to answer a location request for a key it is responsible for** ($key \in I(u)$)**:** in this case, node u has received a Location Request such that $\langle LR, key = \{hash(x), x\}, i(v), flag \rangle$ initiated on node v such that $key \in I(u)$. It has to answer to node v by sending a Location Reply (Reply) $\langle Reply, key = \{i(v), -1\}, \mathcal{C}(x), flag \rangle$.

The routing process is the same whatever the kind of message (LR, RR or Reply) as the routing decision is only based on the key. In every case, the value $flag$ is set to 1 by the node forwarding the message if key belongs to the interval its subtree is responsible for, it is set to 0 otherwise. As detailed later, it is useful for the routing decisions. Remark that node u already knows the location (cluster Id) of its neighbors, of its cluster-head and of the nodes it is responsible for. Thus, if node v is such that $v \in \mathcal{H}(u) \cup \Gamma_1(u)$ or $hash(v) \in I(u)$, node u directly routes toward node v, skipping the localization steps (skipping steps 1 and 2 on Figure 1).

Upon reception of a message M (RR, LR or Reply) containing the key $\{hash(x), x\}$ coming from node u ($u \in \Gamma_1(v)$), node v decides to end routing, forward M or discard M. Note that node u may just forward itself the message and is not necessarily the request initiator. The routing ends when M reaches a node v which either is responsible for the wanted key ($key \in I(v)$) or is the wanted node ($key = \{hash(v), v\}$).

If $key \neq \{hash(v), v\}$, node v forwards M in three cases:

- If $u = \mathcal{P}(v)$ and $key \in I_{tree}(sT(v))$ (the message is coming from node v's parent and the key is in its subtree's interval). See Figure 3(a).
- If $u \in Ch(v)$ (the message is coming from a child of node v), v forwards M:
  - if $key \notin I_{tree}(sT(v))$ (the key is not in its subtree, and obviously neither in the subtree of its child v): the message has to follow its way up the tree. See Figure 3(b).

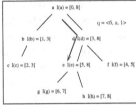
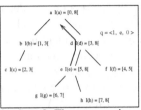
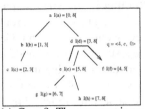

(a) Case 1: The message is go- (b) Case 2: The message is go- (c) Case 3: The message is going down the tree. a is looking ing up the tree. e is looking ing up and down the tree. e is
for the node in charge of 6.   the node in charge of 1.   looking for the node in charge
of 4.

Fig. 3. Different cases of figures of when a message received on node d is forwarded

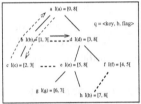
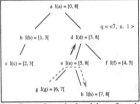
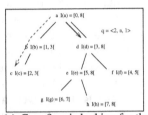

(a) Case 1: The message is (b) Case 2: e is looking for the (c) Case 3: a is looking for the
coming from a neighbor of node key 4. The message is going up key 2. Node d is not in charge
d but does not concern node d. and down the tree on node e but of the researched key. One of its
is heard by node d.   sibling will forward.

Fig. 4. Different cases of figures when a message received on node d is discarded on d. The
dashed arrows represent the possible paths followed by the message.

- if $key \in I_{tree}(sT(v))$ and $key \notin I_{tree}(sT(u))$ $(flag = 0)$ (the key is in
  its subtree but not in the subtree of its child from which it has received the
  request): the message has to be forwarded down its subtree to another child.
  See Figure 3(c).

And node v discards M in all other cases, which means:

- If $u \notin P(v) \cup Ch(v)$ (M is coming from a node which is neither the parent nor a
  child of node v). Figure 4(a).
- If $u \in Ch(v)$ and $key \in I_{tree}sT(u)$ $(flag = 1)$ (M is coming from a child v
  which subtree is responsible for the key). Thanks to the flag, u knows it does not
  need to forward as the message goes up and down the tree via v. Figure 4(b).
- If $u \in P(v)$ and $key \notin I_{tree}(sT(v))$ (M is coming from v's parent but the subtree
  of v is not responsible of the key). u is not concerned by the request, it does not
  forward. One of its siblings will. Figure 4(c).

Algorithm 1 describes the routing operation. When a RR message
$\langle RR, key = \{hash(u), u\}, C(u), flag \rangle$ reaches its final destination v, v updates the location of u in its table. When a Reply message $\langle Reply, key = \{i(u), -1\}, C(x), flag \rangle$
reaches its final destination u, u is thus able to route to $C(x)$ using the hierarchical

routing. When a LR message $\langle LR, key = \{hash(x), x\}, i(w), flag \rangle$ reaches its final destination v (in charge of $hash(x)$), v answers the sender by initiating a Reply message $\langle Reply, \{i(w), -1\}, \mathcal{C}(x), flag \rangle$.

Algorithm 1 Query Forwarding

For all node u, **upon reception of a message** $\langle Type, key = \{hash(x), x\}, X, flag \rangle, X \in \{RR, Reply, LR\}$ **coming from a node** $v \in \Gamma_1(u)$ **and initiated at a node** y:

    if $(u = x)$ **then** Reply sending $\langle Reply, \{X = i(y), -1\}, \mathcal{C}(u), flag \rangle$ and Exit **end**
    ▷ *u is the wanted node. It can answer node y.*
    if $(key \in I(u))$ **then**
    ▷ *u is responsible for storing the key. The message has reached its final destination.*
        if (Type = LR) **then** Send $\langle Reply, \{X = i(y), -1\}, \mathcal{C}(x), flag \rangle$ and Exit **end**
        if (Type = RR) **then** Register the location of node x and Exit **end**
        if (Type = Reply) **then** Route toward the destination cluster X, Exit **end**
    end
    if $(v = \mathcal{P}(u))$ **then**
    ▷ *The message is going down the tree.*
        if $(key \in I_{tree}(sT(u)))$ **then** Set $flag$ to 1 and Forward.
        ▷ $\exists w \in sT(u)$ *such that* $key \in I(w)$. *See Figure 3(a).*
        else Discard. ▷ *See Figure 4(b).*
        end
    else
        if $(v \in Ch(u))$ **then** ▷ *The query is coming up the tree from a child of node u.*
            if $(key \notin I_{tree}(sT(u)))$ **then** Set $flag$ to 0 and Forward.
            ▷ *The query is forwarded up the tree. See Figure 3(b).*
            else ▷$\exists w \in sT(u) \setminus \{u, v\}$ *such that* $key \in I(w)$.
                if $(flag = 0)$ **then** Set $flag$ to 1 and Forward.
                ▷ $key \notin I_{tree}(sT(v))$ *but as* $key \in I_{tree}(sT(u))$, *u has to forward the query to its other children. The query goes up and down. Figure 3(c).*
                else Discard. ▷ *The query goes up and down via v. See Figure 4(c).*
                end
            end
        end
        else Discard. ▷ *See Figure 4(a).*
    end
end

Routing in the physical network. In this section, we detail how we perform our hierarchical routing over our cluster topology by using a reactive routing protocol inside the clusters and a proactive routing protocol between the clusters. Such a proactive routing scheme implies for a node u to know the sequence of clusters to go through from its own cluster $\mathcal{C}(u)$ toward any other one. Algorithm 2 describes our hierarchical routing (We use This function is known by every node as the routing between clusters is proactive.).

Suppose node u needs to reach node v. If node u does not already know how to reach v, it uses the $hash$ function to learn $\mathcal{C}(v)$ before applying the routing rules. The routing process is illustrated on Figure 5. If $\mathcal{C}(v) = \mathcal{C}(u)$, then u initiates a reactive routing within its cluster to reach v. Otherwise, it looks at its routing table for the next cluster $\mathcal{C}(w)$ on the route toward $\mathcal{C}(v)$ and initiates a reactive routing in its cluster to

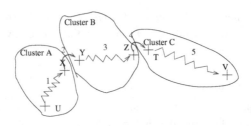

Fig. 5. Node u needs to communicate with node v. It uses successive reactive routing protocols to cross all clusters on its routes toward $C(v)$ and then to reach node v.

look for a node $x \in C(u)$ which is a frontier node of $C(w)$ (x such that $x \in C(u)$ and $\exists y \in \Gamma_1(x) \cap \overline{C(u)}$). Note that $C(u)$ and $C(w)$ may be two neighboring clusters, in this case $C(v) = C(w)$. The message is thus sent to x which forwards it to one of its neighbors y in the neighboring cluster $C(w)$. The routing process is thus reiterated on node y and so on till reaching the final destination. As the reactive routing step is confined in clusters which have low diameters, the induced flooding and its undesirable aspects are limited. Note that it can also be enhanced as in [15].

Algorithm 2 Routing

For a message M **sent by node** $x \in C(x)$ **to node** $y \in C(y)$

    $\mathcal{C}_{current} = C(x)$

    while $(\mathcal{C}_{next} \neq C(y))$

        $\mathcal{C}_{next} = Next\_Hop(\mathcal{C}_{current}, C(y))$

        Reactively route M toward node $u \in \mathcal{C}_{current}$ such that $\exists v \in \Gamma_1(u) \cap \mathcal{C}_{next}$.

        Node u sends M to node v.

        $\mathcal{C}_{current} = \mathcal{C}_{next}$

    end

    ▷ *The message has reached the cluster of the destination node.*

    Reactively route M toward destination node y.

Stretch factor. The *stretch factor* is the difference between the cost (or length) of the route provided by the routing protocol and the optimal one. Flat routing protocols generally provide optimal routes. The length of the routes provided by our proposed routing scheme is equal to twice the length of the route in the tree for locating the target node (Step 1 of the indirect routing), more the length of the final route to the target (Step 2 of the indirect routing). In our proposition, as, the routing process of the second step is performed over the physical topology with flat protocols, the stretch factor of this step is close to 1. Thus, only the stretch factor induced by the first step (routing in the virtual space) is noticeable. Our global stretch factor s is such that $s = 1 + 2 \times l(u, v)$ where $l(u, v)$ is the length of the path in the virtual space (tree) from node u to node v, v being the rendezvous point soring the information needed by u. As this routing scheme is performed within a cluster only, we can bound $l(u, v)$ by the length of the longer path in the tree: $l(u, v) \leq 2 \times Tree\_depth$. Finally, we have $s \leq 1 + 4 \times Tree\_depth$. As already mentioned, $Tree\_depth$ is a low constant (between 3 and 4 hops). Thus,

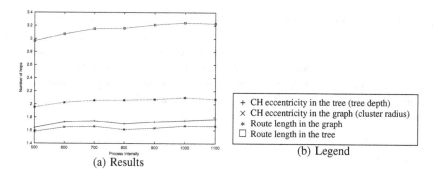

(a) Results   (b) Legend

Fig. 6. Comparisons of distances in the logical and physical topologies

routes provided by our locating/routing proposition may be costly in term of number of hops only for small path in the graph since, as we evolute in a large scale environment, this constant can be expected negligible in front of most of the path lengths. It is not the case only for some routes within a cluster. We evaluated the length and the stretch factor of paths of our indirect routing first step by simulation. We used a simulator we developed. Nodes are randomly deployed using a Poisson point process in a 1×1 square with various levels of intensity λ. In such a process, λ represents the mean number of nodes per surface unit. The communication range R is set to 0.1 in all tests. In each case, each statistic is the average over 1000 simulations. We compared the length of paths between two nodes of the cluster in the graph (physical topology) and in the tree (logical topology) to know the cost for routing in the tree. Results are shown in Figure 6. As we can note, path lengths are quite constant when the number of nodes increase and remains low (between 3 and 4 hops). To reach a rendezvous node v, node u has to use the tree (as it needs to perform the interval routing in the logical space). Using the tree only adds 1 hop in average than using the physical graph (so 2 hops for a round trip) and induce much less traffic overhead and memory requirements than using a classical routing in a cluster. Nevertheless, we remind that the total stretch factor for a node u looking for a node v, actually is equal to the length round trip path from node u to node w responsible for $hash(v)$. It does thus not exactly correspond to this additional hop, but as the direct routing which is used in the second routing step has a negligible stretch factor, it is not far from it. So, we can say that the locating/routing approach we propose does not add much latency, except when a node tries to locate a node in the same cluster than its. However, as we are dealing with a great amount of nodes, the proportion of such communications is low.

5   Conclusion

In this paper, we have proposed a way for locating a node and routing toward it in a large scale clustered wireless multi-hop network. This scheme lies on a hierarchical and indirect routing which only needs $O(1)$ memory size and generates low traffic overhead and latency. It takes advantages of the underlying tree structure to perform an efficient

Interval Routing within clusters. Then, unlike what is usually proposed in the literature, we use a pro-active routing approach between clusters and a reactive one inside the clusters. In future works, we intend to analyze deeper our algorithm concerning refreshing periods of node location registration and re-assignment of the partitions of the logical space over different trees. Then, we plan to compare our proposition to other existing ones as for instance SAFARI [21] which also uses this reverse approach for the hierarchical routing.

References

1. A. Amis, R. Prakash, T. Vuong, and D. Huynh. Max-Min d-cluster formation in wireless ad hoc networks. In *INFOCOM*, Tel-Aviv, Israel, 2000.
2. F. Araujo, L. Rodrigues, J. Kaiser, L. Changling, and C. Mitidieri. CHR: A Distributed Hash Table for Wireless Ad Hoc Networks. In *DEBS'05*, Columbus, USA, 2005.
3. L. Blazevic, S. Giordano, and J.-Y. Le Boudec. Self-organized Terminode routing. *Journal of Cluster Computing*, 5(2), April 2002.
4. B. Chen and R. Morris. L+: Scalable landmark routing and address lookup for multi-hop wireless networks. Mit lcs technical report 837, March 2002.
5. G. Chen, F. Garcia, J. Solano, and I. Stojmenovic. Connectivity-based k-hop clustering in wireless networks. In *HICSS'02*, Hawaii, USA, 2002.
6. Y. P. Chen, A. L. Liestman, and J. Liu. Clustering algorithms for ad hoc wireless networks. *Ad Hoc and Sensor Networks*, 2004.
7. Y. Fernandess and D. Malkhi. k-clustering in wireless ad hoc networks. In *POMC'02*, Toulouse, France, 2002.
8. P. Fraigniaud and P. Gauron. An overview of the content-addressable network D2B. In *PODC'03*, July 2003.
9. P. Krishna, N. H. Vaidya, M. Chatterjee, and D. K. Pradhan. A cluster based approach for routing in dynamic networks. In *ACM SIGCOMM*, pages 49–65, April 1997.
10. J. Li, R. Morris, J. Jannotti, D. S. Decouto, and D. R. Karger. A scalable location service for geographic ad hoc routing. In *Mobicom'00*, pages 120 – 130, 2000.
11. C. R. Lin and M. Gerla. Adaptive clustering for mobile wireless networks. *IEEE Journal of Selected Areas in Communications*, 15(7):1265–1275, 1997.
12. P. Maymounkov and D. Mazires. Kademlia: A peer-to-peer information system based on the XOR metric. In *IPTPS '02*, MIT Faculty Club, Cambridge, USA, 2002.
13. N. Mitton, A. Busson, and E. Fleury. Self-organization in large scale ad hoc networks. In *MED-HOC-NET 04*, Bodrum, Turkey, 2004.
14. N. Mitton and E. Fleury. Distributed node location in clustered multi-hop wireless networks. Technical Report In proceed, INRIA, 2005.
15. N. Mitton and E. Fleury. Efficient broadcasting in self-organizing multi-hop wireless network. In *Ad Hoc Now'05*, Cancun, Mexico, 2005.
16. N. Mitton, E. Fleury, I. Gurin-Lassous, and S. Tixeuil. Self-stabilization in self-organized multihop wireless networks. In *WWAN'05*, Columbus, USA, 2005.
17. E. T. Ng and H. Zhang. Predicting Internet network distance with coordinates-based approaches. In *INFOCOM*, New-York, USA, 2002.
18. D. Niculescu and B. Nath. Ad hoc positioning system (APS). In *GLOBECOM'01*, 2001.
19. N. Nikaein, H. Labiod, and C. Bonnet. DDR-distributed dynamic routing algorithm for mobile ad hoc networks. In *MobiHoc*, Boston, USA, 2000.
20. C. Perkins. *Ad hoc networking*. Addison-Wesley, 2001.

21. R. Riedi, P. Druschel, Y. C. Hu, D. B. Johnson, and R. Baraniuk. SAFARI: A self-organizing hierarchical architecture for scalable ad hoc networking. Technical Report TR04-433, Rice University, February 2005.

22. C. Santivanez, B. McDonald, I. Stavrakakis, and R. R. Ramanathan. On the scalability of ad hoc routing protocols. In *INFOCOM*, New-York, USA, 2002.

23. N. Santoro and R. Khatib. Labeling and implicit routing in networks. *The computer Journal*, 28:5–8, 1985.

24. I. Stoica, R. Morris, D. Karger, M. F. Kaashoek, and H. Balakrishnan. Chord: A scalable peer-to-peer lookup service for Internet applications. In *SIGCOMM'01)*, 2001.

25. J. Van Leeuven and R. Tan. Interval routing. *The computer Journal*, 30:298–307, 1987.

26. A. C. Viana, M. Dias de Armorim, S. Fdida, and J. Ferreira de Rezende. Self-organization in spontaneous networks: the approach of DHT-based routing protocols. *Ad Hoc Networks Journal*, 2005.

27. J. Wu and W. Lou. Forward node set based broadcast in clustered mobile ad hoc networks. *Wireless Communications and Mobile Computing*, 3(2):141–154, 2003.

A Duplicate Address Detection and Autoconfiguration Mechanism for a Single-Interface OLSR Network

Saadi Boudjit, Cédric Adjih, Anis Laouiti, and Paul Muhlethaler

INRIA, France
{saadi.boudjit, cedric.adjih, anis.laouiti, paul.muhlethaler}@inria.fr

Abstract. Mobile Ad hoc NETworks (MANETs) are infrastructure-free, highly dynamic wireless networks, where central administration or configuration by the user is very difficult. In hardwired networks nodes usually rely on a centralized server and use a dynamic host configuration protocol, like DHCP [7], to acquire an IP address. Such a solution cannot be deployed in MANETs due to the unavailability of any centralized DHCP server. For small scale MANETs, it may be possible to allocate free IP addresses manually. However, the procedure becomes impractical for a large-scale or open system where mobile nodes are free to join and leave. Numerous dynamic addressing schemes for ad hoc networks have been proposed. These approaches differ in a wide range of aspects, such as the usage of centralized servers or full decentralization, hierarchical structure or flat network organization, and explicit or implicit duplicate address detection. In [1] we have proposed an autoconfiguration solution for $OLSR$, which can detect and resolve only a single address duplication in the network. In this paper, however, we will present a complete and optimized version of the autoconfiguration solution for $OLSR$ proposed in [1]. This solution detects and resolve duplications whatever the number of address conflicts in the network, and, as in [1], it is based on an efficient Duplicate Address Detection(DAD) algorithm which takes advantage of the genuine optimization of the $OLSR$ routing protocol [3].

Keywords: MANET; Autoconfiguration; OLSR.

1 Introduction

Many fruitful efforts have focused on routing protocols for $MANET$ in recent years. These MANET protocols can be classified into proactive protocols [3] where each node maintains an up-to-date version of the network topology by periodic exchange of control messages with neighboring nodes; and reactive protocols [4] where each node discovers the route to a destination on demand.

Most of these routing protocols assume that mobile nodes in ad hoc networks are configured a priori with a unique address before joining a manet. Because mobile nodes may frequently move from one network to another, it is desirable for them to obtain addresses via dynamic configuration. The IPv6 and ZERO-CONF working groups of the IETF deal with autoconfiguration issues but with

K. Cho and P. Jacquet (Eds.): AINTEC 2005, LNCS 3837, pp. 128–142, 2005.

a focus on wired networks. Automatic address allocation is more difficult in a *MANET* environment than in wired networks due to instability of links, mobility of the nodes, the open nature of the mobile ad hoc networks, and lack of central administration in the general case. Thus performing a DAD (Duplicate Address Detection) generates more complexity and more overhead in ad hoc networks than in wired networks where protocols such as DHCP [7] and SAA [5] can be used.

Recently, a considerable number of dynamic addressing schemes for ad hoc networks have been proposed. These approaches differ in a wide range of aspects, such as address format, usage of centralized servers or full decentralization, hierarchical structure or flat network organization and explicit or implicit duplicate address detection. In this paper we will present a complete and optimized version of the autoconfiguration solution proposed in [1] for the *OLSR* protocol. The autoconfiguration algorithm proposed in [1], works under the assumption that there are only two nodes with a duplicated address in the network. However, our new autoconfiguration mechanism takes into account the case of networks with multiple address duplications. This mechanism is based on an efficient Duplicate Address Detection (DAD) algorithm which takes advantage of the genuine optimization of the *OLSR* protocol.

The paper is structured as follows: section 2 is dedicated to related work, mostly concerning previously proposed autoconfiguration protocols in ad hoc networks. Section 3 gives the main features of the *OLSR* routing protocol. Then section 4 describes the duplicate address detection mechanism which is the core of our proposed autoconfiguration protocol. A formal proof of correctness of this detection algorithm is given. Section 5 proposes different ways to assign an initial IP address to a newly arriving node and to resolve address duplication. The overhead generated by the autoconfiguration algorithm and some simulation results are also provided in section 6. The paper concludes in section 7.

2 Related Work

Numerous studies have been carried out on autoconfiguration protocols and the related issue of duplicate address detection in ad hoc networks. These studies have been proposed within the IETF or published in academic papers.

2.1 Address Autoconfiguration Scenarios and Address Duplications

Before describing algorithms, we first highlight some scenarios where address duplications may occur and which allow to discriminate between the different algorithms.

The first scenario is the simplest: a mobile node joins and then leaves a *MANET*. An unused IP address is allocated to the node on its arrival and becomes free on its departure. In some scenarios, the allocated IP address may be duplicated in the network.

A more complicated scenario is the following: nodes are free to move arbitrarily in the *MANET* and, consequently, the network may becomes partitioned. In the resulting partitions, the nodes continue to use the previously allocated IP addresses. If a new node comes to one of the partitions, it may be assigned an IP address belonging to another partition. Address duplication may occur when the partitions merge later.

Another scenario is when two independent *MANETs* merge. Because the two *MANETs* were configured separately, and the address allocation in each of the *MANETs* is independent of the other, there may be duplicated addresses when the two *MANETs* merge.

Most of the other scenarios, can be thought of as special cases of the three scenarios described above.

2.2   Address Autoconfiguration in Ad Hoc Networks

There are two main approaches to address autoconfiguration in ad hoc networks. The first approach tends to allocate conflict free addresses. The algorithms in this approach are often called conflict-free allocation algorithms. The second approach tends to allocate addresses on a random basis and uses dedicated mechanisms to detect duplicate addresses. When duplications are detected, new addresses with new values are assigned.

The Distributed Dynamic Host Configuration Protocol (DDHCP) [12] is one example of a conflict-free allocation algorithm. In this algorithm, the nodes responsible for allocation try to assign an unused IP address to a new node – unused to the best of their knowledge. Then the new node performs DAD to guarantee that it is an unallocated IP address. DDHCP maintains a global allocation state, so IP addresses which have been used, and addresses which have not yet been allocated, are known. When a new node joins the network, one of its neighbors choses an unused address for it. The same unused IP address in the global address pool could be assigned to more than one node arriving at almost the same time. This is the reason why DAD is still performed by a node after getting an IP address. This algorithm takes into account network partition and merger, and works well with proactive routing protocols.

Dynamic Configuration and Distribution Protocol (DCDP) [13] is another conflict-free allocation algorithm. When a new mobile node joins the *MANET*, an address pool is divided into halves between itself and a configured node. This algorithm takes into account network partition and merge, however conflicts will occur during the merge if two or more of the separately configured *MANETs* taking part in the merge, begin with the same reserved address range.

Another conflict-free allocation algorithm, called the 'prophet allocation protocol' has been proposed for large scale *MANETs* in [14]. The idea is that every mobile node executes a stateful function $f(n)$ to get a unique IP address. $f(n)$ is function of a state value called the *seed* which is updated for each node in the network. In this algorithm, mechanisms are proposed for network partition and merger. The difficulty in this solution is to find a function $f(n)$ which will guarantee the generation of unique IP addresses each time the function is executed by a node.

The algorithms related to the second approach, perform a DAD (Duplicate Address Detection) to ensure the uniqueness of the allocated IP address. The general procedure is that a node generates a tentative address and then performs DAD within its neighborhood (radio range of the node). If the address is unique, the DAD is performed again over the whole network and a unique IP address is constructed. Examples of such approaches include [2], [9] and [10]. DAD mechanisms can also be divided into two categories which differ in when, and how duplicate addresses are detected.

The ADAD (Active Duplicate Address Detection) mechanisms distribute additional control information in the network to prevent address duplication as, for instance, in [9] and [10].

In contrast, PDAD (Passive Duplicate Address Detection) algorithms [11], try to detect duplicates without disseminating additional control information in the network. The idea behind this approach is to continuously monitor routing protocol traffic to detect duplicates rather than sending additional control packets for this purpose. However, in [11] a so-called Address Conflict Notification (ACN) message is introduced for the purpose of conflict resolution, and no less than nine different approaches to detecting duplicated addresses have been presented for proactive link-state routing protocols.

3   OLSR and MPR Technique

This section describes the main features of the *OLSR* (Optimized Link State Routing) protocol [3]. *OLSR* is an optimization of a pure link state routing protocol. It is based on the concept of *multipoint relays (MPRs)* [8]. First, using *multipoint relays* reduces the size of the control messages: rather than declaring all links, a node declares only the set of links with its neighbors that are its *"multipoint relay selectors"*. The use of *MPRs* also minimizes the flooding of control traffic. Indeed only *multipoint relays* forward control messages (Figure 1). This technique significantly reduces the number of retransmissions of broadcast control messages [3] [8]. The two main *OLSR* functionalities, Neighbor Discovery and Topology Dissemination, are now detailed. Then we present *OLSR* "gateway" mechanism used by some nodes to declare reachability to their connected hosts and networks.

3.1   Neighbor Discovery

Each node must detect the neighbor nodes with which it has a direct link.

For this, each node periodically broadcasts *Hello* messages, containing the list of neighbors known to the node and their link status. The link status can be either *symmetric* (if communication is possible in both directions), *asymmetric* (if communication is only possible in one direction), *multipoint relay* (if the link is symmetric and the sender of the *Hello* message has selected this node as a *multipoint relay*), or *lost* (if the link has been lost). The *Hello* messages are received by all 1-hop neighbors, but are not forwarded. They are broadcasted once per refreshing period called the *"HELLO_INTERVAL"* (the default value is 2 seconds). Thus, *Hello* messages enable each node to discover its 1-hop neighbors, as well as its 2-hop neighbors. This neighborhood and 2-hop neighborhood information has an associated holding time, the - *"NEIGHBOR_HOLD_TIME"*, after which it is no longer valid.

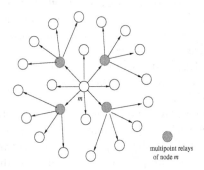

multipoint relays
of node *m*

Fig. 1. Multipoint relays of node *m*

On the basis of this information, Each node *m* independently selects its own set of *multipoint relays* among its 1-hop neighbors in such a way all 2-hop neighbors of *m* have *symmetric* links with $MPR(m)$. This means that the *multipoint relays* cover (in terms of radio range) all 2-hop neighbors (Figure 1). One possible algorithm for selecting these *MPRs* is described in [8]. The *MPR* set is computed whenever a change in the 1-hop or 2-hop neighborhood is detected. In addition, each node *m* maintains its *"MPR selector set"*. This set contains the nodes which have selected *m* as a *multipoint relay*. Node *m* only forwards broadcast messages received from one of its *MPR selectors*.

3.2   Topology Dissemination

Each node of the network maintains topological information about the network obtained by means of *TC (Topology control)* messages. Each node *m* selected as a *multipoint relay*, broadcasts a *TC* message at least every *"TC_INTERVAL"* (the default value is 6 seconds). The *TC* message originated from node *m* declares the *MPR selectors* of *m*. If a change occurs in the *MPR selector* set, the next *TC* can be sent earlier. (e.g. after some pre-specified minimum interval). The *TC* messages are flooded to all nodes in the network and take advantage of *MPRs* to reduce the number of retransmissions. Thus, a node is reachable either directly or via its *MPRs*. This topological information collected in each node has an associated holding time *"TOP_HOLD_TIME"*, after which it is no longer valid.

The neighbor information and the topology information are refreshed periodically, and they enable each node to compute the routes to all known destinations. These routes are computed with Dijkstra's shortest path algorithm [15]. Hence, they are optimal as concerns the number of hops. Moreover, for any route, any intermediate node on this route is a *multipoint relay* of the next node. The routing table is computed whenever there is a change in neighborhood or topology information.

3.3 OLSR "Gateways"

Each node maintains information concerning which nodes may act as "gateways" to associated hosts and networks. These "gateways" periodically generate a *HNA (Host and Network Association)* message, containing pairs of (network address, netmask) corresponding to the connected hosts and networks. The *HNA* messages are flooded to all the nodes in the network by the *MPRs*. These messages should be transmitted periodically every *HNA_INTERVAL*. The collected information is valid for *HNA_HOLD_TIME*. The networks and associated hosts are added to the routing table and they have the same next hop as the one to reach the appropriate "gateway".

3.4 Multiple Interface Declaration

In the full OLSR protocol, a node which has several interfaces, periodically emits a special type message, "Multiple Interface Declaration", in which it lists all its interfaces addresses, along with one of them, which is fixed, and is (arbitrarily) chosen as its main address.

4 Autoconfiguration: Duplicate Address Detection

Our autoconfiguration algorithms are based on two steps. In the first step, an IP address is selected by the arriving node and this latter can join the ad hoc network. Numerous schemes can be used to select this IP address. For instance the node can perform a random selection in a well known pool of addresses; another technique consists of one of the configured neighbors selecting the address on behalf of the arriving node.

After this first step has been performed, the second step takes place. The aim of this step is to detect potential address duplications during an ad hoc session. To perform this task a DAD algorithm is started on this newly configured node. This DAD algorithm allows each configured node to state whether or not its address is duplicated. In such a case a new address can be chosen.

Our DAD algorithms use a special control packet called MAD for "Multiple Address Declaration". This control packet includes a node address and a node identifier (Figure 2). The node identifier is a sequence of bits of fixed length L which is randomly generated. Hence we are using the standard idea that the probability of two nodes having the same identifier is low, and the probability of at least one address collision with N nodes, which is the well known "birthday problem" [6], can be set arbitrarily low by choosing a large enough value of L.

This packet is broadcast in the network, thus all the network nodes must receive this packet. The duplicate address detection algorithm uses the node identifier to detect address conflicts. Hence, a node detects that it is in conflict when it receives an MAD message containing the same address as its own, but with a different identifier. Actually other nodes, not involved in the conflict, will detect the address duplication by receiving two *MAD* messages having the same address, but holding different node identifiers. To spare the channel bandwidth

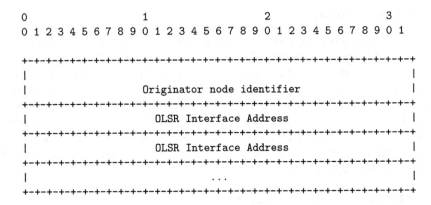

Fig. 2. MAD message

the MAD packet is broadcast using the MPR flooding rules. Actually, applying OLSR relaying optimization rules as they are defined, may not be sufficient to ensure diffusion in some conflictual cases. As an illustration of such possible situations, we give the following example.

Figure 3 shows two conflicting nodes $X1$ and $X2$, in the 2-hop neighbors of node I. Nodes Y and I are not MPRs, then, the "Multiple Address Declaration" messages of the nodes $X1$ and $X2$ can not be propagated throughout the entire network only if the next MPR calculation is done properly. In our scenario, node I coud not calculate its MPR set correctly, because MPR calculation is based on the assumption that there is no address duplication in the 1-hop and 2-hop neighbors. Consequently, node $X1$ and node $X2$ will not detect the address conflict, and the network remains corrupted. To handle such scenarios, a new MPR flooding algorithm for MAD messages diffusion is used. This algorithm takes into account the possibility that the originator adresses of the MAD messages might be duplicated. With this algorithm we can ensure that having two nodes A_1 and A_2 using the same address A but holding different node identifiers ID_{A_1} and ID_{A_2}, and for any value of the distance d between A_1 and A_2, the MAD message of A_1 will be received by A_2 and vice versa the MAD message of A_2 will be received by A_1. This new version of MPR flooding is called *Duplicate Address Detecting MPR Flooding* or *DAD-MPR Flooding*.

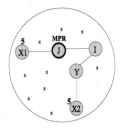

Fig. 3. Address duplicate scenario

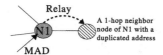

Node N1 detects an address duplication and is
a neighbor node of a node involved in this
conflict.
The MAD message is relayed irrespectively of
the indication of the
MPR flooding algorithm

Fig. 4. Rule 2 added to the OLSR MPR flooding

4.1   Special Rules of the DAD-MPR Flooding Algorithm

The first modification of the classical OLSR MPR flooding, to ensure the MAD message relaying to reach the whole network, is mainly the addition of the rule according to which the MAD duplicate message detection, will be based on the node originator address, the message sequence number, plus the node identifier. We call this rule *Rule 1*.

The second rule is the following: if a given node $N1$ receives a MAD message from a neighbor node $N2$, then, node $N1$ must relay this message irrespectively of the OLSR MPR flooding rules if it detects that one of its 1-hop neighbors has the same address as the one contained in the MAD message but with a different node identifier. The MAD TTL value is set to 1 to avoid the transmission of the MAD message beyond the conflicting nodes. This rule, called *Rule 2*, is explained in figure 4 and illustrated in figure 5 where nodes $A1$ and $A2$ share the same address A. See [1] for more details about *Rule 1* and *Rule 2*.

In figure 5, nodes B and C do not need to choose an MPR to cover respectively their 2-hop neighbors A_2 and A_1 since they have the same address A. Address A is considered as a one hop neighbor. In contrast, A_1 chooses node B as a MPR to reach node C, and node A_2 chooses C as a MPR to reach node B. In this situation, the A_1 MAD messages will reach node C, and the ones generated by A_2 will arrive at node B. But, B and C do not choose each other as an MPR, consequently, A_1 can not receive MAD messages coming from A_2, and A_2 MAD messages will never reach A_1 node.

The *Rule 2* enables MAD relaying for such situations.

Let us now consider the case of multiple conflicts as depicted in figure 6. In this example, each node considers that it has only two neighbors at 1-hop distance and no 2-hop neighbors (i.e the network seen by node A is composed by direct neighbors B and C). None of the nodes present in this network will be elected as an MPR. Hence, MAD messages will not be relayed and never reach other conflicting nodes or at least a neighbor of a conflicting node. In that case the *Rule 2* will not ensure the relaying of MAD messages between nodes in conflict.

We assume that there can be an arbitrary number of nodes with a duplicated address in the network. We also assume that each node in the network picks a globally unique random identifier.

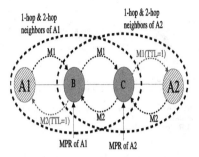

Fig. 5. Relay of MAD messages when duplication is detected nearby one of the nodes invloved in the conflict

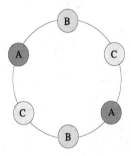

Fig. 6. Case of multiple conflicts

We will add a third rule to the classical MPR flooding mechaninsm to handle multiple conflicts. The property that we will add is extremely simple. We weaken the relaying condition for nodes who are in the 1-hop neighborhood of a node who is sending an MAD message. When these neighbor nodes receive an MAD message, they must relay it irrespectively of the relaying conditions of the OLSR MPR flooding algorithm. We call this rule, *Rule 3*. Actually, this rule covers the *Rule 2*. In fact, when a node $N1$ receives a MAD message from a neighbor node $N2$, and if the node $N1$ detects that one of its 1-hop neighbors, say N, has the same address as the one contained in the MAD message but with a different node identifier, $N1$ may act as if the MAD message was orginated from node N and hence, it applies the *Rule 3*. That is why we merge these two rules into a single rule in section 4.3.

With the previous rules, we will be in the position to prove the correctness of the DAD-MPR flooding algorithm. More precisely in the absence of packet loss an MAD message will finally reach all the nodes in the network.

4.2   Proof of Correctness of the DAD-MPR Flooding Algorithm

Let's denote: A, B, C, D, ... the nodes and '1', '2', ... the addresses. Let's denote 'A{1}' the node 'A' with the address '1' and so on ...

In this part, it is assumed that any conflict with distance ≤ 3 is resolved (it is proved in [1] that such conflicts are resolved).

Lemma 1: In a permanent 4-hop conflict represented by the topology A{1}–B{2}–C{3}–D{4}–E{1}, neither B nor D chooses C as MPR. In other terms, in a 4-hop conflict between two given nodes, the node(s) at the center of the conflict isn't chosen as MPR by the neighbors of the conflicting nodes. A node "in the center of conflict" is defined to be exactly 2-hop away from the both nodes in conflict. At least one such node exists by definition of "4-hop conflict".

Proof: By contradiction.
Assume there is a permanent 4-hop conflict: A{1}–B{2}–C{3}–D{4}–E{1} and C is MPR of B for instance (case with D is symetrical).
  Then:

- The node A originates one MAD message
- The node B retransmits it: because it is a 1-hop neighbor of the MAD message originator (*Rule 3*)
- The node C retransmits it: because it is an MPR of B
- The node D retransmits it: because it detects the conflict and is one hop away from the other node in conflict (*Rule 2*).

which results in the conflict being resolved, which itself is contradictory with the "permanent 4-hop conflict" hypothesis.

Lemma 2: In a permanent 4-hop conflict which can be represented by the topology : A{1}–B{2}–C{3}–D{4}–E{1}, there must be some nodes X and Y such as the topology includes : X{4}–Y–B{2}–C{3}–D{4}–E{1} and Y is MPR of B.

Proof:
Assume there is a permanent 4-hop conflict which can be represented by the topology : A{1}–B{2}–C{3}–D{4}–E{1}. *Lemma 1* shows that C is not MPR of B (nor D, incidentally). And then since D{4} is a 2-hop neighbor of B via C, proper MPR selection in B requires that :

(a) Some node with address 4 is covered by another 1-hop neighbor, chosen as MPR by B.
(b) OR some node with address 4 is a neighbor of B.

  The last case (b) is impossible because ,otherwise, there would be a topology such as X{4}–B{2}–C{3}–D{4}–E{1}, which is a 3-hop conflict [1].
  So (a) must be verified : let us denote X{4} the other 2-hop neighbor of B wit address 4, and Y the 1-hop neighbor to reach it (the MPR chosen by B) [hence a part of the topology is represented by B{2}–Y–X{4}].

---

[1] Recall that, by hypothesis, any conflict with distance ≤ 3 is resolved.

Then the topology includes:

– B{2} – C{3} – D{4} – E{1}
    |
    Y
    |
    X{4}

Which proves the lemma. [Notice here that, the proof is still valid if Y happens to be A].

Theorem: There are no permanent 4-hop conflicts.

Proof: By contradiction.
Assume there is a permanent 4-hop conflict which can be represented by the topology : A{1}–B{2}–C{3}–D{4}–E{1}. Then according to *lemma 2*, there exist nodes X and Y such that : X{4}–Y–B{2}–C{3}–D{4}–E{1} and Y is MPR of B.

Now by noticing that a subgraph of this topology is : X{4}–Y–B{2}–C{3}–D{4}–...(a 4-hop conflict with address 4), we can apply *lemma 2* on this topology again: thus, there exist nodes U and V such that: U{3}–V–Y–B{2}–C{3}–D{4} (and V is MPR of Y) is part of the graph.

Now notice that the topology includes the subgraph of a 4-hop conflict with address 3 : U{3}–V–Y–B{2}–C{3}, but this time with the noteworthy fact that Y is MPR of B{2} (this fact comes exclusively from the first application of *lemma 2*). But this fact is in contradiction with the *lemma 1* applied to the 4-hop conflict between U{3} and C{3} : indeed the *lemma 1* indicates(proves) that Y shall choose neither V, nor B{2} as MPR. The contradiction shows that the hypothesis that "there is a permanent 4-hop conflict" is impossible, hence the theorem.

4.3   Specification of the DAD-MPR Flooding Algorithm

Let us recall the assumptions here.
Each node A periodically sends an MAD message M including:

- The originator address of A, $Orig_A$, in the OLSR message header.
- The message sequence number, $mssn$, in the OLSR message header.
- The node identifier ID_A (a string of bits) in the message itself.

The message is propagated by MPR flooding to the other nodes ; but for DAD-MPR Flooding, the duplicate table of OLSR is modified, so that it also includes the node identifier list in the duplicate tuple. That is, a duplicate tuple, includes the following information:

- The originator address (as in OLSR standard duplicate table).
- The message sequence number (as in OLSR standard duplicate table).
- The list of node identifiers.

The detailed algorithm for DAD-MPR Flooding is the following:

- When a node B receives a MAD message M from node C with originator $Orig_A$, with message sequence number $mssn$, and with node identifier ID_A, it performs the following tasks:
  1. **If** a duplicate tuple exists with the same originator $Orig_A$, the same message sequence number, and ID_A is in the list of node identifiers, **Then**, the message is ignored (it has already been processed). The algorithm stops here.
  2. **Else** one of the following situations occurs :
     (a) A duplicate tuple exists with the same originator $Orig_A$ and the same message sequence number, but ID_A is not in the list of node identifiers: then, a conflict is detected (address $Orig_A$ is duplicated). ID_A is added to the list of node identifiers.
     (b) No duplicate tuple exists with the same originator $Orig_A$, and the same message sequence number $mssn$. A new one is created with the originator address, message sequence number and list of node identifiers containing only ID_A.
  3. The MAD messages should be relayed if one or more of the following rules are met:
     (a) C had chosen this current receiving node ,B, as an MPR(as in normal MPR flooding).
     (b) The node B has a link(symmetric or asymmetric) with the originator address, $Orig_A$, contained in the MAD message M (*Rule 2* and *Rule 3*).

5   Initial Address Assignment, and Resolution of Conflicts

In [1] we have mainly presented two ways to allocate an address to a newly arriving node. The first way is to allocate this node a random address in a well known pool of addresses and then to rely on the DAD algorithm to discover potential conflicts. The second way is to ask for the help of a configured neighbor node to get a valid address. These approaches remain correct whatever the DAD algorithm used.

In [1], we have also proposed a simple rule to solve a detected address conflict between two nodes. In fact, the node with the smallest identifer changes its address by randomly selecting a new one from a well known pool of addresses.

These solutions can be applied to the autoconfiguration mechanism proposed in this paper. See [1] for more details.

6   Simulation

The overhead of the proposed autoconfiguration mechanism for $OLSR$ has been simulated (using a simulator written in the Ocaml programming language) in order to evaluate its performance. An idealized simulation model was choosen in order to evaluate precisely its impact: a given number of nodes are placed in a square area ; there is no mobility, and a unit disk graph[2] is used. There is no MAC, hence no contention or collision, and the transmission delay is uniform. The radio range is a parameter of the simulation.

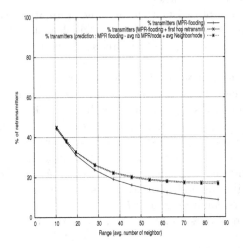

Fig. 7. Overhead generated by MAD messages

The interesting performance metrics include the overhead generated by the diffusion of the MAD control messages: in our simulations, we computed the average cost of one diffusion of MAD message. In MAD diffusion, not all, but only a fraction of the nodes will retransmit the MAD message, hence an interesting parameter is the fraction of the nodes of the network which retransmit on average a MAD message (in our graph, the fraction is expressed in percentage of the nodes of the network). In addition, we know that the MAD-flooding overhead could be compared to MPR-flooding: the overhead of the MAD-flooding should be similar to the overhead of the MPR-flooding, plus the extra transmission of the neighbors of the initial originator of the MAD message - which are not MPR. The average number of the neighbors which are not MPRs of a given node is: avg number of neighbors per node - avg number of MPRs per node. Hence an estimate would be:

> MAD-flooding overhead = MPR-flooding overhead
> + (avg number of neighbors per node - avg number of MPRs per node)

In our results, we evaluated the average cost of one MAD-flooding, the cost of the MPR-flooding, and in order to verify the previous estimate, the estimate of the MAD-flooding cost based on the previous expression and the actual measured cost of MPR flooding was evaluated.

---

[2] In a unit disk graph, each node is identified with a disk of unit radius $r = 1$ in the plane, and is connected to all nodes within (or on the edge of) its corresponding disk.

The figure 7 depicts the result for increasing density: the simulated network comprises 1000 nodes and for each simulation, a different radio range is set in order to modify the density of the neighboring nodes.

The conclusion is that the additional overhead generated by MAD messages remains limited compared to one classical MPR flooding: the cost of the MAD-flooding is similar to the cost of MPR-flooding, with the exception that all the direct neighbors of the originator will retransmit (instead of only the MPR) in the first step.

It also appears that the estimate of the cost of the MAD flooding computed from the measured cost of MPR-flooding is close the actual measured MAD-flooding, showing that the estimate is quite precise.

7 Conclusion

The autoconfiguration procedure that is proposed mainly relies on an efficient and proven duplicate address detection algorithm. A special control message MAD (Multiple Address Declaration) conveys with the address of the node a random identifier. This control message uses the OLSR genuine MPR flooding algorithm to reach all the nodes in the network, however special rules have been added to ensure that even with address duplications, the MAD messages will be propagated throughout the entire network. The proposed autoconfiguration mechanism for $OLSR$ has been simulated, and the simulation results have proved that the additional overhead generated by MAD messages remains limited compared to the classical OLSR MPR flooding.

References

[1] S. Boudjit, A. Laouti, C. Adjih, P. Muhlethaler: *"Duplicate address detection and autoconfiguration in OLSR"*, Proceedings of IEEE SNPD/SAWN 2005, May 2005, Maryland, USA.

[2] S. Boudjit, A. Laouiti, P. Minet, C. Adjih: "OLSR for IPv6 networks", Proceedings of the 3rd Med-Hoc-Net Workshop. Bodrum - Turkey (June 2004).

[3] P. Jacquet, P. Muhletaler, P.Minet, A. Qayyum, A. Laouiti, T. Clausen, L. Viennot: *"Optimized Link State Routing Protocol"*, IETF RFC3626, October 2003.

[4] C.Perkins, E.Belding-Royer, and S.Das: *"Ad Hoc On-Demand Distance Vector(AODV) Routing"*, IETF RFC3561, July 2003.

[5] S. Thomson, T. Narten: "IPv6 Stateless Address Autoconfiguration", IETF RFC 2462, December 1998.

[6] M. Sayrafiezadeh: *"The Birthday Problem Revisited"*, Math. Mag. 67, 1994, pp 220-223.

[7] R. Droms, Ed., J. Bound, B. Volz, T. Lemon, C. Perkins, M. Carney: "Dynamic Host Configuration Protocol for IPv6 (DHCPv6)", IETF RFC 3315, July 2003.

[8] A. Qayyum, A. Laouiti, L. Viennot: "Multipoint relaying technique for flooding broadcast messages in mobile wireless networks", HICSS: Hawai Int. Conference on System Sciences, Hawai - USA, (January 2002).

[9] C. Perkins, J. Malinen, R. Wakikawa, E. Belding-Royer, Y. Sun: "IP Address Autoconfiguration for Ad Hoc Networks", Internet Draft, IETF Working Group MANET, Work in progress, (November 2001).

[10] K. Weniger, M. Zitterbart: "IPv6 Autoconfiguration in Large Scale Mobile Ad-Hoc Networks", Proceedings of European Wireless 2002, Florence - Italy, (Feb 2002).

[11] K. Weniger: "PACMAN : Passive Autoconfiguration for Mobile Ad Hoc Networks", Proceedings of IEEE WCNC 2003, New Orleans - USA, (March 2003).

[12] S. Nesargi, R. Prakash: "MANETconf: Configuration of Hosts in a Mobile Ad Hoc Network", InfoCom 2002, (June 2002).

[13] A. Misra, S. Das, A. McAuley, S. K.Das: "Autoconfiguration, Registration, and Mobility Management for prevasive Computing", IEEE Personal Communication, (August 2001), 24-31.

[14] H. Zhou, L. M.Ni, M. W.Mutka: "Prophet Address Allocation for Large Scale MANETs", IEEE INFOCOM 2003, (March 2003).

[15] S. A.Tanenbaum: "Computer Networks", Prentice Hall, 1996.

On the Application of Mobility Predictions to Multipoint Relaying in MANETs: Kinetic Multipoint Relays

Jérôme Härri, Fethi Filali, and Christian Bonnet

Institut Eurécom*
Department of Mobile Communications
B.P. 193
06904 Sophia-Antipolis, France
{Jerome.Haerri, Fethi.Filali, Christian.Bonnet}@eurecom.fr

Abstract. In this chapter, we discuss the improvements multipoint relays may experience by the use of mobility predictions. Multipoint Relaying (MPR) is a technique to reduce the number of redundant retransmissions while diffusing a broadcast message in the network. The algorithm creates a dominating set where only selected nodes are allowed to forward packets. Yet, the election criteria is solely based on instantaneous nodes' degrees. The network global state is then kept coherent through periodic exchanges of messages. We propose in this chapter a novel heuristic to select kinetic multipoint relays based on nodes' overall predicted degree in the absence of trajectory changes. Consequently, these exchanges of message may be limited to the instant when unpredicted topology changes happen. Significant reduction in the number of messages are then experienced, yet still keeping a coherent and fully connected multipoint relaying network. Finally, we present some simulation results to illustrate that our approach is similar to the MPR algorithm in terms of network coverage, number of multipoint relays, or flooding capacity, yet with a drastic reduction in the number of messages exchanged during the process.

Keywords: Multipoint relay, MPR, mobility prediction, kinetic, broadcasting, manets.

1 Introduction

Mobile Ad Hoc Networks (MANETs) is an emergent concept in view for infrastructure-less communication. These networks rely on radio transmissions, but with the lack of infrastructures, flooding (distributing information to each and every node in the network in an uncontrolled way) happens to be a key part of information dissemination. In wireless networks and particularly when the network is dense, the overhead due to this kind of information dissemination may become prohibitive. Despite its simplicity, flooding is very inefficient and can result in high redundancy, contention and collision. This is the main motivation for many research teams that have proposed more efficient flooding

---

* Institut Eurécom's research is partially supported by its industrial members: Bouygues Télécom,Fondation d'entreprise Groupe Cegetel, Fondation Hasler, France Télécom, Hitachi, ST Microelectronics, Swisscom, Texas Instruments, Thales, Sharp.

K. Cho and P. Jacquet (Eds.): AINTEC 2005, LNCS 3837, pp. 143–156, 2005.

techniques whose goal is to minimize the number of retransmissions while attempting to deliver packets to each node in the network. Different approaches of flooding techniques and broadcasting control protocols exist and are listed in [1, 2].

Multipoint relaying (MPR, [3]) provide a localized way of flooding reduction in a mobile ad hoc network. Using 2-hops neighborhood information, each node determines a small set of forward neighbors for message relaying, which avoids multiple retransmissions and blind flooding. MPR has been designed to be part of the Optimized Link State Routing algorithm (OLSR, [4]) to specifically reduce the flooding of TC messages sent by OLSR to create optimal routes. Yet, the election criteria is solely based on instantaneous nodes' degrees. The network global state is then kept coherent through periodic exchanges of messages. Some studies showed the impact of periodic beacons, which could be compared to increasing the probability of transmission, in 802.11 performances [5], or the effects of beaconing on the battery life [6]. This denotes that these approaches have major drawbacks in terms of reliability, scalability and energy consumptions. The next step to their evolution should therefore be designed to improve the channel occupation and the energy consumption.

In this chapter, we propose to improve the MPR protocol by using mobility predictions. We introduce the *Kinetic Multipoint Relaying (KMPR)* protocol which heuristic selects kinetic relays based on nodes actual and future predicted nodal degrees. Based on this, periodic topology maintenance may be limited to the instant when a change in the neighborhood actually occurs. Our objective is to show that this approach is able to significantly reduce the number of messages needed to maintain the backbone's consistency, thus saving network resources, yet with similar flooding properties as the regular MPR.

The rest of the chapter is organized as follows. In Section 2, we shortly define our motivation for using mobility prediction with MPR. Section 3 describes the heuristic used in order to compute nodes' kinetic degrees. In Section 4, we formally describe our KMPR protocol, and in Section 5, we propose an aperiodic neighborhood maintenance strategy. Finally, Section 6 provides simulation results justifying our approach, while Section 7 draws some concluding remarks and describes some future works.

2   Preliminaries

In this section, we give a short description of mobility predictions and our motivation for using this concept in MANETs. Finally, we provide some related work on this field.

2.1   Mobility Predictions in MANETs

In mobility predictions, a mobile samples its own location continuously or periodically and constructs a model of its own movement. The model can be first order, which provides nodes' velocities, but higher and more complex models providing nodes' accelerations are also possible. The node disseminates its current model's parameters[1] in

---

[1] The model's parameters are assumed to be valid over a relative short period of time depending on the model's complexity.

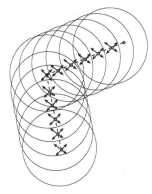 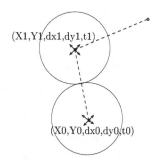

(a) 1 message sent per second ⇒ A total of 12 messages in 12 seconds.

(b) 1 message sent upon any trajectory change ⇒ A total of 2 messages in 12 seconds.

Fig. 1. Illustration of the influence of mobility predictions on the number of location updates

the network. Any changes to the model's parameters is reactively announced by the respective nodes. Every other node uses this information to track the location of this node. Very little location update cost is incurred if the model's prediction is accurate. For example, in Fig. 1, we compare the number of location updates needed with or without mobility predictions when using a linear first order mobility model.

2.2 Average Linear Trajectory Durations in Ad Hoc Networks

A basic assumption in mobility prediction-based techniques is to assume that nodes move following a linear trajectory, then predict to update the neighborhood information when a trajectory change occurs. Therefore, scalability is highly dependent to the number of trajectory changes (or transitions) per unit of time, thereafter called β.

Part of the results obtained in [10] are reproduced in Fig. 2. It shows that even with an average velocity of 20m/s, nodes have an average trajectory duration of $22s$ (Fig. 2(a)) for the Random Waypoint model and $10s$ (Fig. 2(b)) for the City Section model. Figure 2 therefore provides a lower bound on the average trajectory duration, that is $\frac{1}{\beta} \approx 10s$ using extreme values for the configuration parameters of the mobility models. In more realistic situations, this value is rather $\frac{1}{\beta} \approx 30s$. Accordingly, it becomes conceivable to consider predictions to improve ad-hoc protocol the way we will do in this chapter.

2.3 Related Work

Prediction-based protocols are a straight evolution of position-based methods. Indeed, this approach evolved from simple positions, to position and velocity, and finally to trajectory models. The first definition of such protocols has been done under the name *predictive distance-based* protocol [12], and has been cited in possible developments

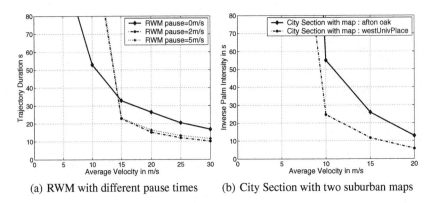

(a) RWM with different pause times   (b) City Section with two suburban maps

Fig. 2. Average nodes' trajectory duration ($\frac{1}{\beta}$) under the Random Waypoint mobility model and the City Section

in the terminode project [11]. Almost at the same time, another study has been performed [13] which illustrated the benefits early unicast and multicast protocols could experience from mobility predictions. It is, however, only in recent months that this model started to get seriously studied. To our knowledge, the authors of [7] have been the first team to analyze it in their proposition of *Kinetic spanning trees* for ad hoc sensor networks. The authors managed to create an auto adaptive shortest path spanning tree to a sink node that was getting rid of periodic beacons inherited from the Bellman-Ford algorithm.

Later, another team studied, under the name *dead-reckoning*, the benefits of predictions for mobile ad hoc networks [16]. They showed that this model was able to deliver superior routing performances than DSR or AODV. They then extended their studies to location services [17]. They conclusions were quite similar, by noticing that the diffusion of predicted future locations of nodes in the network could improve the performances of location services. Recently, the authors in [14] proposed a paper that was analyzing the effect of trajectory predictions on topology management. By they intrinsic behavior, topology control protocols are usually considered as proactive, since they need to maintain a structure between moving nodes. However, the authors managed to show that, by using stochastic prediction-based trajectories, they could create the first totally reactive topology control protocol, in a sense that after an initial organization, the topology is maintained in an event-driven manner, without the need of periodic beacons.

Finally, in recent months many interesting papers has been presented that deal with prediction-based routing protocols. [19] presents an approach that reduces mobility-induced location errors on geographical routing using mobility predictions. [18] in other hand, make use of mobility prediction in order to improve routing protocols. This global interest in mobility predictions greatly justified the motivation we have to dig into that direction, since we firmly believe that such approach would improve any protocol under any configuration.

In this chapter, our objective will be to develop a predicted model adapted to the MPR protocol by modeling nodes predicted degrees, also called *kinetic degree*.

3 Kinetic Nodal Degree in MANETs

We explain in this section the method for modeling kinetic degrees in MANETs. We model nodes' positions as a piece-wise linear trajectory and, as we showed in Section 2.2, the corresponding trajectory durations are lengthy enough to become a valuable cost for using kinetic degrees.

The term "Kinetic" in KMPR reflects the motion aspect of our algorithm, which computes a node's trajectory based on its Location Information [7]. Such location information may be provided by the Global Positioning System (GPS) or other solutions exposed in [8] or [9]. Velocity may be derived through successive location samples at close time instants. Therefore, we assume a global time synchronization between nodes in the network and define x, y, dx, dy as the four parameters describing a node's position and instant velocity [2], thereafter called *mobility*.

Over a relatively short period of time [3], one can assume that each such node, say i, follows a linear trajectory. Its position as a function of time is then described by

$$\mathbf{Pos}_i(t) = \begin{bmatrix} x_i + dx_i \cdot t \\ y_i + dy_i \cdot t \end{bmatrix}, \tag{1}$$

where $Pos_i(t)$ represents the position of node i at time t, the vector $[x_i, \ y_i]^T$ denotes the initial position of node i, and vector $[dx_i, \ dy_i]^T$ its initial instantaneous velocity. Let us consider node j as a neighbor of i. In order to let node i compute node j's trajectory, let us define the squared distance between nodes i and j as

$$D_{ij}^2(t) = D_{ji}^2(t) = \|\mathbf{Pos}_j(t) - \mathbf{Pos}_i(t)\|_2^2$$
$$= \left(\begin{bmatrix} x_j - x_i \\ y_j - y_i \end{bmatrix} + \begin{bmatrix} dx_j - dx_i \\ dy_j - dy_i \end{bmatrix} \cdot t \right)^2$$
$$= a_{ij}t^2 + b_{ij}t + c_{ij}, \tag{2}$$

where $a_{ij} \geq 0$, $c_{ij} \geq 0$. Consequently, a_{ij}, b_{ij}, c_{ij} are defined as the three parameters describing nodes i and j mutual trajectories, and $D_{ij}^2(t) = a_{ij}t^2 + b_{ij}t + c_{ij}$, representing j's relative distance to node i, is denoted as j's linear relative trajectory to i. Consequently, thanks to (1), a node is able to compute the future position of its neighbors, and by using (2), it is able to extract any neighboring nodes' future relative distance.

Considering r as nodes maximum transmission range, as long as $D_{ij}^2(t) \leq r^2$, nodes i and j are neighbors. Therefore, solving

$$D_{ij}^2(t) - r^2 = 0$$
$$a_{ij}t^2 + b_{ij}t + c_{ij} - r^2 = 0, \tag{3}$$

gives t_{ij}^{from} and t_{ij}^{to} as the time intervals during which nodes i and j remain neighbors. Consequently, we can model nodes' kinetic degree as two successive sigmoid functions,

---

[2] We are considered moving in a two-dimensional plane.

[3] The time required to transmit a data packet is orders of magnitude shorter than the time the node is moving along a fixed trajectory.

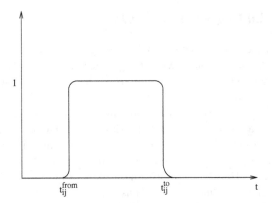

Fig. 3. Double sigmoid function modeling a link lifetime between node i and node j

where the first one jumps to one when a node enters another node's neighborhood, and the second one drops to zero when that node effectively leaves that neighborhood (see Fig. 3).

Considering $nbrs_i$ as the total number of neighbors detected in node i's neighborhood at time t, we define

$$Deg_i(t) = \sum_{k=0}^{nbrs_i} \left(\frac{1}{1 + \exp(-a \cdot (t - t_k^{from}))} \cdot \frac{1}{1 + \exp(a \cdot (t - t_k^{to}))} \right) \quad (4)$$

as node i's kinetic degree function, where t_k^{from} and t_k^{to} represent respectively the time a node k enters and leaves i's neighborhood. Thanks to (4), each node is able to predict its actual and future degree and thus is able to proactively adapt its coverage capacity. Figure 4(a) illustrates the situation for three nodes. Node k enters i's neighborhood at time $t = 4s$ and leave it at time $t = 16s$. Meanwhile, node j leaves i's neighborhood at time $t = 20s$. Consequently, Fig. 4(b) illustrates the evolution of the kinetic degree function over t.

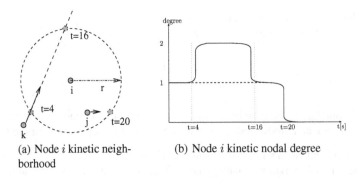

(a) Node i kinetic neigh-borhood

(b) Node i kinetic nodal degree

Fig. 4. Illustration of nodes kinetic degrees

Finally, the kinetic degree is obtained by integrating (4)

$$\widehat{Deg}_i(t) = \int_t^\infty \left(\sum_{k=0}^{k=nbrs_i} \left(\frac{1}{1 + \exp(-a \cdot (t - t_k^{from}))} \cdot \frac{1}{1 + \exp(a \cdot (t - t_k^{to}))} \right) \right) (5)$$

For example, in Fig. 4(b), node i kinetic degree is ≈ 32.

4   Kinetic Multipoint Relays

In this section, we describe our Kinetic Multipoint Relaying (KMPR) protocol. It is mainly extracted from the regular MPR protocol. Yet, we adapt it to deal with kinetic degrees.

To select the kinetic multipoint relays for node i, let us call the set of 1-hop neighbors of node i as $N(i)$, and the set of its 2-hops neighbors as $N^2(i)$. We first start by giving some definitions.

Definition 1 (Covering Interval). *The covering interval is a time interval during which a node in $N^2(i)$ is covered by a node in $N(i)$. Each node in $N^2(i)$ has a covering interval per node i, which is initially equal to the connection interval between its covering node in $N(i)$ and node i. Then, each time a node in $N^2(i)$ is covered by a node in $N(i)$ during a given time interval, this covering interval is properly reduced. When the covering interval is reduced to $\emptyset$, we say that the node is fully covered.*

Definition 2 (Logical Kinetic Degree). *The logical kinetic degree is the nodal degree obtained with (5) but considering covering intervals instead of connection intervals. In that case, t_k^{from} and t_k^{to} will then represent the time interval during which a node $k \in N^2(i)$ starts and stops being covered by some node in $N(i)$.*

The basic difference between MPR and KMPR is that unlike MPR, KMPR does not work on time instants but on time intervals. Therefore, a node is not periodically elected, but is instead designated KMPR for a time interval. During this interval, we say that the KMPR node is active and the time interval is called its activation.

The KMPR protocol elects a node as KMPR a node in $N(i)$ with the largest logical kinetic degree. The activation of this KMPR node is the largest covering interval of its nodes in $N^2(i)$.

Kinetic Multipoint Relaying (KMPR).

The KMPR protocol applied to an initiator node i is defined as follows:

- Begin with an empty KMPR set.
- First Step: Compute the logical kinetic degree of each node in $N(i)$.
- Second Step: Add in the KMPR set the node in $N(i)$ that has the maximum logical kinetic degree. Compute the activation of the KMPR node as the maximum covering interval this node can provide. Update all other covering intervals of nodes in $N^2(i)$ considering the activation of the elected KMPR, then recompute all logical kinetic degrees. Finally, repeat this step until all nodes in $N^2(i)$ are fully covered.

Then, each node having elected a node KMPR for some activations is then a KMPR Selector during the same activation. Finally, *KMPR flooding* is defines as follows:

Definition 3 (KMPR Flooding). *A node retransmits a packet only once after having received the packet the first time from an active KMPR selector.*

5   Adaptive Aperiodic Neighborhood Maintenance

A limitation in per-event maintenance strategies is the neighborhood maintenance. While mobility prediction allows to discard invalid links or unreachable neighbors, it remains impossible to passively acquire new neighbors reaching some other nodes' neighborhood. The lack of an appropriate method to tackle this issue would limit KMPR's ability to obtain up-to-date links and effective kinetic multipoint relays.

We developed several heuristics to help KMPR detecting nodes stealthily entering some other nodes transmission range in a non-periodic way.

- *Constant Degree Detection*— Every node tries to keep a constant neighbor degree. Therefore, when a node i detects that a neighbor actually left its neighborhood, it tries to acquire new neighbors by sending a small advertising message. (see Figure 5(a));
- *Implicit Detection*— A node j entering node i transmission range has a high probability to have a common neighbor with i. Considering the case depicted in Figure 5(b), node k is aware of both i and j's movement, thus is able to compute the moment at which either j or i enters each other's transmission range. Therefore, node k sends a notification message to both nodes. In that case, we say that node i implicitly detected node j and vice versa;
- *Adaptive Coverage Detection*— We require each node to send an advertising message when it has moved a distance equal to a part of its transmission range. An adjusting factor which vary between 0 and 1 depends on the node's degree and its velocity (see Figure 5(c));

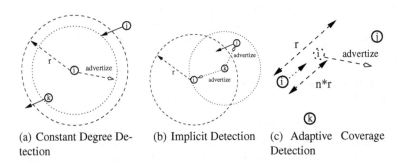

(a) Constant Degree Detection   (b) Implicit Detection   (c) Adaptive Coverage Detection

Fig. 5. Three heuristics to detect incoming neighbors in a per-event basis

6   Simulation Results

We implemented the KMPR protocol under ns-2 and used the NRL-MPR [23] implementation for comparison with KMPR. We measured several significant metrics for Manets: The effectiveness of flooding reduction, the delay before the network receives a broadcast packet, the number of duplicate packets and finally the routing overhead. The following metrics were obtained after the population of 20 nodes were uniformly distributed in a 1500×300 grid. Each node has a transmission range of $250m$. For the initial results, the mobility model we used is the standard Random Mobility Model where we made nodes average velocity vary from $5m/s$ to $30m/s$. Then, we used a Vehicular and a Pedestrian Mobility model [20,21] on the same simulation area, where we made nodes average velocity vary from $5m/s$ to $20m/s$, and $1km/h$ to $7km/h$ respectively. Between each motion change, the Pedestrian mobility model also includes a pause time randomly distributed between $2s$ and $30s$. Finally, we simulated the system for $100s$.

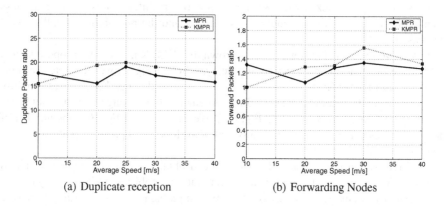

(a) Duplicate reception   (b) Forwarding Nodes

Fig. 6. Illustration of the flooding reduction of MPR and KMPR

Figure 6 illustrates the flooding reduction of MPR and KMPR. Although MPR is slightly more performing than KMPR, we can see that both protocols are close together and have a fairly good flooding reduction, both in terms of duplicate and forwarded packets. Note that the low fraction of relays in Fig 6(b) comes from the rectangular topology, where only a couple of MPRs are used as bridge in the center of the rectangle.

On Figure 7, we depicted the broadcast efficiency of MPR and KMPR. In the simulations we performed, we measured the broadcast efficiency as the time a packet takes before being correctly delivered to the entire network. As we can see, KMPR has a delivery time faster than MPR by 50%. This might comes from two properties of KMPR. Firstly, as described in [15], MPR suffers from message decoding issues. Indeed, MPR often discards correct neighborhood information based on wrong message decoding. It therefore relies on several iterations before being able to obtain correct information about nodes neighborhood. Since MPR nodes are periodically recomputed, the time before which MPR is operational also increases. In KMPR, since we do not rely on periodic retransmissions, we changed the decoding order as suggested in [15]. Secondly,

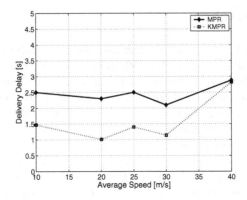

Fig. 7. Illustration of the broadcast efficiency of MPR and KMPR

as we will see in the next figure, KMPR's backbone maintenance is significantly less than MPR. Therefore, the channel access is faster and the probability of collisions is decreased.

In the two previous Figures, we have shown that KMPR had similar properties than MPR in term of flooding reduction and delay. Now, in Fig. 8, we illustrate the principal benefit of KMPR: its *low routing overhead*. Indeed, since KMPR uses mobility predictions and does not rely on periodic maintenance, the routing overhead may be reduced by 75% as it may be seen on Fig. 8(a). We also show on Fig. 8(b) the number of hello messages which drops dramatically with KMPR, yet still preserving the network's consistency.

All Figures we have presented yet have been obtained using the regular Random Waypoint Mobility model. And it is no secret that this mobility model recently drew a lot of criticism mainly on its lack of stability and realism. In order to further analyze KMPR using more realistic mobility scenarios, we also simulated KMPR and MPR us-

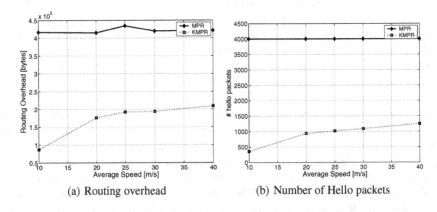

(a) Routing overhead   (b) Number of Hello packets

Fig. 8. Illustration of the network load for MPR and KMPR

ing a vehicular and a pedestrian mobility model [20, 21]. Basically, the macromobility is similar to the RWM, yet nodes are restricted to move on space graphs. Such graphs actually represent streets, roads and sidewalks that car and pedestrians are advised to follow. And to better fit with the increased motion freedom pedestrians usually experience, pedestrian graphs are more dense than the vehicular ones. Finally, whereas the micromobility for pedestrian motion is inexistent in the model we used, the vehicular mobility model also includes a micromobility model called *Intelligent Driver Motion (IDM)* (see [22]), which mainly adjusts the speed to the inter-distance between cars in order to avoid collisions. Cars and Pedestrians velocities have also been adapted to closely fit with real deployment.

In Figure 9, we depict KMPR and MPR's behaviors under a vehicular mobility model. Similar to the previous Figures, Fig 9(a) shows the drastic reduction of KMPR's backbone maintenance, as well as KMPR's improved broadcast delay. The channel access is increased on average up to 60% by KMPR, while the broadcast delay is reduced by 25%. An interesting observation is that unlike Fig 8(b) and Fig. 7, neither the number of hello messages nor the delay remain stable as we increase the average speed. Although not being dramatic, this mainly comes from the IDM. Indeed, as we increase the average speed, we have more chances to get close to another car on a graph edge,

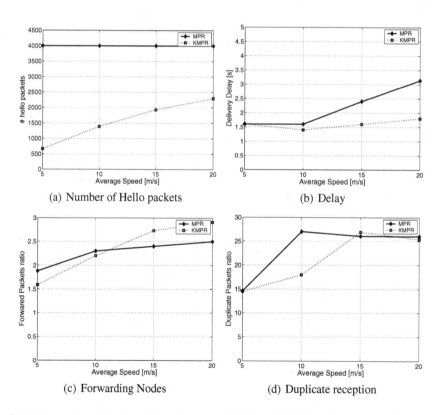

(a) Number of Hello packets

(b) Delay

(c) Forwarding Nodes

(d) Duplicate reception

Fig. 9. Illustration of the broadcast efficiency of MPR and KMPR under vehicular mobility

thus triggering a speed adaptation. Accordingly, KMPR needs to tune its predictions to the new configuration and is consequently less available for broadcasting. In Fig. 9(c) and Fig. 9(d), we can finally see that similarly to Fig. 6, the flooding reduction obtained by KMPR is similar to MPR's, and is thus not influenced by vehicular motions.

Finally, in Fig. 10, we illustrate KMPR and MPR behaviors under a pedestrian mobility model. Unlike vehicular motions, pedestrians move at a much slower speed, yet experiencing more trajectory changes. Actually, there is a direct relationship between the average velocity and the number of sudden trajectory changes. The faster a node moves, the less likely it will abruptly change its trajectory. And this is how the pedestrian motion is modeled here. However, although KMPR has to adapt more often its neighborhood for low speed motions with abrupt motion changes, the number of hello messages remains limited as it can be seen on Fig 10(a). This can be simply explained by the fact that even if a node changed its trajectory, its neighborhood would remain identical, thus limiting the update to a single message. This may also be justified by the pause times included in the pedestrian mobility model, which further increase the time before the occurrence of a motion change. Then, when looking at Fig 10(b), due to the low velocity experienced by pedestrians, both MPR and KMPR have stable delays. Since nodes neighborhood remains much longer stable, both KMPR and MPR may

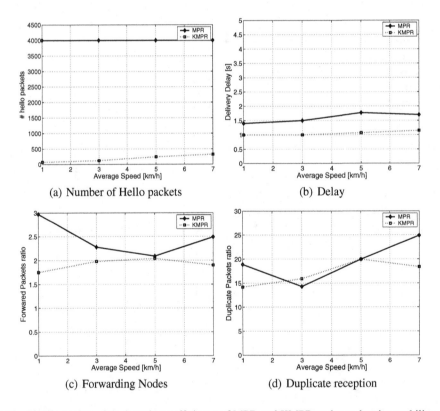

(a) Number of Hello packets

(b) Delay

(c) Forwarding Nodes

(d) Duplicate reception

Fig. 10. Illustration of the broadcast efficiency of MPR and KMPR under pedestrian mobility

improve their backbone for fast delivery. Again, as expected, KMPR's delay always remains 35% smaller than MPR's. Finally, as it is shown in Fig. 10(c) and Fig. 10(d), pedestrian motions do not seem to have a major influence on the flood reduction of KMPR and MPR. As for vehicular motions, both KMPR and MPR's flood reduction remain close together.

7   Conclusion and Future Works

In this chapter, we presented an original approach for improving the well-known MPR protocol by using mobility predictions. We showed that the Kinetic Multipoint Relaying (KMPR) protocol was able to meet the flooding properties of MPR, and this by reducing the MPR channel access by 75% and MPR broadcast delay by 50% under random motions.

We furthermore tested the KMPR protocol under realistic vehicular and pedestrian motions, and could also show that KMPR managed to reduce the channel access by 60% for vehicular motion and even 95% for pedestrian motions. At the same time, the broadcast delay could be reduced by 25% for vehicular motion and 35% for pedestrian motions. And this with similar flood reduction properties than MPR. We consequently illustrated that, after having been studied in other fields of mobile ad hoc networking, mobility predictions are also an interesting technique to improve broadcasting protocols.

In this work, we simulated KMPR and MPR only on a network of 20 nodes. Since KMPR main feature is mobility predictions, we focused our analysis on network mobility and not on network scalability. Yet, the latter would be worth investigating in future works. Moreover, the objective of this work has not been to try to improve the MPR protocol in term of flooding reduction, but instead to reduce its drawback in term of network load. Consequently, in the light of the results we presented, we are now interested in studying the impact of this improved network load on the OLSR protocol.

References

1. Brad Williams and Tracy Camp, "Comparison of broadcasting techniques for mobile ad hoc networks", in *Proc. of the 3rd ACM international symposium on Mobile ad hoc networking & computing (MobiHoc'02)*, pp. 194-205, 2002, Switzerland.
2. X.Y. Li and I. Stojmenovic, "Broadcasting and topology control in wireless ad hoc networks", in *Handbook of Algorithms for Mobile and Wireless Networking and Computing*, Eds: A. Boukerche and I. Chlamtac, CRC Press, to appear.
3. A. Laouiti *et al.*, "Multipoint Relaying: An Efficient Technique for Flooding in Mobile Wireless Networks", *35th Annual Hawaii International Conference on System Sciences (HICSS'2001)*, Hawaii, USA, 2001.
4. T. Clausen and P. Jacquet, "Optimized Link State Routing Protocol (OLSR)", www.ietf.org/rfc/rfc3626.txt, Project Hipercom, INRIA, France, October 2003.
5. Bianchi, G. "Performance analysis of the IEEE 802.11 distributed coordination function", in *Selected Areas in Communications, IEEE Journal on*, Vol. 18 , Issue 3 , pp. 535 - 547, March 2000.

6. C.-K Toh, H. Cobb, and D.A. Scott, "Performance evaluation of battery-life-aware routing schemes for wireless ad hoc networks" in the *2001 IEEE International Conference on Communications (ICC'01)*, pp. 2824-2829, Vol. 9, June 2001.
7. C. Gentile, J. Haerri, and R. E. Van Dyck, "Kinetic minimum-power routing and clustering in mobile ad-hoc networks", in *IEEE Proc. Vehicular Technology Conf. Fall 2002*, pp. 1328-1332, Sept. 2002.
8. R.J. Fontana and S.J. Gunderson, "Ultra-wideband precision asset location system", *IEEE Conf. on Ultra Wideband Systems and Technologies*, pp. 147-150, 2002.
9. C. Gentile and Luke Klein-Berndt, "Robust Location using System Dynamics and Motion Constraints", in *Proc. of IEEE International Conference on Communications*, Paris, 20-24 June, 2004.
10. J. Haerri and Christian Bonnet, "A Lower Bound for Vehicles' Trajectory Duration", to be presented in *IEEE VTC'Fall 2005*, Dallas, USA, September 2005.
11. L. Blazevic *et al.*, "Self-Organization in Mobile Ad-Hoc Networks: the Approach of Terminodes", in *IEEE Communications Magazine*, June 2001.
12. B. Liang and Z. Haas, "Predictive distance-based mobility management for pcs networks", in *Proc. IEEE Infocom Conference* , pp. 1377-1384, 1999.
13. William Su, Sung-Ju Lee, and Mario Gerla, "Mobility Prediction in Wireless Networks", in *Proc. of the IEEE Military Communications Conference (Milcom'05)*, pp. 491 - 495, Vol. 1, October 2000.
14. J. Haerri, and Navid Nikaein, and Christian Bonnet, "Trajectory knowledge for improving topology management in mobile ad-hoc networks", in *Proc. of ACM CoNext Conference 2005*, Toulouse, France, October 2005.
15. J. Haerri, and Christian Bonnet, and Fethi Filali, "OLSR and MPR: Mutual Dependences and Performances", in *Proc. of the 2005 IFIP Med-Hoc-Net Conference*, Porquerolles, France, June 2005.
16. Aarti Agarwal and Samir R. Das, "Dead Reckoning in Mobile Ad-Hoc Networks", *Proc. of the 2003 IEEE Wireless Communications and Networking Conference (WCNC 2003)*, New Orleans, March 2003.
17. V. Kumar and S. R. Das, " Performance of Dead Reckoning-Based Location Service for Mobile Ad Hoc Networks", *Wireless Communications and Mobile Computing Journal*, pp. 189-202, Vol. 4 Issue 2, Mar. 2004.
18. Jian Tang, Guoliang Xue and Weiyi Zhang, " Reliable routing in mobile ad hoc networks based on mobility prediction" in *MASS'04: IEEE International Conference on Mobile, Ad-Hoc and Sensor Systems*, October 25-27, 2004, Fort Lauderdale, Florida, USA.
19. D. Son, A. Helmy and B. Krishnamachari, "The Effect of Mobility-induced Location Errors on Geographic Routing in Mobile Ad Hoc and Sensor Networks: Analysis and Improvement using Mobility Prediction", *IEEE Transactions on Mobile Computing, Special Issue on Mobile Sensor Networks*, To appear 3rd Quarter, 2004.
20. Jerome Haerri, Marco Fiore, Fethi Filali, Christian Bonnet, Carla-Fabiana Chiasserini, Claudio Casetti, "A Realistic Mobility Simulator for Vehicular Ad Hoc Networks", Eurécom Technical Report, Institut Eurécom, France, 2005.
21. Illya Stephanov, "CANU Mobility Simulator", http://canu.informatik. uni-stuttgart.de/mobisim/.
22. Dirk Helbing, "Traffic and Related Self-Driven Many-Particle Systems", in *Reviews of Modern Physics*, Vol. 73, pp. 1067-1081, 2001.
23. NRLOLSR, http://pf.itd.nrl.navy.mil/projects.php?name=olsr

The Architecture of the Future Wireless Internet

Guy Pujolle

LIP6 – Université Pierre et Marie Curie,
8 rue du Capitaine Scott, 75015 Paris, France
Guy.Pujolle@lip6.fr

Abstract. The new wireless Internet is emerging very quickly, so does network architecture. It is encouraging to see the fast development of the new IEEE wireless technologies promising the ultimate Internet service deployment on wireless and mobile infrastructures since they would offer larger bandwidth at cheaper price compared to the telecommunication wireless radio resource. However it is disquieting to see that the TCP/IP protocol stack which is supposed to be the heart of the Internet services deployment is not evolving as fast as the wireless technologies do. Here we come up with the hard question which is the network performance of the TCP/IP architecture over wireless networks. It is probably too early to decide to replace TCP/IP by another protocol stack for wireless network support, but it is important to not ignore the problem and analyse the main drawbacks of TCP/IP in wireless networks and think about a new architecture of network communication over wireless networks. This paper provides a brief survey of the future wireless Internet, and emphasizes on new network architecture to optimize the performance of this network.

1 Introduction

Wireless communications are more and more involved in different user applications, and are already part of our daily lives. The set of all the new Wi-xx networks should form the infrastructure of the future wireless Internet. The new wireless technologies are fortunately growing but unfortunately growing faster than networking technologies. In fact, the basic networking technology is the TCP/IP communication model. This model has been designed at the first place for wired networks. At that time, wireless networks requirements and their applications were not part of the TCP/IP model design. TCP/IP model is facing a crisis in wireless and mobile environment. This crisis is getting more important facing new wireless environments such as sensor networks.

In this paper we describe the future wireless Internet and we illustrate via a Wi-Fi network scenario the difficulty of TCP/IP. We point out the need for a new architecture that includes a smart mechanism which is able to decide which communication protocol is suitable for a certain situation (goal) in the network, and also adapt the parameters of the selected protocols to better react to the present and future state of the network.

The paper is organized as follows. First the Wi-family and the future wireless Internet are introduced, then TCP/IP problems are illustrated via a Wi-Fi network

K. Cho and P. Jacquet (Eds.): AINTEC 2005, LNCS 3837, pp. 157–167, 2005.
© Springer-Verlag Berlin Heidelberg 2005

example. After that, a new autonomic-oriented architecture (AoA) is introduced followed by the description of an implementation of this architecture to improve the TCP/IP architecture.

2 The Wi-Family

The Wi-family is being built around IEEE standard committee sponsorship. A quite large number of committees are now working to set up wireless technologies linked between them by diagonal handovers. The whole family has the intention to build a global solution to the future ambient Internet access.

In the following, a brief survey of the main IEEE working groups is described. Then, the TCP/IP stack optimization difficulties on different wireless networks are pointed out.

The IEEE 802.15 Working Group is working on the PAN environment (Personal Area Networks). This group proposes two general categories of wireless networks, called TG4 (low rate) and TG3 (high rate). The TG4 version provides data speeds of 20 Kbps or 250 Kbps. The TG3 version supports data speeds ranging from 11 Mbps to 480 Mbps. The WiMedia alliance and UWB Forum are the industry associations dedicated to collaboratively developing and administering specs using IEEE 802.15 environment.

The IEEE 802.11 working group proposes a family of specifications for wireless local area networks under the Wi-Fi names.

IEEE 802.16 deals with a group of broadband wireless communication standards for metropolitan area networks. The original 802.16 standard, published in December 2001, specified fixed point-to-multipoint broadband wireless systems operating in the 10-66 GHz licensed spectrum. An amendment, 802.16a, approved in January 2003, specified non-line-of-sight extensions in the 2-11 GHz spectrum, delivering up to 70 Mbps at distances up to 20 miles. 802.16 standards are expected to enable multimedia applications with wireless connections. This technology with a long range will provide a viable last mile technology.

The mission of IEEE 802.16e and IEEE 802.20 is to develop the specification for an efficient packet-based air interface that is optimized for the transport of IP-based services. The goal is to enable worldwide deployment of affordable, ubiquitous, always-on and interoperable multi-vendor mobile broadband wireless access networks that meet the needs of business and residential end user markets.

Both groups are working together to a specification of physical and medium access control layers of an air interface for interoperable mobile broadband wireless access systems, operating in licensed bands below 3.5 GHz. This interface should be optimized for IP-data transport with peak data rates per user in excess of 1 Mbps. The specification will support various vehicular mobility classes up to 250 Km/h in a MAN environment. The focus is on a high spectral efficiency, a sustained user data rates and a large numbers of active users that are all significantly higher than achieved by existing mobile systems.

The IEEE 802.22 standard is intended to enable deployment of interoperable 802 multivendor wireless regional area network products, to facilitate competition in broadband access by providing alternatives to wireline broadband access. This solution should extend the deployment of such systems into diverse geographic areas, including sparsely populated rural areas, while preventing harmful interference to incumbent licensed services in the TV broadcast bands.

Finally, and may be the most important, the IEEE 802.21 group is developing standards to enable handover and interoperability between heterogeneous network types including both 802 and non 802 networks.

The set of all these technologies permit to build the future wireless Internet which is illustrated on Fig. 1.A mobile customer will have to find the best network where he has to be connected on. A sequel of diagonal handovers is necessary to maintain the connectivity of the laptop.

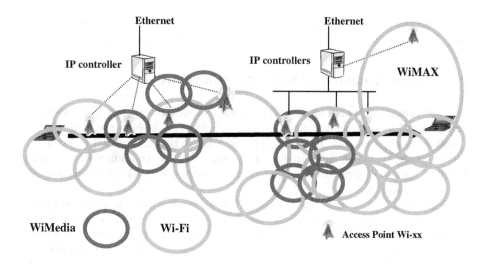

Fig. 1. Architecture of the future Wireless Internet

The TCP/IP stack is intended to be used in all these different wireless networks. However, the TCP/IP stack has to be optimized depending on the characteristics of the network. As an illustration of the difficulties of the TCP/IP model, we measured the energy consumption of a file transfer over a Wi-Fi network. The transmission was at 100 mW level and the receiver was at 2 meters in a direct view so that the quality of the signal is sufficient to avoid retransmission at the MAC layer. We transmit a 100 Mbytes file fragmented into 100 bytes carried out by IP packets. To send one useful bit of the payload on a Wi-Fi wireless network of an IP packet, we needed an energy of approximately 700 nJ. As a cycle of the processor asks for approximately 0.07 nJ, the transmission for one bit of the payload is approximately equal to 10 000 cycles of the processor of the PDA that was used in this experiment.

When measuring the energy consumption for sending just one bit, we obtained 70 nJ. Therefore, we can deduce that the TCP/IP environment is asking on the average 10 times more energy to transmit one useful bit than for the simple transmission of one bit. Note that the transmission of IP packets asks for a large number of signalling packets. This explains partly the high energy consumption. Another part of the energy consumption comes from the number of overhead bits produced by the TCP/IP architecture. A last part of the energy consumption comes from the TCP timers.

Two significant conclusions can be provided from these measurements:

- The TCP/IP protocol over Wi-Fi and more generally over wireless systems is very energy consuming when the segmentation provides small packets.
- When it is necessary to send ten bits for one efficient bit (very compressed ToIP for example), the question that arises is: Is it possible to find another protocol able to improve the energy consumption to send one useful bit?

On this example, we have shown that the TCP/IP stack is not very efficient for wireless networks. We can find other examples where the TCP/IP architecture is not optimal at all on QoS, reliability or security issues.

The sequel of this paper is a proposal for a new architecture able to optimize not only the energy but also different communication performance criteria.

3   AoA-The Autonomic-Oriented Architecture

As user needs are becoming increasingly various, demanding and customized, IP networks and more generally telecommunication networks have to evolve in order to satisfy these requirements. Therefore, a network has to integrate reliability, quality of service, mobility, dynamicity, service adaptation, etc. This evolution will make users satisfied, but it will surely create more complexity in the network generating difficulties in the control process.

Since there is no control mechanism which gives optimal performance whatever the network conditions are, we argue that an adaptive and dynamic selection of control mechanisms, taking into account the current traffic situation, is able to optimize the network resources and to come up with a more important number of user expectations associated with QoS. To realize such functionalities, it is necessary to be able to configure automatically the network in real time. Therefore, all the equipment must be able to react to any kind of changes in the network.

Due to these different issues, an autonomous-oriented architecture (AoA) is a potential solution. The global architecture is shown in Fig. 2. This architecture is composed of 4 planes:

~ The data plane to forward the packets.
~ The control plane to send configuration messages to the data plane in order to optimize the throughput, and the reliability.
~ The knowledge plane to provide a global view of all the information concerning the network.
~ The management plane to administrate the three other planes.

The most important plane in this architecture is the knowledge plane. This plane is supposed to drive the network through the control plane. For this purpose, the knowledge plane will choose the best algorithm to reach the goal decided by the operator. The second action of the knowledge plane is to decide about values of all the parameters of the algorithm. Finally, the knowledge plane has to configure the control plane which itself configures the data plane. Actually, the different control algorithms are

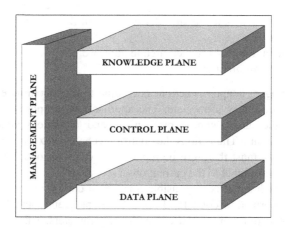

Fig. 2. The Autonomic-oriented Architecture

chosen through a local knowledge and not a global knowledge. The advantages of the global knowledge come from the potential anticipation on the behavior of the control algorithms.

The knowledge plane in our proposal is built with a cognitive multi-agent system. On the contrary, the control plane is built with reactive agents able to react instantaneously. In fact, agents own some features like autonomy, proactivity, cooperation, etc., predisposing them to operate actively in a dynamic environment like IP networks. Agents, by consulting their local knowledge and by taking into consideration the limited available information they possess about their neighbors, select the most relevant control mechanisms to the current situation.

A multi-agent system is composed of a set of agents which solve problems that are beyond their individual capabilities [1]. Multi-agent systems have proven in the past their reliability when being used in numerous areas like: (1) the road traffic control ([2], [3]); (2) biologic phenomena simulation like the study of eco-systems [4] or the study of ant-colonies [5]; (3) social phenomena simulation like the study of consumer behaviors in a competitive market [6]; (4) industrial applications like the control of electrical power distribution systems, the negotiation of brands, etc. By its nature, multi-agent approach is well suited to control distributed systems. IP networks are good examples of such distributed systems. This explains partly the considerable contribution of agent technology when introduced in this area. The aim was mainly to solve a particular problem or a set of problems in networks like: the discovery of topology in a dynamic network by mobile agents ([7], [8]), the optimization of routing process in a constellation of satellites [9], the fault location by ant agents [10], and even the maximization of channel assignment in a cellular network [11].

Our approach consists in integrating cognitive agents [1] in different control equipment and reactive agents within the different routers, firewalls, middle boxes, and so on. The cognitive agents [4] decide about the algorithm to select and the value of the parameters to settle on. The reactive agents decide about the values of network QoS parameters (delay, jitter, loss percentage of a class of traffic, etc.) by adapting the

the activated control mechanisms in order to better fit the traffic nature and volume, and the user profiles.

To achieve the autonomic-oriented architecture, we propose to select the appropriate control mechanisms among:

~ adaptive: the agent adapts its actions according to the incoming events and to its vision of the current system state. The approach we propose is adaptive as the agent adapts the current control mechanisms and the actions undertaken when a certain event occurs. The actions the control mechanism executes may become no longer valid and must therefore be replaced by other actions. These new actions are indeed more suitable to the current observed state;

~ distribution: each agent is responsible for a local control. There is no centralization of the information collected by the different agents, and the decisions the agent performs are in no way based on global parameters. This feature is very important as this avoids having bottlenecks around a central control entity;

~ local: the agent executes actions on the elements of the node it belongs to. These actions depend on local parameters. However, the agent can use information sent by its neighbors to adapt the activated control mechanisms;

~ scalable: the proposed approach is scalable because it is based on a multi-agent system which scales well with the growing size of the controlled network. In order to adaptively control a new node, one has to integrate an agent (or a group of agents) in this node to perform the control.

The model proposed here relies on two kinds of agents: (1) Master agents, that are deliberative or cognitive agents, supervising the other agents, (2) reactive agents responsible for a specific control task within the network equipment. In the control plane, we find different control mechanisms of the node, which are currently activated. Each control mechanism is characterized by its own parameters, conditions and actions, which are monitored and modified by the Master Agents lying in the knowledge plane.

Different agents belong to the control plane (Scheduler Agent, Queue Control Agent, Admission Controller Agent, Routing Agent, Dropping Agent, Metering Agent, Classifying Agent, etc.). Each of these agents is responsible for a specific task within the network equipment. So each agent responds to a limited set of events and performs actions ignoring the treatments handled by other agents lying on the same node or on the neighborhood. This allows to the agents of this level to remain simple and fast. More complex treatments are indeed left to the Master Agent.

If necessary, Master Agents can be grouped in just one centralized unit for the sake of simplicity. The knowledge plane is responsible for controlling the entities of the control plane in addition to the different interactions with the other nodes like cooperation, negotiation, messages processing, etc. A Master agent owns a model of its local environment (its neighbors) that helps him to take its own decisions. The Master Agent chooses the actions to undertake by consulting the current state of the system (neighbors nodes state, percentage of local loss, percentage of its queue load, etc.) and the meta-rules at its disposal in order to have only the most relevant control mechanisms activated with the appropriate parameters. The node, thanks to the two decision levels, responds to internal events (loss percentage for a class of traffic, load percentage of a queue, etc.) and to external ones (message sent by a neighbor node, reception of a new packet, etc.).

The Master Agent owns a set of meta-rules allowing it to decide on actions to perform relating to the different node tasks like queue management, scheduling, etc.

These meta-rules permit the selection of the appropriate control mechanisms to activate the best actions to execute. They respond to a set of events and trigger actions affecting the control mechanisms supervised by that Master Agent. Their role is to control a set of mechanisms in order to provide the best functioning of the node and to avoid incoherent decisions within the same node. These meta-rules give the node the means to guarantee that the set of actions executed, at every moment by its agents, are coherent in addition to be the most relevant to the current situation.

The actions of the routers have local consequences in that they modify some aspects of the functioning of the router (its control mechanisms) and some parameters of the control mechanisms (queue load, loss percentage, etc.). However, they may influence the decisions of other nodes. In fact, by sending messages bringing new information on the state of the sender node, a Master Agent meta-rule on the receiver node may fire. This can involve a change within the receiver node (the inhibition of an activated control mechanism, or the activation of another one, etc.). This change may have repercussions on other nodes, and so forth until the entire network becomes affected.

This dynamic process aims to adapt the network to new conditions and to take advantage of the agent abilities to alleviate the global system. We argue that these agents will achieve an optimal adaptive control process because of the following two points: (1) each agent holds different processes (control mechanisms and adaptive selection of these mechanisms) allowing to take the most relevant decision at every moment; (2) the agents are implicitly cooperative in the sense that they own meta-rules that take into account the state of the neighbors in the process of control mechanisms selection. In fact, when having to decide on control mechanisms to adopt, the node takes into consideration the information received from other nodes.

4 The Agent-Based Autonomic-Oriented Architecture

The autonomic-oriented architecture is composed of mainly two mechanisms: The smart mechanism to select the algorithms and its parameters, and the enforcement mechanism to enforce the decisions of the smart mechanism. For that we use the agent-based scheme described in the previous section, and we use some concepts of the policy-based networking [12] such as the enforcement procedures to implement configurations.

An agent-based platform permits a meta-control structure. It is assumed that, for each network equipment, we associate one or several agents so that the network can be seen as a multi-agent system. The main goal of this system is to decide about the control to use for optimizing some given performance criteria described in the goal distributed by the Master Agent.

Intelligent agents are able to acquire and to process information about situations that are "not here and not now", i.e., spatially and temporally remote. By doing so, an agent may have a chance to avoid future problems or at least to reduce the effects. These capabilities allow the control agents to adapt their behavior according to the traffic flows going through the node.

It is important to note that other works have proposed decision mechanisms able to enforce decisions or policies in the network. These typical architectures enforce high level decisions without considering the problem optimization of parameters related to lower levels of the network. This is a classical top-down approach. In our Autonomous-oriented Architecture, we intend to use the enforcement procedure of policy-based management architecture only from the control plane because this is an interesting concept for automating the enforcement of the smart mechanism decisions. The Autonomous-oriented Architecture considers the optimizing problem related to the control but also the data layers of the network, and enforces the most suitable algorithm and parameters for a given network.

The goal of the Autonomous-oriented Architecture (AoA) is to optimize the SLAs of the different clients. In this implementation, users enter their SLA through a Web service scheme. The manager of the network can also enter the network configurations corresponding to the goals of the network. A Master Agent situated in a Goal Decision Point (GDP) is able to decide about the global goal of the network. This Master Agent can be supported by any kind of centralized servers if any. As soon as defined, the goal is distributed to the different network equipment (the routers in our implementation) that could be named Goal Enforcement Point (GEP). Knowing the goal, the different nodes have to apply policies and define the control mechanisms. A configuration of the routers has to be provided to reach the goal. The configuration affects the software, the hardware and the protocol stack.

The Master Agents plus the agents within the Goal Enforcement points are forming the multi-agent system of the knowledge plane described in the previous section. This knowledge plane is in charge of collecting the different information to be able to decide what the goal is and what the best control algorithms to choose are. This multi-agent system specifies and updates the goal of the network which is about what to optimise in the network and what to be offered by the network.

A Local Policy Decision Point may be implemented in the GEP to decide about the best policy to implement achieving the goal received by the GEP.

As indicated, the configuration can affect the protocol stack. This is why we can introduce a new protocol stack STP/SP (Smart Transport Protocol/Smart Protocol) to better optimize the communication over a wireless link [15].

The choice of the protocol can be seen at two levels: the control and the knowledge. One specific agent in each node at knowledge layer may be defined for deciding the local protocol in cooperation with the other similar agent of the multi-agent system. Each agent has to perform a specific procedure, which is triggered according to the state of the node, to the QoS required, and to any other reason. This constitutes a local level for the decision. Moreover, agents can periodically interact to exchange their knowledge and ask to other agents if they need information they do not have. This constitutes the global level.

Fig. 3 illustrates the AoA implementation for the STP/SP protocol.

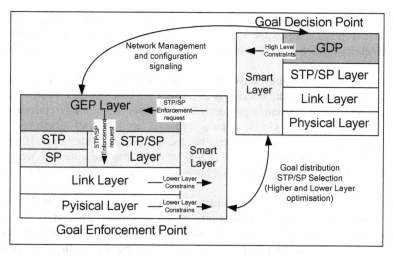

Fig. 3. The Autonomic-oriented Architecture

5 Approach Analysis

In this section, we are interested in a performance evaluation through a simple testbed to understand the pros and the cons of the new autonomic-oriented architecture to support a smart selection of a the most appropriate protocol.

For the STP/SP architecture, we chose only two states for the SP protocol: a protocol using packets as long as possible and a protocol with only short packets (100 bytes). Two kinds of clients were defined:

~ Telephony which induces an IP packet payload of 16 bytes and a throughput of 8 Kbps per call. The IP packet may be either padded to reach 100 bytes or can group several available payloads. In this case the waiting time cannot exceed 48 ms (namely three payloads can be encapsulated in the same IP packet). The response time of the end to end delay cannot be larger than 150 ms and only 1 percent of packets may arrive in late (they are dropped at the arrival but the quality of the voice is maintained).

~ File transfer with 1 billion bytes per file. When available the packets get a 10 000 bytes length and in the other case the file is segmented to produce a series of 100 bytes packet length.

The arrival process of telephone calls is Poisson. The length of telephone calls is 3 minutes on the average and this length is exponentially distributed. The arrival process of the file transfers is Poisson and the average length is 1 billion bytes at a constant rate of 2 Mbps. Traffics introduced by these two applications are identical and equal to 1 Mbps. Namely, idle period and busy periods for the file transfer are 0.5. On the average 125 telephone calls are running.

Two goals were defined: minimizing the energy consumption in the global network and optimizing the number of successful telephone calls.

The model is a tandem queueing system composed of five mobile machines in series. The first queue receives the arriving packets and the queues are FIFO with priority. The service process is dependent on the length of the packets with a rate of 2.5 Mbps.

Results of our simulations show that the lifetime of the networks is more than twice when the length of the packets is as long as possible but 20% of the telephone calls are dropping more than 1 percent of the packets. The energy consumption is divided by more than two. On the contrary, when using 100 bytes length packets, all the telephone calls are running correctly but the lifetime is divided by 2. If we adopt 100 bytes for the telephone calls and 10 000 bytes for the file transfers, we obtain a result in between with 10% of the telephone calls with more than 1% of the packets lost and a 50% shorter lifetime.

When implementing our AoA architecture through an extension of the J-Sim package, and choosing as a goal 1- no dropping of the telephone calls and 2- maximizing lifetimes, we got no telephone call lost and a 20% shorter lifetime than the optimal. This result is due to the fact that as soon as the file transfer traffic is too high, the length of the file transfer packets is automatically reduced to allow the telephone packets to keep an acceptable response time. The reactive agents reply on a too long response time detected by the knowledge plane by shortening the packet length. This permits the short telephone packets not to wait a too long time in the nodes.

Ongoing implementation is building a muti-agent system to enable a real autonomic behavior. This implementation is not just working on a small number of parameters as in the simulation study. In addition, a set of TCP/IP parameters are being identified and a subset of necessary parameters in the network equipment will be specified to build the corresponding STP/SP.

6   Conclusion

This paper brought a new communication paradigm to better support energy consumption, reliability, QoS, and new functionalities in a network using a four-layer architecture. This architecture considers not only the control algorithms provided by the control plane but mainly the knowledge plane able to synthesize all the information of the network. An Autonomic-oriented Architecture is proposed to provide the selection of control mechanisms to optimize the configuration network equipment. This architecture interacts with the network equipment and protocols in order to configure the network with the selected protocols and parameters. An analysis of the proposed architecture intends to confirm that a real time configuration of network equipment in reaction to an ongoing situation in the network brings an important improvement of the network performance.

References

1. Ferber J. - Multi-Agent Systems: An Introduction to Distributed Artificial Intelligence. Addison Wesley Longman, 1999.
2. Bazzan A.L.C., Wahle J. and Klügl F. - Agents in Traffic Modelling - From Reactive to Social Behaviour. KI'99, Bonn, Germany, LNAI 1701, pp 303-307 September 1999.

3. Moukas A., Chandrinos K. and Maes P. - Trafficopter: A Distributed Collection System for Traffic Information. CIA'98, Paris, France, LNAI 1435 pp 34-43, July 1998.
4. Doran J. - Agent-Based Modelling of EcoSystems for Sustainable Resource Management. 3rd EASSS'01, Prague, Czech Republic, LNAI 2086, pp 383-403, July 2001.
5. Drogoul A., Corbara B. and Fresneau D. - MANTA: New experimental results on the emergence of (artificial) ant societies". In Artificial Societies: the computer simulation of social life, Nigel Gilbert & R. Conte (Eds), UCL Press, London, 1995.
6. Bensaid L., Drogoul A., and Bouron T. - Agent-Based Interaction Analysis of Consumer Behavior. AAMAS'2002, Bologna, Italy, July 2002.
7. Minar N., Kramer K.H., and Maes P. - Cooperating Mobile Agents for Dynamic Network Routing. In Software Agents for Future Communication Systems, Chapter 12, Springer Verlag, pp 287-304, 1999.
8. Roychoudhuri R., et al. - Topology discovery in ad hoc Wireless Networks Using Mobile Agents. MATA'2000, Paris, France. LNAI 1931, pp 1-15, September 2000.
9. Sigel E., et al. - Application of Ant Colony Optimization to Adaptive Routing in LEO Telecommunications Satellite Network. Annals of Telecommunications, vol.57, no.5-6, pp 520-539, May-June 2002.
10. White T., et al. - Distributed Fault Location in Networks using Learning Mobile Agents. PRIMA'99, Kyoto, Japan. LNAI 1733, pp 182-196, December 1999.
11. Bodanese E.L., and Cuthbert L.G. - A Multi-Agent Channel Allocation Scheme for Cellular Mobile Networks. ICMAS'2000, USA. IEEE Computer Society press, pp 63-70, July 2000.
12. Verma D. C. - Simplifying Network administration using policy-based management, IEEE Network 16(2), 2002.
13. Merghem L., Gaïti D., and Pujolle G. - On Using Agents in End to End Adaptive Monitoring. E2EMon Workshop, in conjunction with MMNS'2003, Belfast, Northern Ireland, LNCS 2839, pp 422-435, September 2003.
14. Gaïti D. and Pujolle G. - Performance management issues in ATM networks: traffic and congestion control, IEEE/ACM Transactions on Networking, 4(2), 1996.
15. Chaouchi H. and Pujolle G. - STP/SP: the solution for the TCP/IP crisis in the emerging wireless networks, IEEE PIMRC 2004, Barcelona, 2004.

Security Threats and Countermeasures
in WLAN

Dhinaharan Nagamalai[1], Beatrice Cynthia Dhinakaran[2], P. Sasikala[2],
Seoung-Hyeon Lee[1], and Jae-Kwang Lee[1]

[1] Department of Computer Engineering,
Hannam University, 306-791, Daejeon, South Korea
[2] Department of Computer Science and Engineering,
Woosong University, 300-718, Daejeon, South Korea

Abstract. Wireless-the name says it all: "Cut the cord". Today there
is no hotter area of development than wireless networking. Enterprizes
are rapidly adopting wireless local area networks (WLANs). Driving this
rapid adoption is the low-cost and inherent convenience of rapidly deployed connectivity solutions, which are not burdened by a wired infrastructure. WLAN have been in the news quite a lot, recently both from
the perspective of growing popularity and in terms of security vulnerabilities that have been discovered in the wired equivalent privacy (WEP)
security standard, which is supposed to ensure the security of data that
flows over the networks. While wireless LANs are convenient and provide
immediate connectivity, they also impose unique management, security,
and mobility challenges on network. Though "connect anywhere at anytime" promise of the wireless networks is beginning to become popular;
there are a plenty of confusion among prospective users when it comes to
security. Because WLAN impose such significant challenges, should all
802.11 networks be considered hostile to the prospective user? Not necessarily. A well-designed wireless LAN ensures that the network is well
protected and easy to manage, without sacrificing the wireless user's
ability to roam seamlessly. This is a survey paper that focuses on the
catalog of security risks and countermeasures of the wireless network.
We describe the standard mechanisms available for authentication of
users and the protection of the privacy and integrity of the data. A basic
analysis of each security countermeasure is conducted by looking at the
attack techniques addressed by the mechanism. Our analysis takes into
account the perspective of both insiders and outsiders.

1 Introduction

Wireless networking is cable-free, no-strings-attached networking. Wireless
LANs use electromagnetic airwaves to communicate information from one point
to another without relying on any physical connection. A good way to improve
data connectivity in an existing building without the expense of installing a
structure-cabling scheme to every desk is to use WLAN. There are several problems with the physical aspects of wired LAN connections namely loose patch

K. Cho and P. Jacquet (Eds.): AINTEC 2005, LNCS 3837, pp. 168–182, 2005.

cords, locating live data outlets, broken connectors, etc. that generate a significant volume of helpdesk calls. Such problems are reduced in wireless networks. There are number of issues that have to be considered when deploying a WLAN. The most important is the security. The three basic security concepts important to information are confidentiality, integrity and availability [1][2][3]. In most wired LANs the cables are inside the building, so a hacker must defeat physical security measures. But in the radio waves used in wireless networks typically penetrate outside the building, creating a real risk that the network can be hacked from the parking lot or the street. The designers of IEEE 802.11b or Wi-fi (Wireless fidelity) have tried to overcome the security issue by devising a user authentication and data encryption system known as Wired equivalent privacy (WEP)[10][11]. A good wireless networking system should therefore provide a rage of different authentication and data encryption options so that user can be given the appropriate level of security. Another approach is the use of Virtual private network (VPN) hardware[10]. VPN hardware is designed to enable remote users to establish a secure connection to a corporate data network via an insecure medium, namely the Internet. However there are drawbacks to using existing VPN products in a wireless LAN environment. The objective is to deploy and maintain secure, high performance wireless LANs with a minimum amount of time, effort and expense.

The rest of this paper is organized as follows. In section 2 we describe the details of various security risks. Section 3 analyzes and explains the countermeasures needed for every security risk. Finally we conclude our paper in section 4.

2   Security Risks

A complete risk assessment requires a focus on the threats against the three key components of assuring information. That is, the information system should protect against confidentiality, integrity, and availability (CIA) attacks. We start by examining attacks against the confidentiality of communication on the network. We then move into those attacks that actually alter the network traffic, hence destroying the integrity of the information on the network. When looking at confidentiality attacks we start with the least intrusive and work towards more intrusive attacks. In this section we describe chief attack techniques that we use to compare the security technologies available. We chose these attack techniques to be generic enough so that they can be used to evaluate representative security technologies. Security issues, threats and vulnerabilities faced by wireless networks are more or less the same as those faced by the wired networks. But there are few added twists to the wireless network.

2.1   Analyzing Traffic

Analyzing the traffic is a simple technique whereby the attacker can determine the load on the communication medium by the number and size of packets being transmitted. The attacker only needs a wireless card operating in promiscuous (i.e. listening) mode and software to count the number and size of the packets being transmitted. A simple helical directional antenna provides an increased range at which the attacker may analyze traffic. A helical antenna is a simple directional antenna consisting of a horizontal conductor with several insulated dipoles parallel to and in the plane of the conductor. A simple helical antenna made out of a potato chip container, a steel rod, and some washers, an attacker may double the range at which they are receiving transmissions. A helical or spiral antenna, built for less than $100 out of PVC plumbing pipe and copper wire, increases the range by more than double the original distance [1][2].

2.2   Eavesdropping

Due to the radio-based nature of the wireless network, eaves dropping are always possible since the medium of communication is through open air (confidentiality & integrity attack). Intercepting information that is transmitted over the WLAN is generally easier as it can be done from a distance up to kilometers outside the building perimeter without any physical network connection required. The information intercepted can be read if transmitted in clear or easily deciphered if only WEP is used.

Inactive Eavesdropping. In inactive eavesdropping, the attacker passively monitors and has access to the transmission. As already put forward, the directional antenna can detect 802.11 transmissions under the right conditions miles away. Therefore this is an attack that cannot easily be stopped by using physical security measures. Usually we believe that wireless network users would configure their wireless access points to include some form of encryption; however, studies have shown that less than half of the wireless access points in use even have the vulnerable 802.11 wireless security standard, the wired equivalent privacy (WEP) protocol, properly configured and running[8]. Assuming that the session is not encrypted, the attacker can gain two types of information from inactive eavesdropping[1][2]. a) Read data transmitted: The attacker can indirectly read the data transmitted in the session. b) Collect Information: The second information the attacker will be able to gather is information by examining the packets in the session, precisely their source, destination, size number and time of transmission. The information gleaned from this attack is an important precondition for other, more damaging attacks.

Proactive Eaves Dropping. In this attack the attacker monitors the wireless session as described in inactive eavesdropping. However, during proactive eavesdropping, the attacker not only listens to the wireless connection, but also actively inserts messages into the communication medium in order to support them in determining the contents of messages. The preconditions for this attack are that the attacker has access to the transmission and partially access to destination IP address. Since WEP uses a cyclic redundancy check (CRC) to verify the integrity of the data in the packet, an attacker can modify messages (even in encrypted form) so that changing data in the packet (i.e. the destination IP address or destination TCP port) cannot be detected. The attacker's only requirement is to determine the bit difference between the data they want to insert and the original data.

2.3   Unauthorized User Access

This attack is not directed against any individual user, but towards a network as a whole. Once an attacker has access to the network, he can then launch additional attacks or just enjoy usage of network. Due to the physical properties of WLANs, attackers will always have access to the wireless component of the network. In some wireless security architectures this will also grant the attacker access to the wired component of the network. In other architectures, the attacker must use some technique like MAC address spoofing to gain access to the wired component of the network.

2.4   Address Resolution Protocol Attack (ARP)

The Address Resolution Protocol (ARP) maps the Media Address Controller (MAC) address (Layer 2) of a network node to the Internet Protocol address (Layer 3). Altering the mapping of the MAC address to IP address allows an attacker to reroute network traffic through his machine. With the session passing through the attacker's computer the attacker can read plain-text, collect encrypted packets for later decryption, or modify the packets in the session. Routers contain ARP cache poison attacks but a great deal of damage can be done with a successful ARP Cache Poisoning attack. To carry out a successful attack the attacker must have access to the network but nothing else. The attacker sends a forged ARP reply message that changes the mapping of the IP address to the given MAC address. The MAC address is not changed just the mapping [4].

The ARP cache poisoning attack can be used against all machines in the same broadcast domain as the attacker. Hence, it works over hubs, bridges, and switches, but not across routers. An attacker can, in fact, poison the ARP cache of the router itself, but the router won't pass the ARP packets along to its other links. Switches with port security features that bind MAC addresses to individual ports do not prevent this attack since no MAC addresses are actually changed. The attack occurs at a higher network layer, the IP layer, which the switch does not monitor. The ARP attacks can be classified as Denial of service, Man-in the-middle attack and Session Hijacking.

2.5   Man-in-the-Middle (MiM) Attack

The man-in-the-middle attacker can observe this encrypted traffic, but cannot do anything malicious because it is encrypted. Encrypted traffic is not compromised by this attack and the attacker does not gain access to the network. The attacker only gains the ability to target a particular user for the denial of service attack,

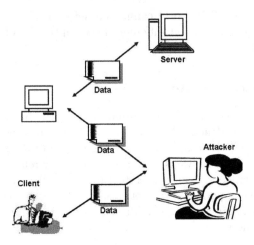

Fig. 1. Man-In-The-Middle Attack

which could be more easily perpetrated by a regular access point disconnected from any network. When encryption is not used, the man in the middle will be able to see the users traffic. This would have also been possible with a network sniffer. Network sniffers can see network traffic of other users, but if that traffic is encrypted, that traffic is useless to any hacker. Therefore, man-in-the-middle type attacks are not a concern.

We explain this attack with an example, when the target has an authenticated session underway. Fig.1. illustrates this type of attack technique. The following steps explain this attack

a) The attacker breaks the session and does not allow the target to reassociate with the access point.

b) The target machine attempts to reassociate with the wireless network through the access point and is only able to associate with the attacker's machine, which is mimicking the access point. Also in step two, the attacker associates and authenticates with the access point on behalf of the target. If an encrypted tunnel is in place the attacker establishes two encrypted tunnels between it and the target and it and the access point.Variations on this attack techniques are based on the security environment. Without encryption or authentication in use the attacker establishes a rogue access point. The target unwittingly associates to the rogue, which acts as a proxy to the actual wireless network.

2.6   Session Hijacking

Session Hijacking is an attack against the integrity of a session. The attacker takes an authorized and authenticated session away from its proper owner. The target knows that it no longer has access to the session but may not be aware that the session has been taken over by an attacker. The target may attribute

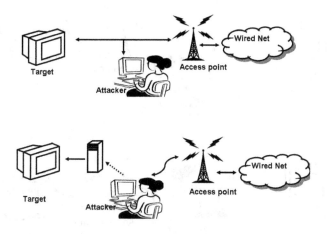

Fig. 2. Session Hijacking

the session loss to a normal malfunction of the WLAN. Once the attacker owns a valid session he may use the session for whatever purposes he wants and maintain the session for an extended time. This attack occurs in real-time but can continue long after the victim thinks the session is over.

To successfully execute Session Hijacking the attacker must accomplish two tasks.

a) He must masquerade as the target to the wireless network. This includes crafting the higher-level packets to maintain the session, using any persistent authentication tokens and employing any protective encryption. This requires successful eavesdropping on the target's communication to gather the necessary information as shown in Fig.2.

b) The second task the attacker must perform is to stop the target from continuing the session. The attacker normally will use a sequence of spoofed disassociate packets to keep the target out of the session[7] as shown in Fig.2.

2.7   Replay Attack

Replay attacks are also aimed at the integrity of the information on the network if not necessarily the integrity of a specific session. Replay attacks are used to gain access to the network with the authorizations of the target, but the actual session or sessions that are attacked are not altered or interfered with in anyway. This attack is not a real-time attack; the successful attacker will have access to the network sometime after the original session(s). In a replay attack (illustrated in Fig.3.) the attacker captures the authentication of a session or sessions as shown in step one. The attacker then either replays the session at a later time or uses multiple sessions to synthesize the authentication part of a session for replay in step two. Since the session was a valid, the attacker establishes an authenticated session without being privy to any shared secrets

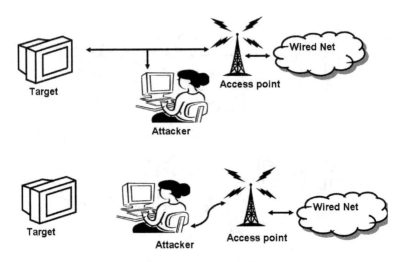

Fig. 3. Replay Attack

used in authentication. Without further security mechanisms the attacker may interact with the network using the target's authorizations and credentials. If the WLAN employs encryption that the attacker cannot defeat the attacker may still be able to manipulate the WLAN by selectively modifying parts of the packet to achieve a desired outcome[5][8].

3   Security Countermeasures and Technologies

The IEEE 802.11 standards group is struggling to solve the problems and the various threats faced by wireless networks. However, it may not be easy as the current technology is already flawed and that cannot be changed. We divide the wireless technology into three broad categories. The first category is authorization. This includes the mechanisms for determining whether or not a client is an authorized user of the WLAN and which authorizations the user should have. It also includes the mechanisms for stopping an unauthorized user from using the WLAN. The second category includes mechanisms for maintaining the privacy of the session once a user is authenticated into the WLAN. Normally, the privacy is maintained by some use of encryption. Stacking another protocol layer on top can circumvent many of the current WLAN security holes. If the traffic on the WLAN is tunnelled through a VPN or an SSH connection, it becomes much harder, if not impossible, for the hacker to capture the traffic. If only tunnelled traffic is allowed, the hacker won't be able to join the network to perform man-in-the-middle attacks either. The problem is that many network administrators do not know how to set up an environment like this, or won't bother with it. The easiest way to set this up is to install a VPN or SSH server right behind the WLAN access point, so that it only allows encrypted traffic to the WLAN direction. In addition to the trouble this of course brings additional costs, as additional procurements in the form of encryption servers are required. One way that has been adopted to use with WLAN is simply accepting that the traffic may be listened on. This is in use at least in some educational facilities in Korea, where the students are simply informed that the WLAN traffic is not encrypted and that they should make sure that they either use a secure way of communicating or their data may be compromised. The problem with this approach is that many end users do not realize how easily they may be revealing critical information. One may think that it's not that bad to download some e-mail that for example only contain jokes sent by friends, but in the process the e-mail account password must be transmitted over the network, and a hacker observing the traffic may capture it as well.

3.1   Authentication

The 802.11 working group is coordinating with the 802.10 groups who are responsible for developing security mechanisms for all 802 LAN standards. The 802.11 standard specifies a mandatory authentication measure for all stations before a connection can be established. There are two authentication mechanisms defined in the 802.11 a) Open System Authentication b) Shared Key Authentication.

Open System Authentication. The first and default authentication method is the open system authentication. The open system authentication works by a station sending an authentication management frame containing the sending station's identity. The receiving station then sends back an acknowledgement with information whether the identity was correct or not. If correct, authentication is complete. This identity is the Service Set Identifier (SSID), which is used to access the network as described above. This basic authentication is rendered useless if Access Points are configured to "broadcast" their SSIDs. This allows any station that isn't configured with an SSID to receive it and then use it to access the wireless network.

Shared RC4 Key Authentication. WEP provides confidentiality for Wireless LANs. WEP provides encryption for all data frames using a 64-bit seed key and the RC4 encryption algorithm. Today most commercial systems use a single secret key shared between all stations. The WEP standard specifies a 40-bit encryption key, however some suppliers are now shipping a 128-bit key version, which is actually only 104-bit. WEP's implementation of shared RC4 Authentication does not offer a high degree of security. An attacker that intercepts a single authentication sequence can then authenticate into the WLAN at will using this key. Many WLANs employ a single key for all users. Regardless, WEP only allows for four total keys, making this vulnerability serious. This security technology offers no protection from a malicious insider. An insider or an attacker masquerading as an insider can authenticate and associate to the WLAN, by virtue of owning the shared secret (key). With access to the WLAN, the attacker has met a necessary precondition for most attacks. An outsider attacker can easily defeat WEP authentication. The attacker will need special tools, but those tools are easily available in the public domain. The skill required to use these tools are minimal. The problems with WEP security are well documented but we include them for completeness.

802.1X. IEEE 802.1X is a standard for port based network access control. The standard can be applied to both wired and wireless networks and provides a framework for user authentication and encryption key distribution. It can be used to restrict access to a network until the network has authenticated the user. It is used in conjunction with one of a number of upper layer authentication protocols to perform the verification of credentials and generation of encryption keys.[6] 802.1x has some serious shortcomings for a wireless network. These come from the reuse of good security mechanisms in an environment for which they were not designed. Reused protocols have been examined more closely than newly developed protocols and therefore are normally more secure. The problem in 802.1x is not the quality of the reused protocols, but the imperfect fit of the wired protocols to a wireless network. For 802.1x to work in a wireless setting, the access point/access controller must allow traffic to the authentication server prior to authentication as shown in Fig.4. Both 802.11a WLAN protocol and 802.1x use state machines to function correctly. The adaptation of 802.1x to 802.11a left the two state machines loosely coupled. Due to the loose coupling between the state machines in the two protocols 802.1x is subject to session hijacking attack from an outsider. The problem of a rogue network connection is much smaller in wired networks than in WLANs. The insider may not have much of an advantage with 802.1x over the outside attacker. Since 802.1x is coupled to wired network authentication (normally through RADIUS) the insider will only have access to their normal resources. If the 802.1x is not coupled to a mechanism for blocking network access like inline authentication then 802.1x only protects network resources from the honest user. The attacker whether insider or outsider, has a platform for launching attacks. Implementing 802.1x security must be coupled with a blocking mechanism so that unauthenticated clients cannot access the network. Using 802.1x to authenticate sessions stops the casual unauthorized user from accessing the WLAN. However, it does not prevent a moderately skilled attacker with few resources from successfully attacking the network[7][3].

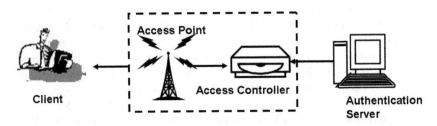

Fig. 4. 802.1X

Extensible Authentication Protocol with Transport Layer Security (EAP-TLS). EAP-TLS is a certificate based authentication protocol and is supported natively in Windows XP. It requires initial configuration by a network administrator to establish the certificate(s) on the user's machine and the authentication server, but no user intervention is required thereafter. The certificates are digital signatures, which are used in conjunction with public key encryption techniques to verify the identity of the client. During an EAP-TLS exchange, the client and authentication server exchange credentials and random data in order to simultaneously synthesize the encryption keys at both ends of the link.[9] Once this has been completed, the server sends the encryption keys to the AP through a secure RADIUS channel and the AP exchanges messages with the client to plumb the encryption keys down to the MAC encryption layer. The problem with mutual authentication for TLS is really just the requirement to distribute client side certificates to each client. Without a good PKI (Public Key Infrastructure) implementation it's not simple to revoke access for any individual certificate either. Mutual authentication stops man-in-the-middle attacks. An attacker cannot fool the client into thinking that he is authenticated into the access point because the client authenticates the access point. Mutual authentication may not stop session hijacking. If each individual packet is authenticated, then it will increase security of the transmission, but at an obvious performance cost. Many replay attacks can be thwarted by mutual authentication. If the authentication includes time or sequence numbers replay attacks will be much more difficult, if not impossible[6].

Tunnelled Transport Layer Security (TTLS). Tunnelled TLS had all of the advantage of TLS and eliminates the need for client side certificate for mutual authentication. What TTLS does is establish the identity of the server side (typically the AP) using normal TLS, much like a web browser establishing the identity of a web server for secure transactions. Once a secure connection is established in this way, the client can use other existing authentication mechanisms such as RADIUS to authenticate. This makes deployment a simpler task.

Protected Extensible Authentication Protocol (PEAP). PEAP is an IETF draft standard and can be used to provide a secure password based authentication mechanism. In a PEAP exchange, only the authentication server is required to have a certificate. After the initial communication with the authentication server, the public key from the authentication server (AS) certificate is sent to the client computer. The client computer then generates a master encryption key and encrypts this key using the AS's public key and sends the encrypted key to the AS. Now that the master key is on both ends of the channel, this key can be used as source material to establish a secure tunnel between the AS and the client over which any subsequent authentication method can be used to authenticate the client computer to the AS. In many cases is it expected that this will be some form of a password based authentication protocol.

Packet Authentication. Packet Authentication is different from the session authentication that the previous paragraphs address. Once an authenticated session is established and the keys are exchanged. Schemes rely on the privacy of an encrypted tunnel and integrity checking on the payload to imply the identity of the sender. This is an effective scheme, however, the addition of packet authentication adds an additional mechanism that an attacker must defeat. We do not believe replay, session hijacking and man-in-the-middle attacks are possible when packet authentication is added to strong session authentication. The receiver must be sure that the individual packets of a session did in fact come from the sender or else the session is subject to man-in-the-middle, replay or session high jacking attacks. These attacks all can succeed because the attacker fools the receiver into believing the packets sent by the attacker are from the target, hence destroying the session integrity of the system. These all rely on breaking an authenticated session. Per-packet authentication adds another layer of defense that an attacker must defeat. She cannot just take over an authenticated session without the ability to authenticate the packets that she generates or modifies. By itself packet authentication does not offer much defense; however, when combined with mutual session authentication it is very effective. This is an example of how properly integrated partial security mechanisms can form a defense-in-depth.

3.2   Encrypted Tunnel or Virtual Private Network (VPN)

Packets are kept private by the use of encryption. Encryption systems are designed to provide a virtual tunnel that the data passes through as it traverses the protected part of the network. If the system is properly designed and correctly implemented, the contents of the payload will be unreadable to those without the proper decryption key. The contents that the receiver decrypts must not only be private, but exactly as the sender intended. In other words correct tunnel will not Only keep the contents private, but also free from modification. This requires the use of a cryptographic integrity checker or checksum.

OSI Network Layer and Endpoints. Two of the key design parameters of VPN are the OSI network layer that is encrypted and the endpoints of the tunnel. Generally, the lower the layer that is encrypted the more secure. Also the longer the tunnel, generally the more secure the tunnel. The drawback is that the more secure these mechanisms the higher the reliance on vendor specific components and the decrease in system performance. Integration with your existing architecture is beyond the scope of this paper but is useful in understanding the array of options.

Endpoints. Encrypted tunnels can have three possible sets of endpoints. The first illustration in Fig.5.shows a tunnel that runs from client to access point. The second runs through the access point but only to an access controller appliance that separates the wired and wireless component Finally end-to-end encrypted tunnels run from the client to the server passing through the wired and wireless

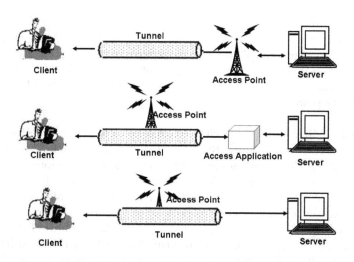

Fig. 5. Endpoint options for Encrypted Tunnels

network segment in the encrypted state. They are decrypted at the destination. These may be used together to form a defense-in-depth.

Encryption Layer. Besides the length of the encrypted tunnel, the other attribute that determines the security of the encryption is the implementation layer. Encrypted tunnels may be implemented at layer 4 (i.e. secure sockets or SSL), the layer 3 (i.e. IPSec or VPN solutions), and/or at the layer 2 (i.e. WEP or AES. Layer 3 tunnels encrypt layers 4 and higher leaving the layer 3 header exposed. Likewise, a layer 2 tunnel encrypts layer 3 data and higher protecting information like the source and destination IP address of the packet. The security of an encrypted tunnel increases when encryption is applied at a lower layer. Thus, a layer 3 tunnel is not as secure as a layer 2 tunnel. For example, spoofing an IP is easier to achieve when a layer 3 tunnel rather than a layer 2 tunnel is implemented because the IP address of the recipient is transmitted in the clear. A layer 2 tunnel decreases the risk of an IP spoofing attack but does nothing to prevent an ARP spoofing attack as the MAC address is still transmitted in plaintext. The rule of thumb for encrypted tunnels is, "when you own a level, you can break security implemented on all the higher levels." This provides a good guide but only when examined in isolation. Combinations of other security mechanisms with encrypted tunnels will increase security.

Encryption Algorithm and Key Size. Another major design feature is the type of encryption to use. In theory, the algorithm and key length combine to make the packets difficult to read. The implementation however, is a key aspect that cannot be ignored. Flaws in implementation can drastically alter the effort required to break the encrypted message negating any theoretical advantage a scheme may have. WEP encryption is a perfect example of how implementation flaws negate the theoretical advantages of the algorithm. While WEP encryption provides minimal protection, triple - DES (Data Encryption Standard) is the current standard and if properly implemented provides adequate security for most applications. The Advanced Encryption Standard (AES) is the newly approved standard, which provides a higher level of assurance while requiring less processor power[8].

IEEE 802.11 Standard or WEP (40 and 104 bit keys). WEP is a layer 2-encryption scheme based on the RC4 stream cipher. It relies on a secret key that is shared by the client and server. WEP uses a no cryptographic checksum of the plaintext to insure integrity. The plaintext and the checksum are encrypted using an initialization vector, the secret key and the RC4 algorithm. The initialization vector and the encrypted payload are then sent to the recipient. [8] As discussed, the WEP 40 bit key size can be attacked by brute force. The 104 bit keys are not currently vulnerable to brute force attacks but regardless of key size WEP is vulnerable to both passive and active eavesdropping. WEP encryption can be defeated passively when the key stream is reused. Because the WEP initialization Vector (IV) is only 24 bits, reuse can occur quite frequently even in a well-implemented version of WEP. The use of 802.11a with a potential for a fivefold increase in bandwidth decreases the time needed to passively defeat WEP to minutes. Active eavesdropping with partially or fully known plaintext are relatively simple attacks to carry out on WEP WLANs[8]. A determined attacker could build a decryption dictionary in a relatively short period of time and thus have real-time access to all message traffic on the WLAN. WEP WLANs that start the IV at 0 when reinitialized or don't change the key stream after every packet make this task simpler still. Attackers can modify a WEP encrypted packet, or create a new packet that meets the WEP authentication standards violating the integrity of WEP sessions. This allows man-in-the-middle and session hijacking attacks to be successful. WEP security is flawed, but does provide some security. WEP provides almost no protection from insiders. With only four keys available, multiple users using the same key are inevitable. Even though WEP can be broken relatively easily by outsiders, the attacker must use special tools and may have to invest days of effort to do so. Neither of these requirements presents much of a hurdle, but discourage the casual attacker looking for an easy target.

Layer 3 Virtual Private Network (VPN) or Strongly Encrypted Tunnel. Strongly Encrypted Layer 3 VPNs leave the IP header data between the VPN client and the VPN concentrator unencrypted while protecting the payload and header information for layers 4 and up. Layer 3 VPNs tend to be more venders independent and can be configured to protect sessions over the portions of the wired network as well as the WLAN. Layer 3 Strongly encrypted Virtual Private Networks (VPN) can provide strong protection against outsider's access to private sessions. Like Layer 2 tunnels, key management is important. The owner of the secret key can eavesdrop on protected sessions. If combined with a public key infrastructure (PKI) this risk can be mitigated. However, PKI

increases the complexity of the infrastructure and can introduce other vulnerabilities to the system. Even without the key, Layer 3 VPNs are more vulnerable
than layer 2 tunnels. In 802.11a WLANs, management packets are not authenticated and an attacker can easily break the connection between the target and
the wireless access point. Without per packet authentication the attacker can
then launch a man-in-the-middle or session hijacking attack.

3.3 Privacy Protection

To protect the privacy of the session we can choose criteria for three major design
parameters: layer, topology, and encryption. Most encryption for WLAN takes
place at layer 2 or layer 3. Most topologies just include the wireless portion of
the network. Normally, an appliance located between the wireless portion of the
network and the remainder of the wired network is the endpoint of the encrypted
tunnel. There are tunnels that extend from the wireless client to the server, like
the security architecture provided by IBM, but they are vendor specific. The
final parameter is the choice of encryption algorithm and key size. WEP and
enhanced WEP are common. Lengthened key size and dynamic rekeying are
common enhancements.

Integrity Checking. Another aspect that must be considered is integrity
checking. Integrity is normally implemented separately from the encryption and
indicates whether or not the packet has been altered from when the sender created it. A cryptographic checksum is a necessity. The question is whether to
protect the message itself or the meaning of the message. The integrity check
mechanism can encrypt the message and authenticate the encrypted message or
it can authenticate the plaintext message and encrypt the authentication and
the message.

4 Conclusion

In this paper we have discussed the various security risks and the countermeasures of WLAN. We believe that a WLAN security architecture must have the
following attributes: mutual authentication; a strongly encrypted layer 2 tunnel;
and strong cryptographic integrity verification. Without these features, not only
is a WLAN vulnerable, but also the entire information infrastructure of which
it is a part is at risk. The major components of security are the technology, the
policies, and the people. A system will not be secure without the right technology, policy, and people. Security technology is only one component, albeit a very
critical component. To develop, run, and maintain a secure network, the administrators must know the value of the information assets and the threats against
them. They must then consider the functionality their organizations need for
mission accomplishment and the resources they have at their disposal. Another
attribute of security to keep in mind is that security is not a state, but a process
of risk management .The administrators must then use risk mitigation strategies

(technologies, policies, user training, etc.) to reduce the risks appropriately. A well designed wireless LAN ensures that the network is well protected both from insiders as well as outsiders.

References

1. W. Stallings.: "Wireless Communication & Networks", Prentice Hall, New Jersey (2001)
2. Ferguson, Niels and Bruce Schneier.: "A Cryptographic Evaluation of IPsec." Counterpane InternetSecurity White Paper (1999)
3. Potter, Bruce.: "802.1x: What it is, how it's broken, and How to fix it", Black Hat 2002, Las Vegas, NV (2002)
4. Krishnamurthy, Prashnt, Joseph Kabara and Tanapat Anusas -amornkul.: "Security in Wireless Residential Networks, IEEE Transactions on Consumer Electronics, Vol 48, No 1,pg 157-166 (2002)
5. S Fratto, Mike.: "Mobile & Wireless Technology Tutorial", Wireless Security in Network Computing,CMP United Business Media (2001)
6. Mishra, Arunesh and William Arbaugh.: " An Initial Security Analysis of The IEE 802.1X Standard", University of Maryland Institute for Advanced Computer Studies Technical Report CS-TR-4328 and UMIACS-TR(2002)
7. Borisov, Nikita, Ian Goldberg and David Wagner.: "Intercepting Mobile Communications: The Insecurity of 802.11", proc of the Seventh International Conference on Mobile Computing and Networking (2001)
8. Cisco Systems. Wireless LAN Enterprise Campus with Multi-Site Branch Offices.www.cisco.com
9. Matthew, A Technical Comparison of TTLS and PEAP, The O'Reilly Network White Paper (2002)
10. B.O'Hara and A.Petrick, IEEE 802.11 Handbook: A Designers Companion, IEEE Press, New York (1999)
11. W. Stallings, IEEE 802.11: Moving closer to Practical wireless LANs, IT Pro, (152), pp. 17-23 (2001)

Securing OLSR Routes

Alia Fourati[1], Khaldoun Al Agha[1,2], and Thomas Claveirole[1]

[1] LRI (Laboratoire de Recherche en Informatique),
Université Paris-SUD XI, 91405, Orsay Cedex, France
{alia, alagha, claveirole}@lri.fr
[2] INRIA
(Institut National de Recherche en Informatique et en Automatique)

Abstract. Ad hoc networks offer novel capabilities in the mobile domain widening the mobile application field and arousing significant interest within mobile operators and industrials. Securing routing protocols for ad hoc networks appeared soon as a challenging scope for industrials and researchers' community, since routing protocols, base of ad hoc networks, haven't defined any prevention measure or security mechanism in their specifications. In this paper, we propose a securing solution providing the integrity of the OLSR routing messages while respecting ad hoc routing protocols characteristics and operation. The solution relies on a distributed securing scheme based on threshold cryptography features and which completely fits to the OLSR operation. Implementation proves finally the successful operation of the security model and show that additional delay is suitable to OLSR routing specifications.

1 Introduction

Routing protocols are the base of ad hoc networks. Unfortunately, they have defined neither prevention measure, nor security mechanism in their specifications. In the OLSR protocol RFC [14], as in the other ad hoc routing protocols RFCs [15][16], this is clearly specified as follows: "currently, special OLSR does not specify any security measures". Securing ad hoc routing protocols appears then as an urgent requirement to promote networks deployment and to widen their application domains.

A survey of ad hoc routing protocols vulnerabilities had lead to classify potential attacks into two main categories. The first category is due to classic attacks on wireless networks and requires cryptographic mechanisms to provide authentication of nodes and integrity of routing messages. The second category of vulnerabilities is due to attacks inherent to ad hoc networks and which occur even when nodes are authenticated.

Consequently, many security solutions and concepts were defined to protect from identified attacks. However, most recent ad hoc routing securing research has focused on providing security services while relying on assumptions which are not fitting to ad hoc operation principles. In [5] for example, the solution relies on the use of a trusted certificate server in the network. This requirement imposes a number of pre-setup restrictions that must be satisfied before establishing the ad hoc network. In

K. Cho and P. Jacquet (Eds.): AINTEC 2005, LNCS 3837, pp. 183–194, 2005.
© Springer-Verlag Berlin Heidelberg 2005

addition, the centralized certificate server represents here a single point of compromise and capture, especially vulnerable in a wireless environment.

Many solutions had focused on distributing security services in ad hoc networks through threshold cryptography techniques. Solutions in [1][2][3] propose to distribute the secret key of the CA over n trusted nodes of the network while basing on threshold cryptography. The restriction here is that an external intervention must initiate the CA by sharing the CA secret key. This scheme is not realizable in the case where an ad hoc would be formed by anonym persons in an unanticipated place.

Even if many security solutions present satisfactory results and reach their initial objectives, we criticize the fact that they often omit to respect ad hoc networks characteristics, and therefore disadvantaging them. In our work, we aim to suit as much as possible to the most essential MANETs features, i.e. auto-organized, dynamic, and ubiquitous characteristics, by avoiding unsuitable assumptions. In fact, no external intervention or pre-setup configurations are required to the operation of our securing scheme.

In this paper, we propose a securing scheme for ad hoc routing protocols, especially designed for the OLSR protocol. In the first section, we introduce the paper and review weaknesses of solutions securing ad hoc routing protocols. In the second section, we first remind ad hoc networks characteristics, and we overview then ad hoc routing protocols vulnerabilities. In the next section, we present our completely distributed and self-organized algorithm securing the OLSR routing protocol. The algorithm is designed in a manner that it fits naturally to the OLSR protocol operation. In fact, one of our main objectives was to not impose initial assumptions implicating restrictions to the autonomous and ubiquitous nature of ad hoc networks. The algorithm is based on threshold cryptography mechanisms to ensure the integrity of routing messages. A fundamental and primary task while securing a routing protocol is to guarantee that routing information is flooded without being altered in order to obtain correct topological tables, and consequently correct routes. In the last section, we finally prove with implementation results the successful operation of the security model and the implication of an acceptable delay suitable to the routing requirements.

2   MANETs Overview and Vulnerabilities

2.1   MANET Characteristics

MANETs (Mobile Ad hoc NETworks) are networks specified as wireless and infrastructureless. They are rapidly deployed and self-organized, offering mobility and connectivity without constraints, anywhere and anytime. Any node of the network is thus constantly free to join and to leave the network. These capabilities qualify ad hoc networks as dynamic, autonomous and ubiquitous.

As a counter part to those ad hoc strong features, appear weaknesses in network communications. In fact, the absence of infrastructure and of centralized services, added to the weak range of wireless transmission mediums, require the forwarding of messages through intermediate nodes in order to guarantee the routing function.

Consequently, all nodes of an ad hoc network operate as routers, which weaken the routing function.

Ad hoc routing vulnerabilities have often been defined and discussed and potential attacks are currently well known [6][7][8]. In this section, we classify them into two main attacks categories: attacks common to mobile and wireless networks in general and those inherent to ad hoc networks and which are due to attacks, called Byzantine.

2.2 Classic Wireless Network Attacks

The attacks common to mobile and wireless networks are the following:

- Sniffing and traffic analysis: They allow intruders to take knowledge of messages forwarding on the network, and even to exploit them dishonestly.
- Impersonation: It allows to a terminal to be identified, or authenticated, toward another one, as a legitimate source by using a false identity.
- Modification, insertion and suppression: Attackers can modify or insert false messages while their forwarding; they can also eliminate them.
- Replay: Attackers can replay sessions by replacing original messages by new false one, and by keeping the same sequence number.
- Deny of Service (DoS): It consists on a volunteer overloading of network connections by sending excessive amounts of data. More generally, DoS attacks aim to deny access to a system, network, application or information to a legitimate user.

This first list of attacks allows the illegitimate exploitation of messages forwarding in the network. If routing messages are targeted, the routing function is then disabled. To avoid this kind of attacks, cryptographic schemes are required to provide basic security services, i.e. authentication and non-repudiation of source nodes, and integrity of routing messages. We note that confidentiality is not required here because routing messages are intended to all the network nodes, so that they can update routing tables.

2.3 Byzantine Attacks

Byzantine attacks concern the cases when authenticated nodes cannot be trusted because they don't act conformably with the protocol specifications [19]. The most important of them are the following:

- Selfishness/Black hole: These attacks take place when nodes omit cooperating in the routing process. It implies that nodes don't forward control messages in order to be diffused in the entire network. As a result, they cause the connectivity rupture of one or several nodes of the network. However, we notice differences between the intentions of both attackers. The goal of selfish nodes is to reap the benefits of participating in the ad hoc network without expending its own resources in exchange. In contrast, the goal of a black hole attacker is to disturb the routing by stopping forwarding routing messages, but still participates in the routing protocol correctly, without regard to its own resource consumption [19].

- Wormhole: This attack requires the participation of at least two nodes, which collude to create one or several tunnels between them. Attackers encapsulate the received messages and exchange them through the created tunnels. This operation deprives intermediate nodes from receiving control messages and causes the rupture of their connectivity.

To prevent these attacks, we usually require IDS (Intrusion Detection Systems). Intrusion detection is defined as the method to identify "any set of actions that attempt to compromise the integrity, confidentiality, or availability of a resource" [20].

3   The OLSR Protocol

The OLSR (Optimized Link State Routing) protocol is a proactive and link state routing protocol developed for ad hoc networks. Since its standardization in October 2003 [14], the protocol arouses an increasing interest in research and military domains. This is revealed by the multiplication of its source code downloads. For further details consult [17]. Its proactive nature implies that routes are continually maintained up-to-date, such that when a node requests for sending a message, an optimal route is already available. In addition, in the OLSR protocol, the availability of connectivity information, i.e. the declaration of the nature of links between neighbors which characterizes link state protocols, is fully re-examined and optimized.

However, the OLSR robustness and popularity are mainly due to its MPR (MultiPoint Relays) principle. The objective of MPR is the optimization of the control traffic flooding [14]. In fact, in the OLSR protocol, the construction of routing tables is carried out through periodic topology control messages (*TC* messages) which must be flooded to all network nodes. MPRs are nodes chosen such that emitted flooding message (*TC* messages), when relayed by the MPR set, must reach all 2-hop neighbors. The MPR set of a node n, denoted MPR(n), represents then the smallest subset of symmetric 1-hop neighbors of n, having symmetric links with all 2-hop neighbors of n. MPR flooding mechanism conducts to the elimination of duplicate transmission and of the minimization of duplicate reception. Indeed, MPRs significantly minimizes the control traffic in term of packet length and control messages number, and optimizes in parallel the bandwidth use. Furthermore, the OLSR protocol is well adapted in an important mobility context.

4   Security Scheme

4.1   Security Requirements and Assumptions

In ad hoc networks, we are faced with two main levels of information to secure. First, there is application data to secure, as it is crucial to achieve regardless to the network nature. Second, there is routing information which is as important and crucial to secure because of the participation of all network nodes in the routing function. In this work, we are interested in securing routing information in ad hoc networks, in particular to assure integrity of routing messages. The objective being to prevent intermediate nodes from altering routing messages while forwarding them, and to ensure then the right routing of messages. We notice that it is not necessary to ensure messages confidentiality since routing messages are intended to all the nodes of the network.

Furthermore, we aim to propose a solution without imposing initial assumptions implicating restrictions to the autonomous and ubiquitous nature of ad hoc networks. This requirement emerged by considering previous solutions which always rely on a necessary third party to their securing scheme operation, e.g. the intervention of an external initiator in the bootstrapping phase of the security system. This prerequisite makes the resulting secured ad hoc network not really auto-organized, nor autonomous, which disadvantage the network capabilities.

Therefore, to better meet the security needs, we aim to assure the following goals:

- Ensuring routing messages integrity while respecting ad hoc networks characteristics, i.e.:
    - Absence of a centralized entity in the network.
    - Absence of an external intervention in the securing system.
- Focusing on only securing the OLSR protocol in order to suit as well as possible to its characteristics and to its requirements. The objective being to obtain an optimal and maximum security level.

4.2 Threshold Cryptography Principles

In threshold cryptography, the secret is shared between several network nodes, in such a way that no single node can deduce the secret without the knowledge of all shares. The principal benefit in using threshold cryptography is to ensure security services by employing encryption without keeping the secret at only one holder, which could be easily compromising.

The idea of Shamir's (k, n) threshold systems consists on sharing a secret key between n parties [13]. Each group of any k participants (share holders), can cooperate to reconstruct the shares and recover the secret. On the other hand, no group of k-1 participants can get any information about the secret.

The interest we have for this technique resides in its flexibility when applied in the ad hoc context. In fact, independently of the fact that Shamir's scheme is perfectly secure, it presents many characteristics suiting to the dynamic and infrastructureless ad hoc nature. The most important are the following:

- The number of shares n could be increased and new shares could be added without affecting the other shares.
- Existing shares can be removed without affecting the other shares, as long as the share is really destroyed.
- It is possible to replace all the shares (or even k) without changing the secret and without revealing any information on the secret by selecting new share holders.
- Some parties can be given more than one share.

As threshold cryptography features reveal strong suiting to ad hoc routing protocols nature, many previous solutions have relied on them to propose distributed key schemes [1][2][3][4][9][10][11][12]. Nevertheless, most of them assume an initial pre-selection of trusted nodes to distribute them secret key shares, which necessarily requires an external intervention to initialize the security system. In our solution, we avoid any external intervention to offer security services in an independent and a transparent way respecting the independent and auto-organized ad hoc networks nature.

4.3 The Security Scheme Operation

On the basis of the assumptions defined in 4.1, we propose a distributed securing algorithm integrated in the OLSR routing protocol, and representing an enhanced version of our proposition in [21]. The algorithm operates as the following:

1. As soon as a node joins an ad hoc network, he begins by listening to the *Hello* messages in order to detect its neighbors. Once all neighbors detected, he can then define its MPR list and sends *Hello* messages declaring its neighbors and its MPR list.
2. If the node is selected as MPR, he has to send *TC* messages announcing MPR selector list and other topology information as in classic operation of OLSR. Before sending *TC* messages, each MPR node generates autonomously a pair of keys (secret and public), and then floods the public key announcing its presence in the network. The flooding of the public key is periodic with the same period as *TC* messages. The flooding is carried out through the MPR flooding mechanism.
3. The following step consists on the selection of 1-hop trusted neighbors through which the signing and the sending of TC messages are distributed. In our scheme, instead of sending a signed *TC* message directly from the originator node, *TC* messages are actually sent through the selected 1-hop neighbors signed with their respective attributed secret shares. The number of shares n is initially predefined. If a source node detects a number k of neighbors inferior to n, he automatically adapts its shares' number to the neighbors one. Thus, instead of distributing its secret to n shares, the MPR node distributes it to k. Moreover, the k 1-hop neighbors are selected according to their trust level. We can reinforce our scheme by relying on trust level algorithms which have previously been defined in literature [11]. The source node sends to each of trusted neighbors a share of its secret key encrypted with its respective public key. When the source node completes sending its shares, he destroys its secret key in order to prevent the key from being compromised.
4. After the secret shares distribution, the source node sends to each of the k 1-hop selected neighbors its *TC* in a unicast message encrypted with the respective neighbor public key.
5. The k 1-hop neighbors flood the *TC* message signed with their shares. The flooding is carried out through MPRs in order to optimize the network traffic. The Figure 1 below illustrates the defined secured flooding of *TC* messages. The source node (S) in the Figure shares its private key with 4 (k=4) 1-hop neighbors. Afterwards, the 4 selected neighbors flood the *TC* message, signed with their

shares (K*i*S*i*). The flooding is optimized as it is relying on the MPR flooding principle. The sharing of the source secret key is completed by threshold cryptography mechanisms.

6. Destination nodes must receive the source *TC* message *k* times - signed with *k* different shares - in order to recover the original *TC* message signature and to verify its integrity with the source node public key. The Figure 2 illustrates the recovery of the source (S) *TC* message by a destination node D. This latter must receive the *TC* 3 times, signed each time by one of the 3 shares attributed to the 3 selected trust neighbors of the source node.

We remind though that *Hello* messages are only diffused to 1-hop neighbors. Consequently, they don't need to be signed. Only *TC* routing messages are secured because they are flooded in all network relayed by improbable intermediate nodes, and that they transmit essential topology information. Source nodes represent nodes selected as MPRs, as only MPRs send *TC* messages.

We assume that even if a public key is modified during the flooding, this malicious operation could be revealed as keys reach destinations through different routes and nodes. The case of isolated nodes is not considered in this work.

Sharing the signature key with several neighbors avoids the modification of routing messages since none of them holds the complete secret which corresponds to the known public key.

Conclusively, the algorithm is completely distributed. Moreover, the system doesn't need at any time the intervention of any external entity, so it is completely auto-organized to ensure the integrity of routing messages.

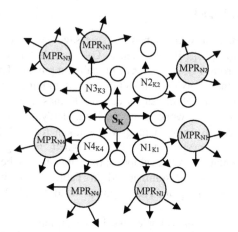

S_K: Source Node with the secret K
Ni_{Ki} : selected 1-hop neighbour having the share K_i of the source secret key K
MPR_{Ni} : MPR of node Ni

Fig. 1. Optimized MPR-based flooding of all shares-signed *TC* messages

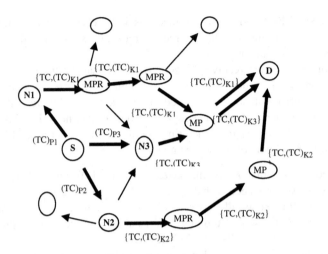

(TC)$_{Pi}$: TC encrypted in unicast messages with public keys Pi of selected trust neighbors
{TC,(TC)$_{K2}$} : TC signed with attributed source secret shares and MPR flooded in the network

Fig. 2. TC recovery

5   Implementation Evaluation

5.1   Implementation Model

An OLSR implementation with signed *TC* messages was realized extending an existing implementation [22]. It is written in C++ and therefore operates on a wide range of machines, from PDAs to conventional computers. This implementation, however, targets only the Linux system so far. In this implementation, public keys are transmitted using new kind of messages called PK messages. These messages are MPR-flooded to the whole network, using the same delay and jitter as *TC* messages. The threshold cryptography implementation was taken from [23]. This library does not implement threshold-cryptography based signed messages mechanism. To complete our implementation, we developed a (*n,n*) threshold scheme. The other main implementation parameters are summarized in Table1. This implementation was tested with 3 PDAs and 1 laptop and can be downloaded from [22].

Table 1. Implementation Parameters

Hello messages diffusion period	2 s
TC messages MPR-flooding period	6 s
Public key flooding period	6 s
Symmetric encryption algorithm	AES 128
Signature algorithm	RSA-SHA1
Threshold cryptography library	Crypto++ Library 5.2.1[23]

In order to test the implementation with a reasonable number of nodes (from 5 to 15), we develop a simulator that create topology configuration of an ad hoc network. This latter consists on generating a graph with wireless connection. The simulator also allows running several clients on the same computer with each client simulating a standalone node. Communications are achieved through UNIX sockets between the clients and a switch process that receives packets and retransmits them to the neighbor nodes. This method has nevertheless few drawbacks: it does not handle mobility and it is not scalable. In fact, since each node is considered as a single process by the system, having too much of them slows it down and makes simulations irrelevant. Also, simulations are done on one computer and because communications were done by Unix sockets, the MAC layer was removed. As a consequence, the propagation of *TC* messages can be computed only by mean of number of hops and not by second.

The simulations were realized with random topologies containing 15 nodes and having a diameter varying between 1 and 4 hops. Among used topologies, two examples are illustrated on Figure 3.

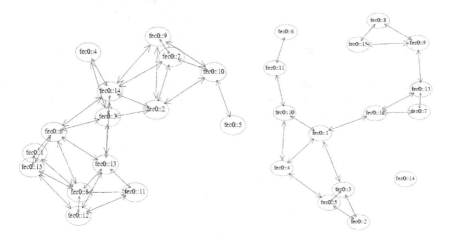

Fig. 3. Examples of random topologies used for simulations with the implementation (Nodes on the Figure are represented by their IPV6 address)

Five simulations were realized: one using the classical OLSR protocol, and four using respectively 2, 3, 4 and 5 shares for *TC* messages. Each simulation consists on 40 tests carried out through different random topologies. These topologies are organized as the following: 10 have a diameter of 1 hop, 10 have a diameter of 2 hops, and so on. Each test is run for 60 seconds, and then the output of each client is analyzed. The output represents the average of *TC* propagation time from each source to all destinations. This average is computed considering all *TC* of all nodes and in all scenarios.

5.2 Implementation Results

Figure 4 shows the main effect of the introduced security features: more delay for *TC* arrival. However this delay stills acceptable comparing to the period generation of *TC* messages. In fact, for these tests and as mentioned in the implementation model, the MAC layer were removed and replaced by Unix sockets. The *TC* propagation is computed by considering that each link on the network forwards packet in a one unit time (1 second); on Figure 4 the average of *TC* propagation reaches a maximum value of 1.8 hops that remains quite inferior to the *TC* generation period.

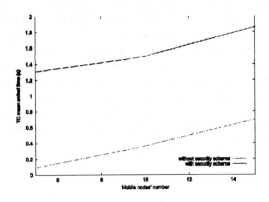

Fig. 4. *TC* mean arrival time without security scheme vs. with security scheme

In the Figure 5, we observe influence of shares' number on the *TC* mean arrival time. We simulate different configurations where mobiles' number varies from 5 to 15. The results are illustrated on Figure 5.

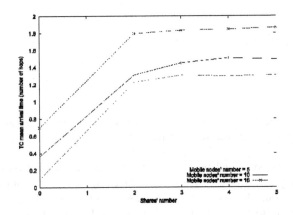

Fig. 5. OLSR *TC* mean arrival time according to shares' number

The Figure 5 presents a progress of the *TC* mean arrival time while increasing the number of shares. We observe that the number of shares does not affect the *TC* mean arrival which remains the same with the different values of shares. In fact, all shares are MPR flooded by using the shortest paths. Thus, incrementing the shares number does not affect *TC* messages propagation.

When the number of shares is equal to zero, the security scheme is not applied and simulations shows classic OLSR *TC* mean arrival time.

Implementation results prove the efficiency of our security scheme. We particularly notice an additional delay which is acceptable for *TC* messages propagation, and that remains inferior to the *TC* sending period. We finally observe that the routing function is not affected by the introduced securing operations.

6 Conclusion and Future Works

In this paper, we propose a solution securing the OLSR routing protocol. The solution is designed in a manner that it fits well to ad hoc networks characteristics. It is based on a completely auto-organized and distributed algorithm realized through threshold cryptography mechanisms. Moreover, we avoid assumptions restricting ad hoc networks to be completely autonomous and ubiquitous. Finally, the implementation results show acceptable and satisfactory additional delay suitable to the OLSR routing specifications, as well as the successful operation of the securing algorithm.

Nevertheless, some aspects of the algorithm have to be enhanced. For example, trust level algorithms have to be improved to better suit to the ad hoc context, and especially to the OLSR protocol. A more crucial problem to resolve remains that of isolated nodes.

References

[1] Zhou L., and Haas Z.J., "Securing Ad Hoc Networks", *IEEE Networks Magazine*, vol. 13, no. 6, 1999, pp. 24-30.

[2] Kong J., Zerfos P., Luo H., Lu S., and Zhang L., "Providing Robust and Ubiquitous Security Support for Mobile Ad-Hoc Networks", *The 9th International Conference on Network Protocols (ICNP'01)*, Riverside, California, USA, November 11-14, 2001.

[3] Luo H., Zerfos P., Kong J., Lu S., and Zhang L., "Self-securing Ad Hoc Wireless Networks", *IEEE Symposium on Computers and Communications (ISCC)*, Taormina, Italy, July 1-4, 2002.

[4] Papadimitratos P., and Haas Z.J., "Secure routing for mobile ad hoc networks", *Communication Networks and Distributed Systems Modeling and Simulation Conference (CNDS 2002)*, San Antonio, Texas, USA, January 27-31, 2002.

[5] Sanzgiri K., Dahill B., Levine B.N., Shields C., and Belding-Royer E.M., "A Secure Routing Protocol for Ad Hoc Networks", *The 10th IEEE International Conference on Network Protocols (ICNP)*, Paris, France, November 12-15, 2002.

[6] Fourati.A., Badis H., and Al Agha K., "Security Vulnerabilities Analysis of the OLSR Routing Protocol", *The 12th International Conference on Telecommunications (ICT 2005)*, Cape Town, South Africa, May 3-6, 2005.

[7] Macker J., and Corson S., "Mobile Ad hoc Networking (MANET): Routing Protocol Performance Issues and Evaluation Considerations", *IETF RFC 2501, IETF Network Working Group*, January 1999.

[8] Buttayan L., and Hubaux J.P., "Report on Working Session on Security in Wireleww Ad Hoc Networks", *ACM SIGMOBILE Mobile Computing and Communications Reviews*, vol. 6, no. 4, 2002.

[9] Khalili A., Katz J., and Arbaugh W.A., "Toward Secure Key Distribution in Truly Ad-Hoc Networks", *2003 Symposium on Applications and the Internet Workshops (SAINT-w'03)*, 2003.

[10] Luo H., Zerfos P., Kong J., Lu S., and Zhang L., "Self-securing Ad Hoc Wireless Networks", *The 7th International Symposium on Computers and Communications (ISCC'02)*, Taormina, Italy, July 1-4, 2002.

[11] Makhuntot R., and Phonphoem A., "Security Enhancement of the Private Key Distribution method in Wireless Ad Hoc Networks", *The 7th National Computer Science and Engineering Conference (NCSEC 2003)*, Burapha University, 2003, p. 282-286.

[12] Mukherjee A., Deng H., and Agrawal D. P., "Distributed Pairwise Key Generation Using Shared Polynomials For Wireless Ad Hoc Networks", *IFIP TC6 / WG6.8 Conference on Mobile and Wireless Communication Networks (MWCN 2004)*, Paris, France, 2004, p. 215-226.

[13] Shamir A., "How to Share a Secret", *Communications of the ACM*, Vol. 22, 1979, pp. 612-613.

[14] Clausen T., Jacquet P., Laouati A., Minet P., Muhltahler P., Qayyum A., and Viennot L., "Optimized Link State Routing Protocol", *IETF RFC 3626*, 2003.

[15] Perkins C. E., and Royer E. M., "Ad hoc on-demand distance vector routing", *IETF RFC 3561*, 2003.

[16] Johnson D. B., Maltz D. A., and Hu Y.C., "The Dynamic Source Routing Protocol for Mobile Ad Hoc Networks (DSR)", *IETF Draft 10*, 2004.

[17] http://www.olsr.org, last visit on October 9, 2005.

[18] Huang Y., and Lee W., "A Cooperative Intrusion Detection System for Ad Hoc Networks", *The ACM Workshop on Security of Ad Hoc and Sensor Networks (SASN '03)*, Fairfax VA, October 2003.

[19] Awerbuch B., Curtmola R., Holmer D., Nita-Rotaru C., and Rubens H., "Mitigating Byzantine Attacks in Ad Hoc Wireless Networks", *Technical Report Version 1*, March 2004.

[20] Zhang Y., and Lee W., "Intrusion Detection in Wireless Ad Hoc Networks", *International Conference on Mobile Computing and Networking (Mobicom)*, Boston, USA, 2000.

[21] Fourati A., and Al Agha K., "A completely distributed and auto-organized algorithm securing the OLSR routing protocol", *IFIP Open Conference on Metropolitan Area Networks (IFIP MAN'05)*, HCMC, Viet Nam, April 11-13, 2005.

[22] http://qolsr.lri.fr, last visit on October 9, 2005.

[23] http://www.eskimo.com/~weidai/cryptlib.html, Free C++ class library of cryptographic schemes, last visit on October 9, 2005.

SPS: A Simple Filtering Algorithm to Thwart Phishing Attacks

Daisuke Miyamoto, Hiroaki Hazeyama, and Youki Kadobayashi

Nara Institute of Science and Technology,
8916-5 Takayama, Ikoma, Nara, Japan
{daisu-mi, hiroa-ha, youki-k}@is.naist.jp

Abstract. In this paper, we explain that by only applying a simple filtering algorithm into various proxy systems, almost all phishing attacks can be blocked without loss of convenience to the user. We propose a system based on a simple filtering algorithm which we call the *Sanitizing Proxy System (SPS)*. The key idea of SPS is that Web phishing attack can be immunized by removing part of the content that traps novice users into entering their personal information. Also, since SPS sanitizes all HTTP responses from suspicious URLs with warning messages, novice users will realize that they are browsing phishing sites. The SPS filtering algorithm is very simple and can be described in roughly 20 steps, and can also be built in any proxy system, such as a server solution, a personal firewall or a browser plug-in. By using SPS with a transparent proxy server, novice users will be protected from almost all Web phishing attacks even if novice users misbehave. With a deployment model, robustness and evaluation, we discuss the feasibility of SPS in today's network operations.

1   Introduction

In this paper, we argue that by applying only a simple filtering algorithm into secure proxy servers and browsers, we can thwart almost all phishing attacks. We propose an architecture based on the simple filtering algorithm which we call *Sanitizing Proxy System* (SPS).

A phishing attack is a scam to deceive novice users into disclosing their personal information in various ways [1, 2]. Recently, the number of phishing attacks has grown rapidly. According to trend reports by Anti-Phishing Working Group (APWG) [3], the number of reported phishing attacks was 15,050 in June 2005, increasing from 6,957 in October 2004. The Gartner Survey reported 120 million consumers lost 929 million dollars through phishing attacks[4].

Generally, a phishing attack is composed of two phases: attraction and acquisition. First, *email spoofing* [5] attracts users using a 'spoofed' email, as if it were sent by a legitimate corporation. To acquire the users' personal information, the spoofed email leads users to execute the attached crime-ware, such as a key-logger or a redirector, or to access a 'spoofed' Web site, the so-called "phishing site". The acquisition method using a phishing site is defined as *Web spoofing* [6].

K. Cho and P. Jacquet (Eds.): AINTEC 2005, LNCS 3837, pp. 195–209, 2005.
© Springer-Verlag Berlin Heidelberg 2005

Web spoofing can be categorized by techniques of stealing personal information, downloading crime-ware, cross site scripting (XSS), and deceit. Downloading and XSS cases employ technical subterfuges. On the other hand, in the deceit case only social engineering, that is, the misbehavior of users is employed.

Although countermeasures against technical subterfuges have been studied and practiced [7, 8], there is no sufficient solution against deceit cases of Web spoofing. Ideally, to avoid the deceit cases, every user should distinguish between phishing sites and legitimate sites by themselves, and should pay attention to phishing attacks while browsing Web sites. Considering the growth rate of phishing attacks, however, many novice users are likely to disclose their personal information into phishing sites. Although several proposals have tried to alert novice users about Web spoofings by warning messages [9, 10, 11, 12, 13, 14, 15, 16, 17, 18], the effectiveness of these proposals for novice users is suspect [19]. Several application firewalls [7, 8] try to protect users from the deceit cases with URL filtering and/or stateful anomaly detection, they, however, cannot completely rule out phishing attacks because of a trade-off for maintenance costs, error rate and convenience of users.

Our proposal, Sanitizing Proxy System (SPS), focuses on blocking the deceit cases of Web spoofing, and on alerting novice users of phishing attacks. The key idea of SPS is *removing part of the content which enables novice users to input their personal information*. The characteristics of SPS are summarized in the following points:

- *Usability of the filtering algorithm*: SPS employs a filtering algorithm, *two-level filtering*. The two-level filtering is composed of *strict URL filtering* and *HTTP response sanitizing*. Combining two filtering methods, SPS prevents novice users from sending their personal information to phishing sites while allowing novice users to browse other contents of phishing sites.
- *Flexibility of the rule-set*: On filtering HTTP responses, SPS distinguishes between verified legitimate sites and other suspicious Web sites based on a rule-set written by the operator of SPS.
- *Simplicity of the filtering algorithm*: Two-level filtering is a very simple algorithm, which can be described in roughly 20 steps. Hence, it is easy to apply the SPS functions into existing proxy implementations, browser plug-ins, or personal firewalls. In section 3, we show SPS based on two different open-sourced proxy implementations to proof the simplicity and availability of the two-level filtering algorithm.
- *Accountability of HTTP response sanitizing*: SPS prevents novice users from disclosing their personal information to phishing sites by HTTP response sanitizing, that is, by removing malicious HTTP headers or HTML tags from HTTP responses. Through the sanitized Web pages, SPS alerts novice users that the requested Web page contains suspicious parts which are under the threat of phishing attacks at the time.
- *Robustness against both misbehavior of novice users and evasion techniques*: An SPS built-in proxy server can protect novice users from almost all deceit cases of Web spoofing, regardless of novice users' misbehavior and evasion

techniques by phishing attacks. Furthermore, applying SPS both in application firewalls and in browser plug-ins can block any Web spoofing phishing attacks.
- *Feasibility of deployment*: It is feasible to deploy SPS into today's Internet operational environment in spite of the maintenance cost of the rule-set.

The rest of this paper is organized as follows: In section 2, we describe related work pertaining to phishing attacks. In section 3, we specify the detail of SPS design, and discuss the feasibility of SPS. In section 4, we verify the function of SPS and evaluate the processing overhead of SPS. Finally, section 5 concludes our paper.

2   Related Work

There are many research contributions to characterize or model phishing attacks [1, 2, 20, 21, 22]. Generally, a phishing attack is separated into two distinct phases, attraction and acquisition. In many cases, phishers attract users by *email spoofing* [5]. Email spoofing attracts and leads users to one of the acquisition tricks. One trick attracts users and suggests executing an attachment crime-ware, such as a virus or Trojan. The other leads users to access a phishing site. The acquisition trick, which employs some phishing sites, is called *Web spoofing* [6]. Web spoofing can be categorized into three types; downloading, cross site scripting (XSS), and deceit. Downloading and XSS use such technical subterfuges as the acquisition trick. Downloading attracts users to download and install some of crime-ware, such as the key-logger or redirector. Once users install these crime-ware, phishers can steal a users' personal information arbitrarily. XSS exploits the vulnerabilities of a legitimate site to forward personal information to a phishing site.

While technical means of Web spoofing have been discovered so far, most Web spoofing only employs social engineering, that is, misbehavior and carelessness of users. In the deceit case, a spoofed email convinces users to access a URL leading to a phishing site. The phishing site is well-designed to be a look-alike of the targeted legitimate site; attracted users easily believe the phishing site is the legitimate site, and the users are likely to disclose or input their personal information to the phishing site, without verifying URL or SSL certification. Merwe et al. pointed out a responsibility shift from businesses to users, and suggested the necessity of education both for the users and for the businesses [20]. Many consortia or security vendors have published guidelines against phishing attacks[1, 2, 23].

On technical issues, sufficient countermeasures have not been proposed yet, although several countermeasures against phishing attacks have been studied and practiced in both the users and the businesses [1, 2]. Several spam-filtering methods are effective for email spoofing because email spoofing can be included in the spam email category. Focusing on the email spoofing itself, a lightweight trust architecture has been proposed and evaluated [24, 25]. Although several security tools [9, 10, 11, 12, 13, 14, 15, 16, 17, 18] support users to alert the phishing attacks or to confirm URL and SSL certification, Wu et al. indicate that these tools are not effective for protecting novice users [19]. By testing three security toolbars, 34% of test users disclosed their personal information to emulated phishing sites. The reasons why 34% of test users were attracted to the phishing sites are as follows: (i) there are some users who ignore warning messages when the phishing site attracts them good enough, and (ii) there are poorly designed legitimate sites which is hard to distinguish from phishing sites.

Application Firewalls have the possibility to block phishing attacks[7, 8]. An application firewall filters several application layer protocols. The coverage of the application firewall is broad and it includes emails, Web applications, instant messaging, p2p applications, and etc. By using signatures or rule-sets, application firewalls can filter spoofed emails or virus-mails, protect servers from XSS attacks, or prevent users from downloading crime-ware.

It is difficult, however, to thwart deceit cases of Web spoofing by application firewalls. In URL filtering, there is a trade-off for convenience, safety and maintenance cost. Strict filtering, which allows users to access to only legitimate Web sites, diminishes the convenience of the user, because users cannot browse anything about filtered pages. Also, strict filtering hinders administrators who want to catch up to the new phishing sites, although most URLs will be meaningless sooner or later because of the average life-span of the phishing sites [3]. On the other hand, loose filtering, which denies well-known illegitimate Web sites, can reduce maintenance costs and improve the convenience of users, in spite of sacrificing safety. URL filtering methods alone cannot prevent users from disclosing their personal information while allowing them to browse Web pages.

Some application firewalls try to deal with the case of deceit in Web spoofing by combining URL filtering and stateful anomaly detection. Stateful anomaly detection have the possibility of detecting new varieties of phishing attacks by tracking and analyzing every TCP session in real time. Combined with loose URL filtering, stateful anomaly detection can ease the maintenance cost of the rule-set, and improve security while preserving the convenience of the user. Stateful anomaly detection is, however, likely to be hindered by high false positives and false negatives. By improving the accuracy of the detection algorithm, the detection algorithm becomes complex and requires a high frequency CPU and large amounts of memory to analyze all of the TCP flow in real time. ASIC, FPGA, or a Network Processor may accelerate throughput, but these special hardware requires simple algorithms to be written in limited steps.

3   SPS: Sanitizing Proxy System

SPS focuses on the case of deceit in Web spoofing. The key idea of our proposal requires *removing part of the content which enables novice users to input their personal information*. If HTML contents does not contain any input form, a user never inputs, that is, never discloses his or her account name, password or PIN code to phishing sites. In this section, we first analyze the behaviors of novice users and of phishers to formulate requirements for SPS. Next, we describe the details of the SPS design including assumptions and requirements, and discuss the feasibility of SPS from several aspects.

3.1 Assumptions

We assume behaviors of novice users and of phishers as follows:

- Novice users cannot verify a Web site by themselves, or cannot distinguish phishing sites and legitimate sites.
- Novice users may be credulous to access a phishing site easily.
- Novice users may ignore warning messages about phishing attacks, if those attacks are highly attractive to the user.
- Novice users may disclose their personal information without being aware of phishing attacks.
- Phishers deceive novice users in various ways, with changing URLs or tricks in the short term.
- Phishers try to hijack the names of legitimate Web sites which have been verified by some trustworthy third parties.
- Phishers try to evade countermeasures against phishing attacks.

3.2 Requirements

According to our assumptions in 3.1, the requirements for SPS are as follows:

1. *Alerting novice users that they have visited the phishing site.* The alert will help novice users to understand why they cannot input their personal information. Without any alerts, novice users will be confused and wonder what is happened to their browser.
2. *Forcing novice users not to disclose their personal information into phishing sites.* SPS should prevent novice users from disclosing their personal information to phishing sites.
3. *Distinguishing between phishing sites and legitimate sites.* Because SPS must allow novice users to input their personal information into legitimate sites, a distinguishing mechanism between phishing sites and legitimate sites is necessary.
4. *Being independent from misbehavior of novice users.* The SPS must not be controlled by novice users, because phishing is caused by novice users' careless mistakes.
5. *Being as simple as possible.* A simple algorithm facilitates developers to incorporate it into various platforms, and it will have opportunities to be applied in various environments, in user-side applications, or in an Internet Service Provider (ISP) / Security Service Provider (SSP) solutions. Because of tunability and availability, the filtering algorithm for SPS should be as simple as possible.
6. *Making the SPS robust against phishers.* Phishers will try to escape from SPS filtering by various evasion techniques. The architecture of SPS should be designed as robust as possible against evasion techniques.

Algorithm 1. SPS main routine

1: **procedure** *SPS_PROXY* main routine
2: **for all** httpRequest from client **do**
3:     **send** httpRequest to server
4:     **receive** httpResponse from server
5:     **if** server.URL == VALIDATED **then**
6:         **send** httpResponse to client
7:     **else**
8:         **apply** *HTTP_RESPONSE_SANITIZING* to httpResponse
9:         **send** httpResponse to client
10:     **end if**
11: **end for**
---

7. *Being easy to deploy.* In order to compete with the rapid growing of phishing attacks, the quick deployment of SPS must be considered. Hence, the benefits produced by SPS should overwhelm the operational costs.

3.3   Two-Level Filtering Algorithm

In this section, we describe the design of a two-level filtering, the key component of SPS. The two-level filtering is composed of *strict URL filtering* (SUF) and *HTTP response sanitizing* (HRS). To filter out the suspicious URLs, the SUF delineates URLs to safe or suspicious ones at first. Then, HRS removes any input forms on every HTTP responses from the suspicious URLs.

Strict URL filtering
Although the SUF is similar to the URL filtering of application firewalls [7, 8], SPS employs the SUF to tell if each HTTP response came from one of the suspicious URLs. SPS does not use the SUF for an actual blocking method as an application firewall do. Because the SUF does nothing but to check the source of the HTTP response, the algorithm can be described simply, as shown in Algorithm 1.

The SUF rule-set is a set of prioritized allow/deny rule, combined with a varying degree of URL specification, as shown in Figure 1, The rule-set contains the top directory URL of each validated Web site. It also includes URLs which are under the threat of phishing attacks in validated Web sites. By using this rule-set, SPS distinguishes safe and validated URLs from other suspicious URLs.

HTTP response sanitizing
After delineating valid URLs from suspicious URLs by SUF, SPS applies HRS to every HTTP responses from suspicious URLs. In the context of XSS, a sanitizing method is often employed. Sanitizing the XSS converts special characters to safe characters, since special characters may lead to unexpected actions. For example, sanitizing XSS replaces ")" to ">".On the contrary, HRS replaces malicious HTTP headers and HTML tags with warning messages. Table 1 shows the HTTP

Algorithm 2. HTTP Response Sanitizing

```
1: procedure HTTP_RESPONSE_SANITIZING
2: if httpResponse.httpStatusCode == MALICIOUS_HTTP_HEADER then
3:    httpResponse.httpStatusCode ← HEADER_SANITIZED_MESSAGE
4: end if
5: if httpResponse.htmlContent includes MALICIOUS_HTML_TAGS then
6:    replace MALICIOUS_HTML_TAGS to TAG_SANITIZED_MESSAGES
7: end if
8: return httpResponse
```

```
filter{
    url "http://validurl.com/suspicious/allowed_directory/" {
        priority 1;
        allow;
    }
    url "http://validurl.com/suspicous/" {
        priority 2;
        deny;
    }
    url "http://validurl.com/" {
        priority 3;
        allow;
    }
    url "http://validurl.net/" {
        priority 3;
        allow;
    }
    default {
        deny;
    }
}
```

Fig. 1. Pseudo SPS Rule-set

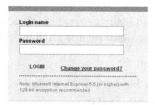   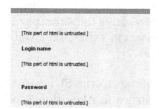

Fig. 2. Malicious Parts Before Sanitiza- **Fig. 3.** Malicious Parts After Sanitization
tion

headers and HTML tags which have the possibility of being used by phishing
attacks. Usually, these headers and tags are used to construct input forms for
user accounts or passwords. Therefore, a Web page using these headers and
tags require users to input their personal information, such as account names
or pass-phrases. To prevent phishing attacks completely, it is better to remove
these headers and tags from Web pages or from HTTP responses if possible.

Table 1. Malicious HTTP Headers

401 Unauthorized
402 Payment Required
407 Proxy Authentication Required
WWW-Authenticate

Table 2. Malicious HTML Tags

`<script>`	`<isindex>`	`<form>`
`<input>`	`<textarea>`	`<select>`
`<applet>`	`<object>`	`<embed>`

HRS alerts novice users about the malicious parts by a sanitized the Web page, as in Figures 2 and 3. The warning messages on sanitized pages are useful for troubleshooting on browsers. It also helps to educate novice users as to what is vulnerable or what type of pages are suspected as phishing sites. Because it replaces any malicious HTTP headers and malicious HTML tags, HRS is robust to possible evasion attacks as long as no phisher hijacks legitimate URLs, as discussed below:

- *Evasion techniques by Authentication Request Headers*
  HRS sanitizes the malicious HTTP Header listed on Table 1, which pops up a dialog box on the browser and asks the user to input his/her ID and password. Phishers cannot escape HRS by Authentication Request Headers.
- *Evasion techniques by CGIs*
  HRS removes `<form>`, `<input>`, `<textarea>`, `<select>` and `<isindex>` tags (as shown in Table 2) which compose an input form by only HTML. This process is applied to dynamically generated contents as well as static contents. Hence, phishers cannot evade HRS by using CGI.
- *Evasion techniques by Active Components*
  HRS replaces the `<script>` tags which can dynamically generate documents, which in turn may contain malicious tags. Phishers cannot escape from HRS even with dynamic scripts
- *Evasion techniques by Dynamic Scripts*
  HRS replaces the `<script>` tags which dynamically generate malicious tags as shown in Table 2 on a Web page. Phishers cannot escape HRS by dynamic scripts.

In addition to the SUF algorithm, HRS can be written in a simple pseudo code as shown in Algorithm 2., which requires only 8 steps.

The effects of the two-level filtering

SPS is usable for novice users and maintainers, because the two-level filtering can ease the trade-off for convenience, safety, and maintenance costs. While preventing novice users from phishing attacks, SPS does not prohibit novice users from browsing any Web pages, except forbidding them to disclose their personal information to unknown URLs. Therefore, SPS can improve both the convenience and safety of users. Also, because SPS uses the rule-set for strict filtering, the maintainers of SPS are no longer bothered by registering new phishing sites on the rule-set to protect users.

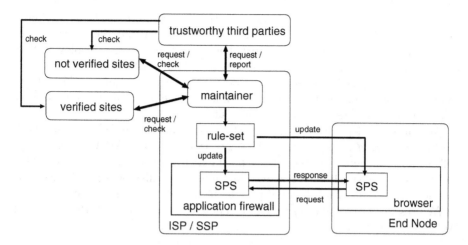

Fig. 4. Deployment Model of SPS

Considering Algorithms 1. and 2., the two-level filtering algorithm is a very simple algorithm, which can be described in roughly 20 steps. We tried to implement SPS in two different open-sourced proxies; tinyproxy [26] and privoxy [27]. We implemented SPS by adding only 46 lines to the 37,449 lines of privoxy;213 lines to the 17,621 lines of tinyproxy. Hence, SPS has the portability to various proxy implementations with only small changes.

Applying SPS into proxy servers is also efficient. As shown in Figure 4, the SPS can be applied both in a proxy server and in a local proxy as a browser plug-in or personal firewall. According to APWG reports, 94.5% of the port number employed by phishing attacks is 80, the well-known HTTP port [28]. Therefore, applying SPS into proxy servers can prevent novice users from almost all phishing attacks. Of course, this proxy server solution is affected by the proxy setting on the browser used by a novice user. If SPS is implemented as a transparent proxy server, it can be protect novice users from phishing attacks while remaining completely independent from the misbehavior of novice users. Also, combining the SPS with application firewalls by ICAP [29] or built-ins can block not only deceit cases, but also other technical subterfuge cases of phishing attacks. Furthermore, SPS can be implemented in the form of browser plug-in. SPS browser plug-in can even sanitize SSL/TLS-encrypted HTTP responses.

3.4 Deployment Issues

In this section, we explain the deployment of SPS. We previously explained the key idea of SPS and its variations of SPS in section 3.3. When deploying SPS in real network operations, we should consider the trade-off of users' convenience, safety, and maintenance costs.

Essentially, HRS has no false positives or false negatives, and has robustness against evasion techniques. The false positives and false negatives of SPS are caused by the accuracy of the rule-set for SUF. Also, the accuracy of the rule-set affects convenience of users. If some legitimate sites have not been registered in the rule-set by a loose operation, users will complain that the maintainer of SPS has to register the legitimate sites. Therefore, the success of deploying SPS depends on the cost of achieving a sophisticated and accurate rule-set regarding legitimate sites. If such a sophisticated and accurate rule-set can be achieved and shared by both users and businesses, it would be helpful novice users in order to find a safe and trustworthy legitimate site.

We consider a deployment model of SPS as shown in Figure 4. A double-check mechanism among ISPs, CDNs, SSPs, or consortia such as APWG or OWASP, will prevent phishers from registering phishing sites to the rule-set. Also, SSPs, and consortia have already started searching new phishing sites. Hence, the double-check mechanism potentially exists in today's network operations.

We present preliminary *supply and demand* analysis of users, ISPs/SSPs, and Web sites as follows:

- *Users*: Novice users already know their responsibility in dealing with their personal information carefully, however, they do not know or do not operate protecting methods against phishing attacks. Potentially, novice users need a solution to protect them from phishing attacks.
- *ISPs or SSPs*: ISPs or SSPs already know their responsibilities to protect their users or customers from phishing attacks. To reduce maintenance costs and to minimize incidents, they explore a light-weight solution to block phishing attacks completely.
- *Web sites*: Web sites already know their responsibility to eliminate the vulnerabilities of their Web sites; they also ask their customers to pay attention to spoofed emails. Since Web sites may lose their business opportunities due to the customer's fear for phishing attacks, they want a solution to protect them from the influences of phishing attacks.

According to the potential needs, the existing double-check mechanism, the effectiveness of the SPS, and the availability of the SPS, we conjecture that SPS can be deployed in real network operations without major difficulty.

3.5   Robustness Against Evasion Attacks

In this section, we explore the evasion attacks against the SPS and describe the robustness of SPS.

- *Deceiving Novice Users*: Phishers may deceive novice users into changing their proxy settings on browsers. Even then, a transparent SPS proxy, mentioned in section 3.3, can prevent novice users from phishing attacks with evasion techniques.
- *Deceiving the Rule-Set Maintainers*: Phishers may try to deceive the maintainer of SPS on some ISPs into listing phishing sites as valid URLs. According to the deployment model, it is hard for phishers to register their phishing sites into the rule-set because of the double-check mechanism.

- *Rank Pollution*: Phishers may try to pollute rankings on several search engines. If rank pollution has occurred, users who find the URL of legitimate sites with search engines would be lead to phishing sites. SPS always sanitizes suspicious URLs, therefore, SPS can prevent novice users from phishing attacks with rank pollution.
- *DNS Cache Poisoning*: Phishers may try to evade SPS by pharming hosts files or by pouring dirty DNS records. SPS can protect novice users from pharming hosts files on novice users, because SPS resolves by its own hosts files or DNS servers. The pharming on the SPS proxy server never occurs unless a phisher cracks the SPS proxy server.

  SPS, however, cannot prevent novice users if the cache records on DNS servers are spoofed. Although DNS cache poisoning is one of the problems on the Internet, it is a security issue of DNS, therefore, it is outside of the scope of this paper.

According to these discussions, the SPS is robust against evasion attacks except for DNS cache poisoning.

4   Evaluation

4.1   Functional Verification

To verify that the SPS meets design requirements, we tried browsing the various emulated phishing sites in several client environments. We prepared 5 client environments (Table 3) and 20 emulated phishing sites. These emulated phishing sites are made from Web pages of legitimate enterprises: Banks, Brokerage, Online payments and Online shops. Notice that the emulated phishing sites built in the test-bed network, are not allowed to be persistently accessible.

As the results of verifications, SPS could identify the phishing sites, sanitized all suspicious part of each HTML content mentioned in section 3.3, and alerted its presence to test users in all environments in Table 3.

4.2   Measurement of Processing Overhead

In this section, we evaluate the processing overhead of SPS. The processing overhead is generated by the SPS, when the SPS checks URLs and sanitize the contents. Also, the processing overhead is an important index for deployment,

Table 3. Client environments

Operating System	Browser
Windows XP Professional	Internet Explorer 6.0
Windows XP Professional	Mozilla Firefox 1.0.6
Mac OS X 10.4.2	Internet Explorer 5.2.3
Mac OS X 10.4.2	Safari 2.0
Sun OS 5.9	Mozilla 1.4

because novice users are not satisfied as long as the processing overhead is too big. To evaluate processing overhead, we measure the time spent to download content in various cases, and compare the response speed with and without SPS.

Our test environment is comprised of a client host, server host, and SPS host. We prepare GNU Wget [30], a well-known HTTP client program, in the client host running Linux 2.4.22, the 1.7GHz Celeron processor and 256MB RAM. We also prepare Apache [31], another well-known HTTP server, in the server host running FreeBSD 4.11, the 800MHz Pentium III processor and 256MB RAM. The SPS host runs FreeBSD 4.11, the 1.8GHz Pentium4 processor and 512MB RAM.

First, we show the base case that there is only one valid URL and the content size is 10Kbytes. Next, we show the performance overhead of increasingly valid URLs and the content size. Moreover, we show the performance overhead of the content compression. Content compression is sometimes a useful HTTPextension

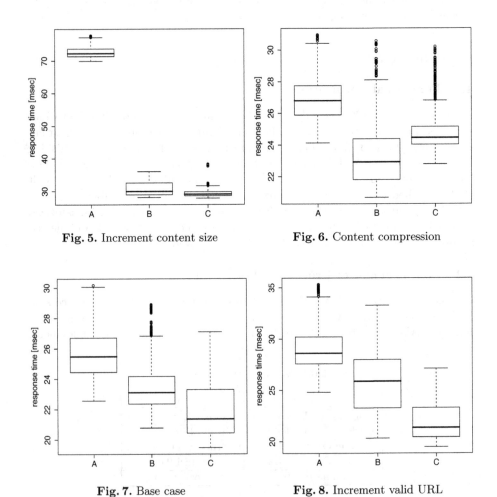

Fig. 5. Increment content size

Fig. 6. Content compression

Fig. 7. Base case

Fig. 8. Increment valid URL

to save bandwidth between a client and a server, but is not suitable for SPS, because SPS must decompress transactions to sanitize contents.

Base case

Figure 7 describes the base case of processing overhead. 1% of top and 1% of bottom data are omitted for accuracy. We assume that pattern A shows the processing overhead for phishing site, so that when the client uses SPS, a requested URL is invalid and sanitizing is also needed. Also, we assume that pattern B shows the processing overhead for the legitimate sites, so that the client uses SPS and the requested URL is valid. Pattern C shows that the client does not use SPS.

The processing overhead for the phishing site is 3.59 millisecond, which is obtained by subtracting from the average of pattern A to the average of pattern C. We consider the 3.59 millisecond of overhead not to be critical, because it is not sensible, and there are too many factors of long delays on the Internet. On the other hand, the processing overhead for the legitimate site is 1.45 millisecond, which is obtained by subtracting from the average of pattern B to the average of pattern C. We may assume that the processing overhead can be reduced by increasing the legitimate sites, such that SPS does not need to sanitize.

The performance overhead of the increasing legitimate sites

To evaluate the performance overhead of the increasing legitimate sites, we change the number of valid URLs from 1 to 100. The result is described in Figure 8. Interestingly, the processing overhead is similar to the base case, even if the valid URLs are increased. In this case, the average of overhead for the phishing site is 6.99 millisecond and the average of overhead for the legitimate site is 3.90 millisecond. In summary, the processing overhead may not be affected sensitively by the increment of valid URLs.

The performance overhead of the increasing content size

In order to evaluate the performance overhead of the increasing content size, we change the content size from 10Kbytes to 100Kbytes. The result is described in Figure 5. The processing overhead for phishing is much higher than the base case, and the average of overhead for phishing site is 43.37 millisecond, otherwise the average of overhead for the legitimate site is 1.38 millisecond.

It can be assumed that SPS needs processing time to sanitize rather than validate the URL, and the overhead is affected sensitively by content size.

The performance overhead of content compression

To evaluate the performance overhead of the content compression, we use the HTTP contents encoding scheme [28]. The result is described in Figure 6. It is interesting to compare pattern B and C, because the processing overhead is reduced by the SPS. Otherwise the client decompresses the content by him or herself in pattern C, but in pattern B, the client does not need to decompress because the SPS sends the content to the client after decompression. Also, the processing overhead for the phishing site is not much higher than the base case, and the loss is only 0.43 millisecond. We may assume that the performance overhead of content compression is not high.

As a result, we conclude from the measurements that processing overhead for legitimate sites is low enough, even if the valid URLs are increased. The processing overhead for phishing sites is high when the contents are too big. This is not a problem because novice users never want to browse phishing sites quickly.

5   Conclusion

We contribute a simple filtering algorithm and an architecture based on a simple filtering algorithm as a robust countermeasure against Web phishing attacks. In this paper, we verified and evaluated our proposal, and showed the results of verifications and evaluations. Our proposed SPS is sufficiently robust and feasible on the Internet.

If SPS is deployed and the double-check mechanism works well, then, a sophisticated and accurate rule-set, that is, a list of verified legitimate sites will be produced. This list will be useful for rating Web sites that tell users which site is safe, trustworthy, and legitimate. Also, the result of such rating of legitimate sites will be available for constructing a countermeasure against Rank Pollution. This is one of future work on against phishing attacks.

References

1. Kumar, A.: Phishing - A new age weapon. Technical report, Open Web Application Securitry Project (OWASP) (2005)
2. Tally, G., Thomas, R., Vleck, T.V.: Anti-Phishing: Best Practices for Institutions and Consumers. Technical report, Anti-Phishng Working Group (2004)
3. Anti-Phishing Working Group: Phishing Activity Trends Report - June, 2005 (2005)
4. McCall, T., Moss, R.: Gartner Survey Shows Frequent Data Security Lapses and Increased Cyber Attacks Damage Consumer Trust in Online Commerce (2005)
5. Drake, C.E., Oliver, J.J., Koontz, E.J.: Anatomy of a Phishing Email. In: Proceedings of CEAS '04. (2004)
6. Felten, E.W., Balfanz, D., Dean, D., Wallach, D.S.: Web spoofing: An internet con game. In: Proceedings of NISSC '97. (1997)
7. Blue Coat Systems, Inc.: Spyware Prevention - Blue Coat Systems. Inc. - proxy servers (2005)
8. F5 Networks, Inc.: Trafficshield Application Firewall (1998-2005)
9. Dhamija, R., Tygar, J.: The Battle Against Phishng: Dynamic Security Skins. In: Proceedings of SOUPS 2005. (2005)
10. Dhamija, R., Tygar, J.: Phish and HIPs: Human Interactive Proofs to Detect Phishing Attacks. In: Proceedings of HIP '05. (2005)
11. Kirda, E., Kruegel, C.: Protecting Users Against Phishing Attacks with AntiPhish. In: Proceedings of 29th COMPSAC 2005. (2005)
12. Chou, N., Ledesma, R., Teraguchi, Y., Boneh, D., Mitchell, J.C.: Client-side defense against web-based identity theft. In: Proceedings of NDSS '04. (2004)

13. Wenyin, L., Huang, G., Xiaoyue, L., Min, Z., Deng, X.: Detection of Phishing Webpages based on Visual Similarity. In: Proceedings of WWW 2005. (2005)
14. CoreStreet: SpoofStick (2005)
15. malwareremover: Phishing Sweeper (2005)
16. Herzberg, A., Gbara, A.: TrustBar: Protecting (even Naïve) Web Users from Spoofing and Phishing Attacks. Cryptology ePrint Archive, Report 2004/155 (2004)
17. Netcraft: Netcraft Toolbar (2005)
18. Chou, N., Ledesma, R., Teraguchi, Y., Boneh, D., Mitchell, J.C.: SpoofGuard: Preventing online identity theft and phishing. Cryptology ePrint Archive, Report 2004/155 (2004)
19. Wu, M., Miller, R., Garfinkel, S.: Do Security Toolbars Actually Prevent Phishing Attack? In: Proceedings of SOUPS 2005, poster session. (2005)
20. Van der Merwe, A., Loock, M., Dabrowski, M.: Characteristics and Responsibilities Involved in a Phishing Attack. In: Proceedings of ISICT 2005. (2005)
21. Jakobsson, M.: Modeling and Preventing Phishing Attacks. In: Proceedings of FC '05, Pishing Panel. (2005)
22. Jakobsson, M.: Distributed Phishing Attacks. In: Proceedings of DIMACS Workshop on Theft in E-Commerce. (2005)
23. Anti-Phishing Working Group: Anti-Phishing Working Group: Resources (2005)
24. Adida, B., Hohenberger, S., Rivest, R.L.: Fighting Phishing Attacks: A Lightweight Trust Architecture for Detecting Spoofed Emails. In: Proceedings of DIMACS Workshop on Theft in E-Commerce. (2005)
25. Hohenberger, S.: Separable Identity-Based Ring Signatures: Theoretical Foundations For Fighting Phishing Attacks. In: Proceedings of DIMACS Workshop on Theft in E-Commerce. (2005)
26. Kaes, R.J., Young, S.: Tinyproxy (2004)
27. Privoxy Developers: Privoxy (2001-2004)
28. Fielding, R., Gettys, J., Mogul, J., Frystyk, H., Masinter, L., Leach, P., Berners-Lee, T.: Hypertext Transfer Protocol – HTTP/1.1. RFC 2616, Internet Engineering Task Force (1999)
29. Elson, J., Cepra, A.: Internet Content Adaptation Protocol(ICAP). RFC 3507, Internet Engineering Task Force (2003)
30. Free Software Foundation, Inc.: GNU Wget (2005)
31. The Apache Software Foundation: The Apache Software Foundation (1999-2005)

On the Stability of Server Selection Algorithms Against Network Fluctuations

Toshiyuki Miyachi[1], Kenjiro Cho[1,2], and Yoichi Shinoda[3]

[1] School of Information Science,
Japan Advanced Institute of Science and Technology,
Nomi, Ishikawa 932–1292 Japan
[2] Internet Initiative Japan Inc. Research Laboratory,
Chiyoda, Tokyo 101–0051 Japan
[3] Center for Information Science,
Japan Advanced Institute of Science and Technology,
Nomi, Ishikawa 932–1292 Japan

Abstract. When a set of servers are available for a certain client-server style service, a client selects one of the servers using some server selection algorithm. The best-server algorithm in which a client selects the best one among the available servers by some metric is widely used for performance. However, when a network fluctuation occurs, the best-server algorithm often causes a sudden shift of the server load and could amplify the fluctuation. Reciprocal algorithms in which a client selects a server with a probability reciprocal to some metric are more stable than the best-server algorithm in the face of network fluctuations but their performance is not satisfactory.

In order to investigate trade-offs between the stability and the performance in server selection algorithms, we evaluate the existing algorithms by simulation and visualize the results to capture the stability of the server load.

From the simulation results, we found that the performance problem of the reciprocal algorithms lies in selecting high-cost servers with a non-negligible probability. Therefore, we propose a 2-step server selection scheme in which a client selects a working-set out of available servers for efficiency, and then, probabilistically selects one in the working-set for resiliency. We evaluate the proposed algorithm through simulation and show that our method is adaptive to environments, easy to load-balance, scalable, and efficient.

1 Introduction

There are many client-server style services on the Internet. When a set of servers are available for a certain service, a client selects one of the servers using some server selection algorithm.

In many cases, the best-server algorithm in which a client selects the best one among the available servers by some metric is widely used. However, the best-server algorithm distributes the load unevenly to different servers so that it often places high loads on a few servers while the rest of the servers are lightly loaded.

Skewed load distribution itself is not a problem because it also allows to solve the high load of a server by either adding another server near the congestion point or replacing the server with more powerful one.

K. Cho and P. Jacquet (Eds.): AINTEC 2005, LNCS 3837, pp. 210–224, 2005.

However, the best-server algorithm has another problem; a network fluctuation can trigger clients to shift to another server at a time, which in turn could lead to further network fluctuations. The best-server algorithm often causes a sudden shift of the server load in the face of network fluctuations, and it is difficult to manage by server placement.

On the other hand, the uniform algorithm randomly selects servers and evenly distributes the load among available servers. But the performance of the uniform algorithm is much worse than the best-server algorithm.

There are other algorithms which falls in between the best-server algorithm and the uniform algorithm but their performance is not as good as the best-server algorithm.

In this paper, we first evaluate the existing server selection algorithms in terms of stability and performance, and then, propose a new algorithm which is resilient to network fluctuations and efficient in performance.

2   Existing Server Selection Algorithms

There are several algorithms to select a server to use among a set of available servers. Here, we describe three typical server selection algorithms.

2.1   Best-Server Algorithm

The best-server algorithm measures a metric such as hop count and round-trip time(rtt) to servers and selects one as the best server.

The performance of the best-server algorithm is optimal in static environment since each client selects the best performing server for receiving the service.

However, wide-area networks are never static and conditions keep changing. With the best-server algorithm, clients' requests often concentrate on a small number of servers with lower cost. When a network fluctuation occurs and metrics to these servers are changed, the clients reselect the best server. Many clients tend to select the same server, which often results in amplifying network fluctuations.

2.2   Uniform Algorithm

The uniform algorithm selects a server out of available servers randomly, regardless of the metric.

The load of servers also becomes uniform. Therefore, this approach does not have the problem of skewed server load.

However, the performance of the uniform algorithm is poor because it does not take the access cost into consideration. Server placement also becomes difficult with the uniform algorithm since the performance is usually dominated by the worst performing server located far away.

2.3   Reciprocal Algorithm

The reciprocal algorithm selects a server with the probability proportional to the reciprocal of the access cost. There are several reciprocal functions that can be used.

The loads of servers are distributed to some extent as each client uses all the servers with certain probabilities.

On the other hand, the performance is much worse than the best-server algorithm because servers with high access costs are probabilistically used.

2.4   A Case Study: DNS

DNS [1][2] is a service to translates host names to IP addresses. It is a distributed database and has a hierarchical tree structure. When a client sends a query to the local DNS server to resolve a host name, the local DNS server resolves the name by traversing the name hierarchy and returns the answer to the client. If there are multiple authoritative servers available to resolve a name, the local DNS server selects one out of the authoritative servers. It is a common practice to have a few authoritative name servers for both availability and performance. The characteristic of the server selection in DNS is that the number of authoritative servers for a name is relatively small, usually from 2 to 13.

There are several DNS implementations employing different server selection algorithms.

  BIND. The Berkeley Internet Name Domain system(BIND) [3] is the most widely
      used implementation. BIND (both version 8 and 9) maintains smoothed rtt(srtt) as
      a server metric, and the srtt is decayed while the server is not used. It is a variant of
      the reciprocal server algorithm.
  DJBDNS and Microsoft Windows Internet Server.
  DJBDNS [4] and Windows Internet Server employ the uniform algorithm.

Although the heuristics employed by BIND work well for many cases, there are rooms to improve. The load distribution of 2 nearby servers is too skewed, and it is not suitable for applications requiring a large number of servers since it needs to keep the states of all servers [5].

3   Evaluation of the Existing Server Selection Algorithms with Simulation

In order to evaluate the behavior of the existing server selection algorithms, we use simulations in which network fluctuations occur in a complex topology.

To create unbalanced server load, the simulation topology is generated by a topology generator based on a scale-free network model [6]. We adopt the following rules in the topology generator.

  – The first node is placed without any edge.
  – Thereafter, a node is placed by choosing an existing node as a peer node to connect
    to with the probability proportional to the number of edges. This rule implements a
    scale-free model.
  – After placing every ten nodes, a new server is placed at the node with the largest
    number of edges. (If the node already has a server, the next one is selected.)

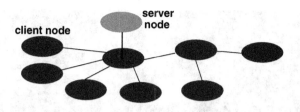

Fig. 1. Concept of the simulation topology

- Whenever the number of clients exceeds 20 for a server, the server is split into 2 servers. (When making the topology, we use the best-server algorithm) When a server splits, the node connected with the server also splits. The edges of the original node are randomly re-assigned to the 2 new nodes.
- After placing every hundred nodes, a new edge is randomly created by choosing 2 nodes with the probability proportional to the number of edges. If the 2 nodes already have a direct edge, 2 nodes are re-selected. This rule is to create a loop in the topology.

Figure 1 shows the concept of the topology.

The link cost between nodes is initialized to 10 and the link cost between a server and its bounded node is 15.

To construct a large-scale topology, we place 500 nodes for the simulation using the above rules. Due to the server-split rule, the final topology has 510 nodes and 60 servers.

In order to simulate network fluctuations, the following 2 steps are performed 50 times in the simulation.

1. Choose a server randomly, and change the cost between the server and the bounded node to a random value in the range from 1 to 40.
2. Restore the cost to the original value.

We designed and implemented a simulator with these rules since there is no apprpriate simulator for evaluating server selection algorithms in a complex topology. The simulator is written in the C language and consists of about 3000 lines of the code.

To visually understand the bias in server load, we draw the topology graphs with colored server loads.

3.1   Visualization of Server Loads

We made topology graphs with colored server loads, in order to easily understand the impact of the server selection algorithms to server loads. In particular, we are interested in observing the shift of the server loads when a network fluctuation occurs or when a new server is placed to distribute server loads.

We use Tulip [7], a graph visualization tool, for graph layout and rendering. Tulip comes with several layout algorithms including 3D layouts.

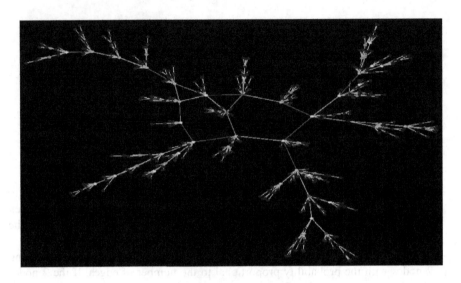

Fig. 2. The simulation topology

Figure 2 shows the simulation topology. In this topology, blue nodes show clients, and server nodes have other colors varying with the load of servers. In the topology making phase, each node selects servers 100 times whenever a new client is added. The server load is shown as the number of clients for each server. A green server has less than 600 clients, a yellow one has 600 to 1099 clients, an orange one has 1000 to 1599 clients, and a red one has more than 1600 clients.

With these topology graphs, it becomes intuitive to observe server loads by color as well as the movement of the server loads.

3.2   Best-Server Algorithm

Figure 3 and 4 shows the results of the simulation. Figure 3 shows the time-series average and maximum cost. In each step, each client selects a server 100 times. The average cost from each client to the selected servers is computed in each step, and then, the average cost over all the clients is computed. The maximum is the one with the largest average cost in the step. The average cost from a client to a server of the best-server algorithm is about 22. This is optimal in the simulation where the link cost between nodes is 10 and the link cost between a server and a node is 15 that is fluctuated between 1 to 40. The maximum cost from a client to a server in each step is between 50 and 60, which is the lower bound of the maximum cost.

Figure 4 shows the time-series server loads. It can be observed that, when there is an upward spike, there is a downward spike of the same size in the same step. For example, server 5's load becomes 2000 to 0 at step 63 because the cost between server 5 and the bounded node is increased from 10 to 38, and server 1's load increases from 1400 to 3400. That is, all the clients of server 5 move to server 1. It illustrates the effect of a network fluctuation to the server load.

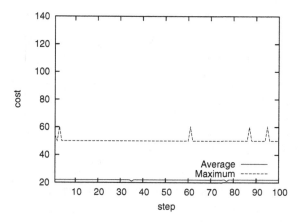

Fig. 3. Average and maximum costs of best-server algorithm. The average cost is about 22 that is optimal in our simulation.

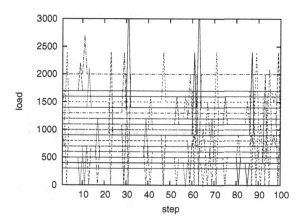

Fig. 4. Server load of best-server algorithm. When a network fluctuation occurs, the load of an affected server shifts to another server.

For the same reason, it is easy to predict a similar load shift when a new server is placed and becomes the best server for many clients. Therefore, it is difficult to control the distribution of server load by placing new servers with the best-server algorithm.

These results illustrate the unstability of the best-server algorithm that we have observed on the actual Internet.

3.3   Uniform Algorithm

Figure 5 and 6 show the simulation results of the uniform algorithm. The average cost is about 77 that is 3.5 times higher than the value of the best-server algorithm. It shows poor performance of the uniform algorithm.

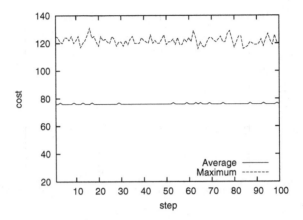

Fig. 5. Average and maximum costs of uniform algorithm. The average cost of the uniform algorithm is about 3.5 times higher than that of the best-server algorithm.

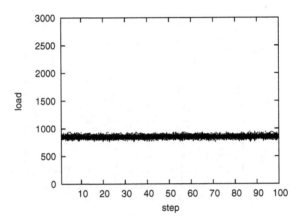

Fig. 6. Server load of uniform algorithm. All the servers have similar loads, and they are not affected by network fluctuations.

The servers' loads are almost flat. It means there are small influences by network fluctuations.

These results are also what we expected.

3.4 Reciprocal Algorithm

In this simulation, we evaluate two reciprocal functions, $1/cost$ and $1/cost^2$.

Using function:$1/cost$. The simulation results are shown in Figure 7 and 8. The average cost is 67.5 that is 3 times higher than the value of the best-server algorithm and the performance improvement is only 14% compared with the uniform algorithm.

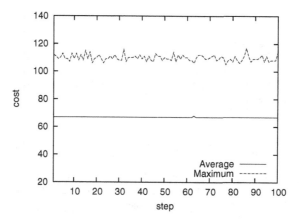

Fig. 7. Average and maximum costs of reciprocal algorithm($1/cost$). The average cost of the reciprocal algorithm($1/cost$) is 3 times higher than the best-server algorithm.

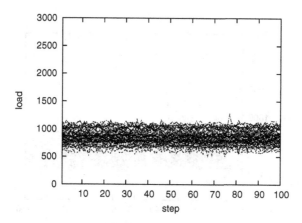

Fig. 8. Server load of reciprocal algorithm($1/cost$). The loads are relatively stable but the range of the loads is 2-3 times wider than that of the uniform selection.

As for the stability, when a network fluctuation occurs, the influence on server loads is larger than the uniform algorithm but it is much smaller than the best-server algorithm.

Using function:$1/cost^2$. The behavior of the reciprocal algorithm with $1/cost^2$ is shown in Figure 9 and 10.

The behavior is closer to the best-server algorithm because the probability of using a server with a lower cost is much higher than the $1/cost$ function.

The average cost is 55.5 that is 2.5 times higher than the value of the best-server algorithm.

The influence on server loads by a network fluctuation is smaller than the best-server algorithm, but there are some spikes in Figure 10. When a network fluctuation occurs

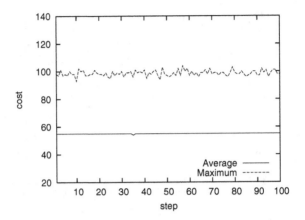

Fig. 9. Average and maximum costs of reciprocal algorithm($1/cost^2$). The average cost of the reciprocal($1/cost^2$) is 2.5 times higher than that of the best-server algorithm, and better than the reciprocal($1/cost$).

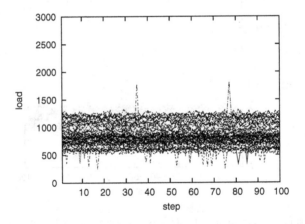

Fig. 10. Server load of reciprocal algorithm($1/cost^2$). The server load of the reciprocal algorithm($1/cost^2$) is not so skewed. The range of the server loads is wider than $1/cost$. The server load shifted by network fluctuations is absorbed by multiple servers.

near a server with high load, its load is distributed to multiple servers. Therefore, the probability of amplifying a fluctuation is smaller than the best-server algorithm.

3.5   Summary of the Simulation

We summarize the results of the simulation.

The best-server algorithm always uses the best server. The uniform algorithm chooses one server regardless of the performance. The reciprocal algorithm changes the probability of selecting a server with server's cost.

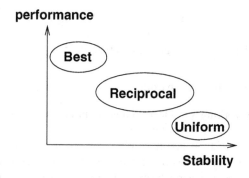

Fig. 11. The relationship of performance and stability with the 3 algorithms

Figure 11 shows the relationship of the performance and the stability of the 3 algorithms.

The best-server selection selects the best server to use, therefore the performance is always the best. However, when a network fluctuation occurs, it often amplifies the fluctuation. And it is difficult to distribute server loads by placing new servers as each client always selects the best server.

The performance of the uniform selection is the lowest among the algorithms examined but it is the most stable. However, the performance can be hardly improved even if a new server is placed at clients' concentration point.

The behavior of reciprocal algorithm is positioned in the middle of that of the best-server algorithm and that of the uniform algorithm. It selects a server with a high cost to some extent, and thus, the performance is not satisfactory.

Considering these results, we can avoid amplifying a network fluctuation if we probabilistically select one server out of multiple servers. However, the performance of the existing probabilistic algorithms is not good enough.

In the next section, we discuss this problem and propose a new algorithm.

4   Two-Step Server Selection Algorithm

The existing reciprocal algorithms do not have satisfactory performance, because they probabilistically use those servers which have fairly high costs. The target server should be selected from servers with small costs but a single server should not be selected to avoid amplifying a network fluctuation.

The problem lies in using a single algorithm for all available servers. A single probabilistic algorithm includes 2 different functions; selecting good servers and load-balancing among good servers. There are suitable algorithms for each function, but the existing algorithms relies on a single algorithm for both functions. Thus, the performance is inevitable to deteriorate, especially when poorly performing servers exist.

Thus, we propose a 2-step server selection algorithm that separates selecting good servers from load-balancing. In the first step, we select a small number of good servers as a working-set. In the second step, we probabilistically select a server to use out of the working-set, which allows us to use only good performing servers. This 2-step selection offers scalability, flexibility and efficiency.

We describe the details of our algorithm in this section.

4.1   Working-Set Selection Algorithm

The objective of the working-set selection algorithm is to efficiently select a small set of good performing servers out of a large number of available servers.

Our insight is that the time granularity of the working-set selection can be much coarser than that of the target server selection. That is, we can reduce the cost of the working-set selection by increasing the probe interval. It is desirable to frequently probe all servers to quickly adapt to environmental changes. However, it is costly to probe the access costs of all servers, especially when the number of servers is large.

Our algorithm probes better performing servers more frequently since better performing servers are more likely to be selected for the working-set. The algorithm reduces the probe cost by increasing the probe interval of poorly performing servers. The disadvantage of this method is that it takes longer to detect sudden improvements of poorly performing servers. However, detecting improvement is less critical than detecting deterioration so that we believe it is a sensible design choice.

Our algorithm sorts servers by their costs, and adjusts the probe interval of each server to a value proportional to the server rank. Therefore, the probe interval becomes a linear function of the rank of the server.

When there are N servers, the rank of the best performing server is 1 and the rank of the worst performing server is N. Let i be the rank of a server. Then, the probe frequency of server i is

$$q(i) = \frac{C}{i}$$

Here C is a scaling factor. As there are N servers, the total number of the probes per unit time $Q(N)$ is

$$Q(N) = \sum_{i=1}^{N} q(i) = C \sum_{i=1}^{N} \frac{1}{i}$$

As the total number of server N increases, the total probes $Q(N)$ increases much more slowly so that the algorithm can handle a large number of servers.

The top W servers are selected for the working-set. As an optimization, we can probe servers in the working-set when the servers are actually used so that separate probes may not be necessary for these servers.

4.2   Target Server Selection Algorithm

The objective of the target server selection algorithm is to distribute the load to multiple servers in order to reduce the influence by a network fluctuation. Our algorithm selects one server out of the working-set using a reciprocal algorithm.

The servers in the working-set are the top W servers ranked by the working-set selection. A simple reciprocal algorithm is used to select a target server within the working-set to quickly adapt to cost changes of servers in the working-set.

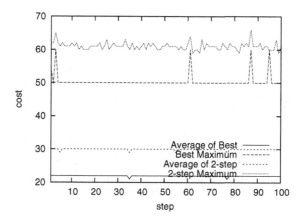

Fig. 12. The maximum and average values of the 2-step algorithm compared with the best-server algorithm. The average of the 2-step algorithm is only 36% higher than that of the best-server algorithm.

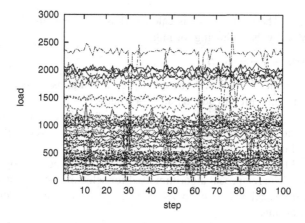

Fig. 13. The server load of the 2-step algorithm. The server loads are skewed as in the best-server algorithm. But when a network fluctuation occurs and the load of a server shifts, its load is distributed to multiple servers. It is more stable than the best-server algorithm.

The target server selection relies on the working-set selection to keep a rough set of best-performing servers. Just a rough set is required because, if the performance of a server in the working-set degrades, this server is excluded by the target server selection in a short-term and eventually taken out of the working-set by the working-set selection.

4.3   Result of the Simulation

In the simulation, we use round-trip time(rtt) as a cost metric. Real rtt includes server processing time, network delay and other factors, and represents the required time for a client to actually receive a service. To filter out instantaneous outliers in rtt, we use

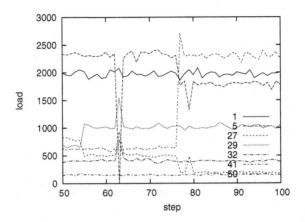

Fig. 14. The loads of 7 servers. The graph illustrates that multiple servers absorb the shifted load caused by a network fluctuation.

smoothed rtt(srtt). The value of srtt is calculated as Exponentially-Weighted Moving Average (EWMA) by the following formula.

$$srtt = \alpha \times srtt + (1 - \alpha) \times rtt$$

Here, α is the weight of EWWA. Although it is common to use α between 0.7 to 0.8, we use 0.6 in our simulation to quickly adapt to changes.

The size of the working-set can be small in our method since, if the performance of a server degrades, it will be automatically replaced by the next server by the working-set selection algorithm.

We evaluated the 2-step algorithm with $W = 4$ on the same topology used for the other algorithms. The reciprocal algorithm used for selecting a target server is $1/cost$. We chose $1/cost$ instead of $1/cost^2$ to distribute the server load more evenly among the servers in the working-set.

Figure 12 shows the performance of the 2-step algorithm compared with the best-server algorithm. The average cost is about 30. It is only 36% lower than the best-server algorithm, 51.1% better than the reciprocal algorithm using $1/cost$, and 40% better than the other using $1/cost^2$.

Figure 13 shows changes of server loads. When network fluctuations occur, the load of the affected server is distributed to multiple servers.

To illustrate the impact of network fluctuations, Figure 14 shows the load of the 7 servers with the lowest srtt from a certain client at step 74. At 63 step, the load of server 5 decreased sharply because rtt between server 5 and the bounded node was changed from 10 to 38. By this fluctuation, clients connected to server 5 moved to other servers. However, the load of the other servers did not change as much as server 5 because the load was distributed to multiple servers.

At step 77, the cost of server 50 is changed to 3 from 10. Server 50 was put into the working-set for many clients. As a result, many clients moved to server 50 from multiple servers and the load of server 50 increased.

Although the skew in the server load distribution is observed, it is not so large and is useful to control the load by server placement. It is difficult to manage large shifts of server load in a small time-scale but skews in a large time-scale allow to manage the server load by placing a new server near a high-load server. The proposed 2-step algorithm can suppress fluctuations of server load in different time scales while maintaining preferential server selection.

5   Validity of Simulation

In this section, we discuss the generality of the topology and the fluctuations used in our simulation.

5.1   Topology

The simulation topology is made by the rules described in Section 3. In this topology, a small number of nodes have a large number of edges and become hubs. Many nodes have only one edge and become leaf nodes. Servers are placed at the hubs so that the distance from clients to the nearest server is fairly short on average. We believe that the properties of real network topologies are well reflected into our simulation topology. Although we have not fully investigated into the generality of our simulation topology, it is at least good enough for our evaluation of server selection algorithms.

Our simulation uses 510 clients and 60 servers. The scale of the topology is large enough to capture the complex behavior of a wide-area network.

5.2   Changing Metric of Servers

To simulate network fluctuations, we changed the cost of edges in the topology in different simulation steps.

In real networks, network fluctuations are caused by various factors such as the load of servers and routers. Our simulation does not take these factors into account but the simulated fluctuations are enough to capture the impact of different types of fluctuations. The simulation results and its visualization allow us to observe and predict the behavior of clients in the face of network fluctuations.

6   Conclusions

The best-server algorithm is widely used because of its simplicity and good performance but it could lead to amplifying network fluctuations.

In this paper, we have evaluated the existing server selection algorithms by simulation, and visualized the server load to capture their behavior in the face of network fluctuations. Then, we have proposed a 2-step algorithm that is adaptive to network fluctuations and still provides the performance comparable with the best-server algorithm.

There are two important properties of a server selection algorithm. One is preferential server selection in which clients prefer good performing servers. From operational point of view, it allows operators to control the server load distribution by server placement. The other is that a server selection algorithm should not propagate network fluctuations. If the performance of two servers are similar, clients should use both of the servers equally.

Our proposed method improves the performance by separating working-set selection from target server selection. The proposed 2-step algorithm selects a small working-set out of available servers, and probabilistically selects a target server in the working-set.

We showed that the 2-step algorithm improves adaptability to metric changes of servers, load-balancing, scalability, and efficiency.

In future work, we will investigate the behavior of server selection algorithms using different network topologies and service types. We will also investigate the impact of server placement from an operational point of view.

References

1. P.Mockapetris.: Domain names - concepts and facilities. RFC1034, IETF, November 1987.
2. P.Mockapetris.: Domain names - implementation and specification. RFC1035, IETF, November 1987.
3. ISC BIND, http://www.isc.org/
4. DJBDNS, http://www.djbdns.org/
5. Ryuji Somegawa, Kenjiro Cho, Yuji Sekiya and Suguru Yamaguchi.: The Effects of Server Placement and Server Selection for Internet Services. IEICE Trans. on Commun. Vol.E86-B No.2. February 2003. p.542–551.
6. A.-L. Barabási, R. Albert, and H. Jeong,: Mean-field theory for scale-free random networks Physica, 272, 173-187 (1999).
7. Tulip Software, http://www.tulip-software.org/

Application-Level Versus Network-Level Proximity

Mohammad Malli, Chadi Barakat, and Walid Dabbous

Projet Planète, INRIA-Sophia Antipolis, France
{mmalli, cbarakat, dabbous}@sophia.inria.fr

Abstract. We motivate in this paper the need for application-level proximity. This proximity is a function of network characteristics that decide on the application performance. Most of existing protocols rely on the network-level proximity as for example the one based on the delay (e.g., the delay closest peer is the best peer to contact). In this paper, we study how much the two proximity definitions differ from each other. The work consists of running extensive measurements over the PlanetLab overlay network and comparing different proximity definitions. Our major observation is that the delay proximity is not always a good predictor of quality and that other network parameters are to be considered as well based on the application requirements. Particulary, the best peer to contact is not always the delay closest one. This can be explained by our other observation, that of the slight correlation of network characteristics with each other.

1 Introduction

The emerging widespread use of Peer-to-Peer (P2P) and overlay networks argues the need to optimize the performance perceived by users at the application level. This amounts to defining a proximity function that evaluates how much two peers are close to each other. The characterization of the proximity helps in identifying the best peer to contact or to take as neighbor[1].

Different functions are introduced in the literature to characterize the proximity of peers, but most of them [7, 6, 13, 15] are based on simple metrics such as the delay, the number of hops and the geographical location. We believe that these metrics are not enough to characterize the proximity given the heterogeneity of the Internet in terms of path characteristics and access link speed, and the diversity of application requirements. Some applications (e.g., transfer of large files and video streaming) are sensitive to other network parameters as the bandwidth and the loss rate. Therefore, the proximity should be defined at the application level taking into consideration the network metrics that decide on the application performance. We propose to do that using a utility function that models the quality perceived by peers at the application level. A peer is closer than another one to some reference peer if it provides a better utility function, even if the path leading to it is longer.

---

[1] This work was supported by Alcatel under grant number ISR/3.04.

K. Cho and P. Jacquet (Eds.): AINTEC 2005, LNCS 3837, pp. 225–239, 2005.

The question we ask ourselves is whether the delay proximity is a good approximation of the application-level proximity. We try to answer this question with extensive measurements carried out over the Planetlab platform [12]. To this end, we consider a large set of peers and we measure path characteristics among them. We focus on the delay, the bottleneck bandwidth, the available bandwidth and the loss rate. Then, we consider two typical applications: a file transfer running over the TCP protocol, and an interactive audio service. For both applications, we evaluate the degradation of the performance perceived by peers when they choose their neighbors based on the delay proximity instead of the application-level one. The evaluation is done using the utility functions introduced in [8, 3].

Our main observation is the following. Delay, bandwidth and loss metrics are slightly correlated, which means that, in our setting, one cannot rely on one of these metrics in defining proximity when the application is more sensitive to the others. For example, if one uses the delay to decide on the closest peer to contact for a file transfer, the application performance deteriorates compared to the optimal scenario where neighbors are identified based on the predicted file transfer latency. Furthermore, if one contacts the delay closest peer for an interactive audio service, the speech quality is not as high as that obtained when the peer to contact is the one providing the best predicted speech rating. The same result extends to the other neighbors beyond the closest one.

The paper is structured as follows. Next we present our measurement setup. In Section 3, the correlation of the different measured network characteristics is studied. Section 4 illustrates the difference in performance between the delay based proximity and the application based one for the two typical applications we consider. The paper is concluded in Section 5.

2   Measurement Setup

Our experiment consists of real measurements run in February 2005 over the Planetlab platform [12]. Although it is widely used, this platform was proved to be appropriate for measurements [17]. We take 127 Planetlab nodes spread over the Internet and covering America, Europe, and Asia. Forward and reverse paths between each pair of nodes are considered, which leads to 16002 measurements. In the following, we call a Planetlab node a peer. All our results concerning peers are averages over the 127 peers.

We measure the end-to-end characteristics of the paths connecting peers using the *Abing* tool [10]. This tool is based on the packet pair dispersion technique [9]. It consists of sending a total number of 20 probe packet-pairs between the two sides of the measured path. It has the advantage of short measurement time on the order of the second, a rich set of results (e.g, bandwidth in both directions), and a good functioning over Planetlab. The measurement accuracy provided by this tool on Planetlab is quite reasonable compared to other measurement tools [11].

For each unidirectional path between two peers, we measure the round-trip time RTT, the available bandwidth ABw, the bottleneck bandwidth (or capacity)

BC, and the packet loss rate P. P is estimated as the ratio of the number of lost and sent packets. BC is the speed of the slowest link along the path. For any of these metrics, say X, we denote by $X(p_i, p_j)$ the value of the metric associated to the path starting from peer p_i and ending at peer p_j.

3   Network Proximity Definitions

Different definitions were studied in the literature for characterizing the proximity among peers, and hence for selecting the appropriate peer to contact. These definitions can be classified into two main approaches static and dynamic. The difference between these approaches lies in the metric they consider. Static approaches [7, 16] use metrics that change rarely over time as the number of hops, the domain name and the geographical location. Dynamic approaches [6, 16, 13] are based on the measurement of variable network metrics. They mainly focus on the delay and consider it as a measure of closeness of peers; the appropriate peer to contact is often taken as the closest one in the delay space. The focus on the delay is for its low measurement cost (i.e., measurement time , amount of probing bytes). However, its use hides the implicit assumption that the path with the closest peer (in term of delay) has the minimum (or relatively small) loss rate and the maximum (or relatively large) bandwidth.

While we believe that the delay can be an appropriate measure of proximity for some applications (e.g., non greedy delay sensitive applications or those seeking for geographical proximity), it is not clear if it is the right measure to consider for other applications whose quality is a function of diverse network parameters. Greedy applications and multimedia ones are typical candidates for a more enhanced definition of proximity. To answer this question, we use our measurements results and study the correlation among path characteristics. We want to check whether (i) the characteristics are correlated with each other, and (ii) how much a proximity-based ranking of peers using the delay deviates from that using other path characteristics. As we will see in this section, there is a clear low correlation among path characteristics which motivates the need for an enhanced model for proximity. The closest peer in terms of delay is far from being optimal in the bandwidth space or loss rate space, and vice versa.

3.1   Delay vs. Bottleneck Capacity

Take a peer p and let p_d be the closest peer in terms of delay and p_b the best peer in terms of bottleneck capacity. First, we want to study how much the bottleneck capacity of the path connecting p to p_d, $BC(p, p_d)$, deviates from the largest one measured on the path between p and p_b, $BC(p, p_b)$. Figure 1 draws the complementary cumulative distribution function (CCDF) of the ratio $BC(p, p_d)/BC(p, p_b)$. The curve is calculated over all peers. For a value x on the x-axis, the corresponding value on the y-axis gives the percentage of peers having on their path to the nearest peer a bottleneck bandwidth larger than x times the maximum bottleneck bandwidth.

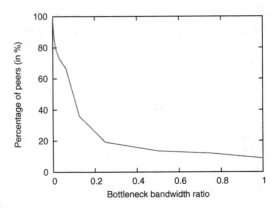

Fig. 1. CCDF of BC on the delay shortest paths

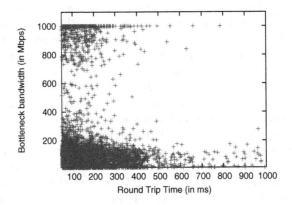

Fig. 2. Variation of BC when RTT increases

The figure shows that only (i) 8.8% of peers have the maximum BC on their path to the nearest peer, (ii) 12% have more than 75% the maximum BC, and (iii) 19.2% have more than 25% the maximum BC. This indicates that selecting the best peer in terms of delay leads in most cases to a bottleneck capacity far from the optimal. Applications having a high bandwidth requirement could suffer from this choice.

Now, we generalize our results to the other peers than the closest one. We plot in Figure 2 the bottleneck capacity versus the round trip time for the 16002 paths. Each point in the figure represents one path. The Figure shows that BC does not decrease uniformly when RTT increases. Furthermore, the correlation coefficient between these two variables is small and equal to -0.128^2.

Figure 3 plots the average bottleneck capacity for all peers of rank r in the delay space, r varying from 1 to 126. In other words, for each peer among the 127 peers, we take its neighbor of rank r in the delay space, we measure its bottleneck

---

^{2} One would have expected a coefficient closer to -1 than to 0.

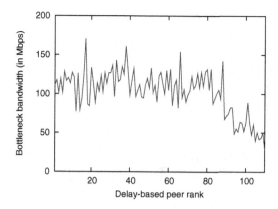

Fig. 3. Variation of BC with the delay-based rank

capacity, then we average this bandwidth over the 127 peers. Again, the figure shows a slow decrease of the bottleneck capacity with the delay-based peer rank. The delay closest peers are far from yielding the best bottleneck capacity.

3.2   Delay vs. Available Bandwidth

We repeat the same analysis but this time for the delay and the available bandwidth. For a peer p, we denote by p_a the best peer in terms of available bandwidth. Figure 4 shows the CCDF of the ratio $ABw(p, p_d)/ABw(p, p_a)$. This should illustrate how far is the available bandwidth on the delay shortest path from the optimal available bandwidth. The figure is plotted over the 127 peers.

We can see that only (i) 12% of peers have the maximum ABw on their path with the nearest peer, (ii) 19.2% have more than 75% of the maximum ABw, and (iii) 45.6% have more than 25% of the maximum ABw. Even though these numbers are better than in the bottleneck bandwidth case, the delay is still far

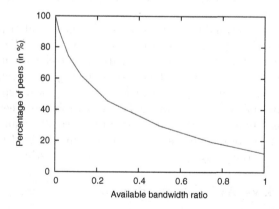

Fig. 4. CCDF of ABw on the delay shortest paths

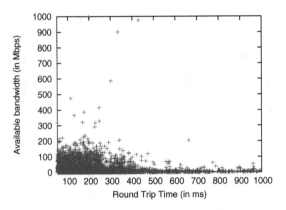

Fig. 5. Variation of ABw when RTT increases

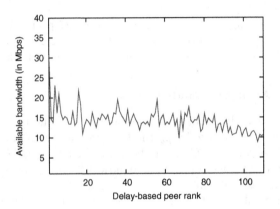

Fig. 6. Variation of ABw with the delay-based rank

from being the proximity metric to use to detect the peer with the maximum available bandwidth.

Figure 5 plots the available bandwidth versus the round trip time for the total 16002 paths. There is no strong correlation between *ABw* and *RTT*. In our setting, the two variables are lightly negatively correlated with a coefficient equal to −0.096. Similar result can be observed in Figure 6 where we plot the average available bandwidth for peers of rank r in the delay space, r varying from 1 to 126. We notice that looking at farther and farther peers in the delay space does not lead to an important decrease in the available bandwidth, and so there is a high chance of having the optimal peer from bandwidth point of view located far away (in the delay space) from the peer requesting the service.

3.3   Bottleneck vs. Available Bandwidth

The bottleneck capacity provides an indication on the maximum performance one can achieve. The available bandwidth indicates how much the network is

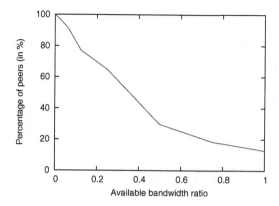

Fig. 7. Percentage of peers having more than the x ratio of maxABw on their maxBC paths

loaded. It is linked to the bottleneck capacity, but since Internet paths are differently loaded, there should be no reason to think that these two characteristics can replace each other when defining proximity for applications sensitive to the bandwidth. This is what we analyze in this section.

For a peer p, we plot in Figure 7 the percentage of peers having a ratio $ABw(p, p_b)/ABw(p, p_a)$ larger than x, x between 0 and 1. In other words, we check the difference between the available bandwidth on the path having the maximum bottleneck bandwidth and the maximum available bandwidth. The figure shows that (i) 12.8% of peers have the maximum ABw on the path having the best BC, (ii) 18.4% of peers have more than 75% the maximum ABw, and (iii) 64.8% of peers have more than 25% the maximum ABw. Clearly, selecting the peer with the maximum bottleneck capacity is not equivalent to selecting the one with the maximum available bandwidth, and the error is not negligible.

Then, we study how these two characteristics behave over all peers. We plot in Figure 8 the available bandwidth versus the bottleneck bandwidth for the total 16002 paths. A positive correlation can be seen, which when computed, yields a coefficient equal to 0.475. Figure 9 plots the available bandwidth averaged over all peers having the rank r in the decreasing-order bottleneck bandwidth space. Clearly, the farther a peer in the bottleneck space, the smaller the available bandwidth. But, in spite of this correlation, we suggest not to replace these two metrics in the proximity definition when the application requires one of them. Both need to be considered simultaneously for the proximity definition to be efficient.

3.4   Delay vs. Loss Rate

Applications are sensitive to the loss rate. We want to check in this section how well a definition of proximity based on delay satisfies the loss rate. We find that all peers have a null loss rate ($P = 0$) on their paths with at least one other

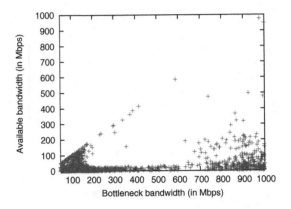

Fig. 8. Variation of ABw when BC increases

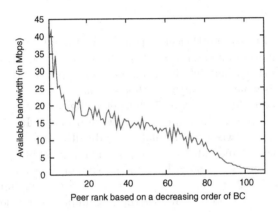

Fig. 9. Variation of ABw with the BC-based rank

peer. To check whether the nearest peer results in the minimum loss rate (i.e., zero), we plot in Figure 10 the cumulative distribution function (CDF) of the loss rate on the path connecting a peer p to its nearest peer p_d. The distribution is computed over the 127 peers. We can see that 87.4% of peers have the minimum loss rate ($P = 0$) on their path to the nearest peer. However, as long as we move away from a peer in the delay space, the loss rate jumps to values on the order of several percents, then it increases slowly. This is illustrated in Figure 11 where we plot the packet loss rate on the path connecting a peer to its neighbor of rank r in the delay space, r changing from 1 to 126. The figure is averaged over the 127 peers.

In our setting, the path with the nearest peer has a minimum loss rate. It seems that it is located in a non-congested neighborhood. Now, when it comes to selecting more than one peer for a certain service sensitive to the loss rate, taking the delay as a metric of proximity stops being efficient, and the loss rate has to be considered as well.

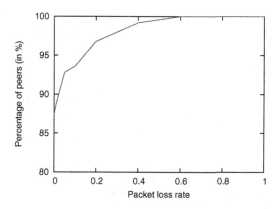

Fig. 10. CDF of loss rate on the delay shortest paths

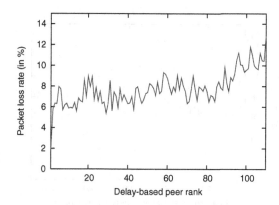

Fig. 11. Variation of P with the delay-based rank

3.5   Load vs. Loss Rate

Finally, we check the correlation between the network load ($\rho = 1 - ABw/BC$) and the loss rate. Surprisingly, we find these metrics to be lowly correlated, with a coefficient of correlation equal to 0.0277 in our setting[3]. As in the delay case, we want to check whether the loss rate is satisfied if one takes the load as a proximity metric. Let p_ρ be the peer with the minimum load. We plot in Figure 12 the distribution of $P(p, p_\rho)$ computed over the 127 peers. The figure shows that (i) 66.14% of peers have the minimum loss rate ($P = 0$) on their lowest loaded path, (ii) 79.52% of peers have a loss rate smaller than 0.1, and (iii) 92.12% of peers have a loss rate smaller than 0.5. We complete the analysis by plotting in Figure 13 the packet loss rate as a function of the peer rank in the load space. The line is an average over the 127 peers.

---

[3] One would have expected these two metrics to be strongly correlated with a coefficient of correlation closer to 1 than to 0.

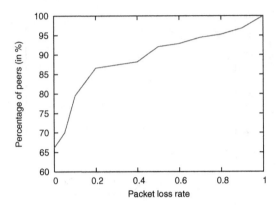

Fig. 12. Percentage of peers having less than the x loss rate on their minimum loaded paths

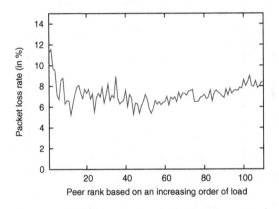

Fig. 13. Variation of P with the load-based proximity

The observation we can make from these figures is that load and loss rate are not highly correlated, and so they need to be both considered simultaneously for an efficient proximity definition (if the application requires both of them).

4   Impact of Proximity Definition on Application Performance

The weak correlation among path characteristics pointed out by our measurement results motivates us to compare between the application-level proximity and the network-level proximity. Basically, the proximity can be determined at the network-level by measuring a network parameter as the delay or the bandwidth. But, it can also be determined at the application-level by estimating some utility function that models the application quality such as the transfer time for the file transfer application and the speech rating for interactive audio applications. Such utility function decides on the proximity at the application level.

As an example of network proximity definition, we consider the delay based one given its wide use. We check whether the delay proximity is a good approximation of the application-level proximity. To this end, we consider two typical applications, which are: (i) file transfer over TCP protocol, and (2) interactive audio service.

4.1   File Transfer over TCP

Firstly, we take the case of file transfer over the TCP protocol. This case can be encountered in the emerging file sharing P2P applications or in the replicated web server context. Applications using TCP are known to form the majority of Internet traffic [4]. For such applications, the optimal peer to select is the one allowing the transfer of the file within the shortest time. We call *latency* the transfer time.

The latency of TCP transfers is known to be a function of diverse network parameters including the available bandwidth, the loss rate, and the round-trip time [14, 8]. The optimal ranking of peers from the standpoint of a certain peer is the one providing an increasing vector of transfer latency. This ranking defines the application-level proximity among peers. Any other ranking results in a different vector and yields a latency increase. We evaluate in this section the degradation of the TCP latency when the delay proximity is used instead of the application-level proximity to perform the ranking of peers from the best to the worst.

To predict the TCP transfer latency, we consider the function PTT (Predicted Transfer Time) that we compute in [8]. This function is the sum of a term that accounts for the slow start phase of TCP and another one that represents the congestion avoidance phase. The function considers the case when a TCP transfer finishes in the slow start with no losses. We omit the window limitation caused

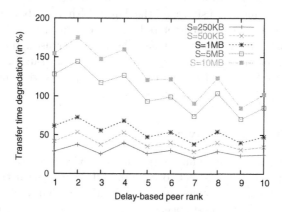

Fig. 14. Transfer time degradation when ranking peers is based on the delay proximity

by the receiver buffer to allow a better understanding of the impact of path characteristics.

The latency of a TCP transfer depends on the file size. Short transfers are known to be dominated by the slow start phase which is mainly a function of the round-trip time. Long transfers are dominated by the congestion avoidance phase where the available bandwidth and the loss rate figure in addition to the round-trip time. This difference in the sensitivity to network parameters makes interesting the problem of peer ranking for applications using TCP.

The degradation of TCP latency between the delay proximity and the optimal one is computed as follows. Take a peer p and denote the peer having the rank r in the delay space by $p_d(r)$, i.e., the peer having the r-th smallest RTT on its path to p. Denote by $p_o(r)$ the peer having a rank r with the optimal definition based on PTT. Let $PTT(x, y)$ denote the transfer latency between peer x and peer y. We define the *degradation* at rank r as:

$$degradation(r) = \frac{PTT(p, p_d(r)) - PTT(p, p_o(r))}{PTT(p, p_o(r))}. \tag{1}$$

Then, we average the degradation at rank r over all peers p (of number 127). With this degradation function we are able to evaluate how well on average ranking peers based on the delay proximity performs at the application level with respect to the optimal case.

We plot in Figure 14 the transfer time degradation as a function of the rank r for the delay proximity and for different file sizes. The closest 10 peers are considered. The figure shows that the degradation becomes larger when the file size increases, but this degradation decreases for larger r (when we get farther from the peer requesting the transfer). For large files, the degradation can be as high as 150%. For small files, the degradation is small and uniform. This discrepancy between small and large files is due to the different sensitivities the slow start and congestion avoidance phases have to network parameters. Indeed, short transfers are sensitive to the delay, and since the ranking is based on the delay, the degradation is small and we are close to the optimal case. Long transfers are more sensitive to the bandwidth (i.e., bandwidth greedy) and since the bandwidth is uncorrelated with the delay (see previous section), the degradation is large. However, when the rank increases, the transfer time increases in the optimal case and becomes closer to the transfer time obtained when peers are ranked based on the delay. This explains why the delay policy performs better for large rank and file sizes.

4.2   Interactive Audio Service

We consider now the case of an interactive audio service, where a set of replicated servers are distributed in the network to provide clients with the same audio communication. To serve a client, a central unit is in charge of identifying the server that can provide the best speech quality. As in the case of file transfer, we do not consider the load at end points and subsequently, we ignore the issue of

Table 1. R intervals, quality ratings, and the associated MOS

R interval	Quality of voice rating	MOS
$90 < R < 100$	Best	$4.34 - 4.5$
$80 < R < 90$	High	$4.03 - 4.34$
$70 < R < 80$	Medium	$3.60 - 4.03$
$60 < R < 70$	Low	$3.10 - 3.60$
$50 < R < 60$	Poor	$2.58 - 3.10$

load balancing. This allows to focus on the impact of network path parameters on proximity characterization.

The speech quality suffers mainly from packet loss and delay. The optimal ranking of peers with respect to a certain peer is the one providing a decreased order of the speech quality. Ranking peers in this way defines the application-level proximity. Any other ranking yields a lower speech quality. Our aim is to evaluate how delay-based ranking of peers deviates from the optimal one. This can tell us if the delay-based proximity is a good predictor of the application-level proximity from the standpoint of interactive audio applications.

Speech quality can be characterized subjectively using the Mean Opinion Score (MOS) test. It can also be determined with the E-model, defined by ITU-T G107 [1], which predicts the subjective quality using objective measures (e.g. delay, loss rate). The E-model expresses the audio quality as a rating factor R that accounts for the different transmission parameters having an impact on the conversation. The R-factor calculated by the E-Model ranges from 100 (the best case) to 0 (the worst case). The mapping from the R-factor to the subjective quality and to the MOS score is illustrated in Table 1.

Many papers, as [3], apply the E-model to evaluate the impairments of IP telephony applications and provide expressions for the rating factor R. In this paper, we use the analytical model obtained in [3] and we consider the case of the well known G.711 codec [2]. This model is a reduction of the ITU-T's E-Model.

Let $R(p_i, p_j)$ denote the speech quality rating between peer p_i and peer p_j. According to the E-model proposed in [3], it can be written in the following reduced form:

$$R(p_i, p_j) = R_0(p_i, p_j) - I_d(p_i, p_j) - I_e(p_i, p_j), \qquad (2)$$

where $R_0(p_i, p_j)$ is the intrinsic quality of the used codec, $I_d(p_i, p_j)$ is the impairment caused by the end-to-end delay, and $I_e(p_i, p_j)$ is the impairment caused by the end-to-end packet loss.

The end-to-end packet loss process is mainly composed of a network component and another one introduced by the de-jitter buffer (i.e., the buffer used at the end points to compensate jitter). The end-to-end delay is composed of the codec delay component, the delay introduced by the de-jitter buffer , and the network delay component. The codec delay component of G.711 codec is equal to $10 \cdot N$ where N is the number of $10ms$ voice frames packed into a single IP packet. We take $N = 2$ and subsequently set the codec delay to $20ms$ (i.e. encoding and packetization delay). We assume that the de-jitter buffer introduces a

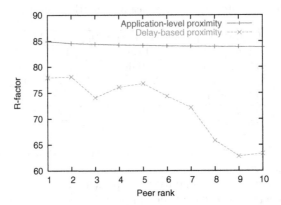

Fig. 15. Speech quality rating when ranking peers is based on the delay and on the optimal proximity

$50ms$ delay and a 2% packet loss. These typical values are considered due to the unavailability of a standard reference model for de-jitter buffer implementations. For the network delay component, we take the half of the measured RTT by assuming that the path is symmetric due to the difficulty to measure accurately the one-way delay.

Figure 15 shows the speech quality rating for the r-th closest peer in the delay space and in the optimal space. Recall that the optimal ranking of peers corresponds to a decreased order of the R-factor. The delay ranking of peers is obtained by the increased order of the RTT. The closest 10 peers are considered. We observe in this figure that the application-level proximity provides an optimal ranking with an R-factor around 85, which represents a high audio quality over the 10 closest peers. On the other hand, the delay-based proximity provides a ranking with an R-factor changing between 80 and 60 over the 10 closest peers. This represents a medium audio quality for the 7 delay closest peers and a low quality for the remaining three others.

We conclude that the delay-based proximity is a poor predictor for the interactive audio quality. This stems from the non consideration of the loss rate.

5   Conclusion

We introduce in this paper a new notion of proximity that accounts for path characteristics and application requirements. With extensive measurements over the Planetlab platform, we motivate the need for this notion of proximity by showing that path characteristics are not highly correlated, and so a proximity in one space, say for example the delay, does not automatically lead to a proximity in another space as the bandwidth one. Thus, the proximity needs to be defined as a function of the metrics impacting the application performance. In a future work, we will focus on the deployment of this new definition of proximity and on the evaluation of its gain with other application types.

References

1. ITU-T Recommendation G.107: The E-model, a computational model for use in transmission planning, 2000.
2. ITU-T Recommendation G.711: Pulse Codec modulation (PCM) of voice frequencies, 1988.
3. R. Cole, and J. Rosenbluth: Voice over IP Performance Monitoring, ACM SIGCOMM Computer Communication Review, v. 31 , p. 9 - 24, Sigcomm'01, 2001.
4. Cooperative Association for Internet Data Analysis, *http://www.caida.org/*.
5. L. Ding, and R. Goubran: Speech Quality Prediction in VoIP Using the Extended E-Model, IEEE Globecom, 2003.
6. F. Dabek, R. Cox, F. Kaashoek, and R. Morris: Vivaldi: A Decentralized Network Coordinate System, Sigcomm'04, 2004.
7. B. Gueye, A. Ziviani, M. Crovella, and S. Fdida: Constraint-Based Geolocation of Internet Hosts, IMC'04, 2004.
8. M. Malli, C. Barakat, and W. Dabbous: An Efficient Approach for Content Delivery in Overlay Networks, CCNC'05, 2005.
9. J. Navratil, L. Cottrell: ABwE: A Pratical Approach to Available Bandwidth Estimation, PAM'03, April 2003.
10. J. Navratil, L. Cottrell: available at *http://www-iepm.slac.stanford.edu/tools/abing/*, 2004.
11. J. Navratil: available at *http://www.slac.stanford.edu/ jiri/PLANET*, 2004.
12. An open, distributed platform for developing, deploying and accessing planetary-scale network services, see *http://www.planet-lab.org/*.
13. L. Tang, and M. Crovella: Virtual Landmarks for the Internet, IMC'03, 2003.
14. N. Cardwell, S. Savage, T. Anderson: Modeling TCP Latency, Infocom'00, 2000.
15. E. Ng and H. Zhang: Predicting Internet network distance with coordinates-based approaches, IEEE Infocom, 2002.
16. V. Padmanabhan and L. Subramanian: An Investigation of Geographic Mapping Techniques for Internet Hosts, SIGCOMM'01, 2001.
17. L. Peterson, V. Pai, N. Spring, and A. Bavier: Using PlanetLab for Network Research: Myths, Realities, and Best Practices, technical report, 2005.

An Adaptive Application Flooding
for Efficient Data Dissemination
in Dense Ad-Hoc Networks

Yuki Oyabu, Ryuji Wakikawa, and Jun Murai

Keio University, Dept. of Environmental Information,
5322 Endo Fujisawa Kanagawa, Japan, 252-8520

Abstract. Recent years, trend of researches in ad-hoc networks are routing, multicasting and optimized flooding. Focusing on data flooding in
ad-hoc networks, proposed scenarios of common researches are limited
due to lack of scalability. If there are lots of communication originators, it
is obvious that wireless resources are inapplicable. In this study, our typical scenario is that a mobile ad-hoc network is deployed in a city area
and many shops, such as shopping malls and restaurants, disseminate
their commercial advertisements to the network. This paper points out
the lack of scalability issues considering existing IP flooding schemes and
proposes a novel application flooding scheme supporting automatic data
size adaptation. In this scheme, size of flooded data is dynamically sampled depending on network congestion level so that data can be delivered
in a scalable manner to all nodes using MANET.

1 Introduction

A Mobile Ad-hoc Network (MANET) is dynamically formed among neighboring
nodes with wireless connections. MANET has distinct characteristics compared
to legacy Internet. Especialy, the following four characteristics should be considered for a stable and scalable communication in MANET. 1) Since nodes use
only wireless in MANET, then total available bandwidth is much smaller than
with usual connectivities. 2) No fixed infrastructures such as access routers,
network switches and access points are necessary to form MANET. Therefore,
every node must operate routing and forwarding to establish communications.
These operations develop considerable overhead to nodes which have less computation, battery and network resources. 3) Data dissemination to neighboring
nodes are relatively simple by using flooding because MANET is a multi-hop
wireless network. However, if numbers of data are flooded, total overhead of
network performance is inavoidable due to highly flooding overheads. 4) Reachability between end-nodes changes frequently by each node's arbitrary mobility.
Therefore, a connection oriented session such as TCP is inoperationable when
reachability is frequently interrupted. From these four characteristics, it is obvious that a establishing stable communication using MANET requires difficulty
compared to a communication in the Internet.

K. Cho and P. Jacquet (Eds.): AINTEC 2005, LNCS 3837, pp. 240–253, 2005.
© Springer-Verlag Berlin Heidelberg 2005

In the trend researches in MANET, several researches such as routing, multicasting and optimized flooding have been introduced for MANET. There are standardized routing protocols such as AODV [10], OLSR [3] and TBRPF [8]. Indeed, MANET can be deployed in practical operations in real world and can be crucial technology for the next future Internet. Although there are many applications and service ideas on MANET, deployment of MANET application has not been truly succeeded. It is because typical MANET applications have special missions such as emergency and relief operations, military exercises and combat situation. These services have important tasks and features, however they are only used in specific situations. Therefore people is not familiar with these applications in daily life. To attain MANET deployment, we introduce an adaptive application flooding scheme for MANET applications which provide lifestyle-oriented service and can be used generally in our life.

The rest of this paper is organized as follows: Section 2 introduces an assumed scenario and an application. In Section 3, we discuss considerable problems when our scenario is achieved by using MANET. Section 4 compares and evaluates existing delivery mechanisms and Section 4.2 introduces common standardized application flooding scheme. Section 5 proposes our adaptive application flooding scheme. We introduce implementation guideline in Section 6 and give our future work with conclusion in Section 7.

2   Application Scenarios

As described in Section 1, our scenarios target daily life situations rather than emergency or military operations. A typical application scenario is a commercial advertisements dissemination from shops to user's cell phone in a city area. This study assumes that MANET is basically formed among user's cell phone or PDAs in a city area. Shops disseminate timely information such as commercial advertisements, today's lunch menu to as many nodes as possible. The application will bring an impact to user's life because of rich information availability in a city area.

Figure 1 illustrates an image of the typical application in our assumed MANET. In Figure 1, the solid arrow represents data path and the dotted circle shows transmission radius (i.e. radio range). A data delivery method is similar to classical flooding scheme [6]. An originator first floods data to its transmission

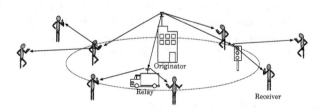

Fig. 1. Example of Applications

radius. Neighbor nodes receive the data and re-transmit the data to their transmission radius. This circle is repeated until the data is delivered to all nodes in MANET. In our research, we assume the following three types of nodes: receivers, originators and relays. MANET is formed among them automatically in a city.

- Receiver
  Receiver is a device which receives commercial data from originators. As a receiver of commercial advertisements, user's devices such as cell phone and PDA are assumed. These devices equips wireless interfaces to join MANET and may have a display and speaker to play a received commercial advertisement. These devices automatically form MANET and stay the MANET unless the devices are away from the city.
- Originator
  Originator is a device which sends commercial data to other nodes in MANET. These devices can be installed as a transmission nodes for MANET considering an installation in public space such as restaurants, department stores, or any shops. Originators are required to send the data to as many nodes as possible in MANET. For this requirement, the originator sends rich informtions to neighboring receivers and may decide to deliver dense light informations to other nodes. However, for far receivers, the originator can decrease the information's content according to network condition. For example, an originator sends video contents to nearby receivers and sends texts contents to far receivers.
- Relay
  Relay is a device which receives commercial data from a neighbor relay or an originator and re-transmits the data to other nodes. The relay can propagate the data unless hop count specified for the data is expired. Not only mobile nodes, we also assume nodes at fixed infrastructures such as traffic lights, electric pole and semi-mobile nodes such as parked vehicles. Unlike mobile nodes, these fixed and semi-mobile nodes have more computational capacity and battery resources. Thus, we expect fixed and semi-mobile nodes are responsible to deliver data to nodes located far away from the originator.

3   Problem Statement

In this study, we expect that many nodes should be able to receive flooded data regardless of the network size and the number of nodes whereas wireless and battery resources are limited. However, the node density in a city area is expected to be higher. Based on these remarks, the following two problems arise: "Dense Mobile Ad-hoc Networks" and " Desired Data Quality". They are described in the following section.

3.1   Dense Mobile Ad-Hoc Networks

As described in Section 2, a density of nodes is high due to a number of users in a city and a number of fixed nodes installed in fixed infrastructures. For the dense

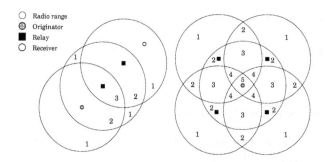

Fig. 2. Network traffic into a single distribution area

MANET, data congestion and data propagation delay drawbacks are frequent drawbacks.

With wireless medias such as IEEE 802.11b, a wireless transmission effects all the neighbor nodes located in the transmission radius. For example, in Figure 2, a node sends data to MANET. Each circle represents the transmission radius of a node. On the left, unicast is used during communications whereas on the right multicast or flooding is used. The numeric value in each transmission radius represents how many times the same packet is received in the range.

Figure 2 shows that a packet will be received more than 2 times in a same area where an originator and relay are located when a unicasting is used. On the other hand, when multicasting or flooding is used, a same packet can be received the number of 1hop-relay-nodes(N) times at a neighboring area of an originator. If continuous data such as video or audio are sent, network will be congested due to duplicated transmission radiuss. Therefore, we can conclude that more congestion and delivery delay are occurred when the number of nodes and flows are increased. It is the critical problem for real services in a city area. Thus, without applying scalability to the application, some of nodes can not receive all advertisements sent to the city MANET. We must consider a scheme not only for efficient data distribution, but also for controlling network traffic.

3.2   Desired Data Quality

Considering the fact that a receiver will receive anonymous commercial data periodically, delivered data quality must be carefully considered. For example, a user is walking in a city and looks for a nicer restaurant for lunch. Intimate information of restaurants located in 5km far is often useless for the user. Intimate information is, for instance, a today's menu, seat availability, wine lists, a map and a direction, etc. In this case, detailed information of nearby restaurants are preferred. Therefore, according to a distance between a receiver and an originator, desired data quality must be changed. In addition, changing data quality is also advantage in a dense network. If each content size is larger, more congestion and propagation delay occur. However, if a content size can be varied according to network conditions, data can be disseminated efficiently to a larger number of nodes in MANET.

4   Delivery Mechanism Consideration

For data distribution to multiple receivers, we select an application based flooding. In this section, the other existing delivery mechanisms are introduced and evaluated.

4.1   Various Delivery Approaches

There are three different flooding mechanisms such as "IP multicasting", "application forwarding" and "IP forwarding".

Application Flooding: It is required that each application (e.g. routing daemon) must send a flooded packet to all neighbors with a hop-limit set to one. The received neighbors process the flooded packet and will re-send it to their own neighbors if necessary. Flooded packets are thus repeatedly forwarded by an application until it has reached all nodes inside a mobile ad-hoc network. Most of MANET routing protocols use application flooding for protocol message exchanges. Application flooding schemes are put into some categories such as area-based flooding [9] and neighbor-knowledge-based flooding [4]. When using flooding, a distribution range must be limited by setting Time To Live (TTL). Each protocol does not require nodes to exchange any control packets for creating and sustaining the distribution tree.

IP Multicasting: As same as legacy multicasting, IP multicasting requires nodes to exchange control packets so as to create and to sustain a multicast tree. By creating a multicast tree, multicast mechanisms avoid network overhead of packet, collision and excessive retransmission. However, for stable multicast communication, a multicasting tree must be re-created depending on each node's frequent moving. As a result, multicast mechanisms bring more control packets to keep the multicast tree updates. Additional management overhead is not negligible under MANET due to limited wireless resources. There are several multicast mechanisms proposed for MANET such as a mesh-based multicasting [11], a state maintenance based multicasting [5] and a location-aided multicasting [7].

IP Flooding: In IP flooding, flooded packets can be forwarded by IP layer. The forwarding is totally blind from applications. An application on each MANET node receive the flooded packet, but the flooded packet is forwarded to next neighbors at IP layer at the same time. The application should not need to re-send the flooded packet.

In IP flooding mechanisms, a relay makes decision whether received packets are retransmitted or not, according to a Time To Live (TTL) of the packets. In IP flooding schemes, TTL is the only information to make the decision. On the other hand, IP multicast mechanisms have more information such as subscription and group management. However, IP multicasting requires considerable overhead to manage the group and the membership specially in MANET environments due to node's frequent moving. Therefore, we take a position that the received data is processed for retransmission at an application layer with rich information to decide the further operation of the flooding such as stopping flooding, decreasing data size, etc.

Moreover, IP flooding and IP multicasting cannot modify received data quality to a desired data quality at intermediate nodes because of IP blindly forwarding. These function may implement at the IP layer, but there is less information to decide an appropriate data quality at the IP layer. Data quality adaptation has advantage to avoid network congestion where many flows are launched at MANET. If a network is congested, intermediate nodes start sampling data and decrease the total overhead of flooding. This adaptation can not be achieved by IP multicasting and IP flooding.

4.2   Simplified Multicast Forwarding Scheme

The Simplified Multicast Forwarding (SMF) is standardized for a common application flooding scheme at the IETF MANET working group. SMF is designed for applications to disseminate data to all nodes in MANET. SMF considers interoperability with a conventional IP multicast, however it does not support group and membership management due to complexity. SMF is fully independent from MANET routing protocol and can operate on any types of MANETs. Thus, we decided to use SMF as a base flooding protocol in this research. SMF has two basic functions such as "Duplicate Packet Detection" and "Reliable Node Selection".

The duplicate packet detection is an essential function of flooding. Without duplicate packet detection, necessary packets keep re-sending in a network and causes flooding storm. This feature enables to avoid retransmitting and re-processing redundant packets which were already received and forwarded. Several researches aimed at duplicate packet detection have been introduced [6]. Assigning a unique identifier to each packet is a possible solution for duplicate packet detection. Each node detects duplication by comparing the received identifier and recorded identifiers of previous processed packets. Along with the identifier, several fields are introduced such as a packet sequence number, and a hash calculation on the entire packet. In our research, we will use a packet sequence number for a packet identification because the hash based calculation requires more computation complexity.

Selecting reliable relay nodes is also one of the common requirements for flooding. Purpose of this scheme are following two. To select reliable nodes as relay nodes is one of the purposes. Considering that a relay nodes or receivers are mobile node, their resources such as computation capacity and battery are more limited. For this reason, if a node which has few resources is selected as an originator, it can not retransmit packets properly. Avoiding unnecessary retransmission of packets is also a key requirement of flooding. If plenty of nodes are randomly deployed within transmission radius of the node and all of them retransmit packets, unnecessary network congestion will occur. Unnecessary network congestion is illustrated in Figure 3(a). In Figure 3(a), the node located

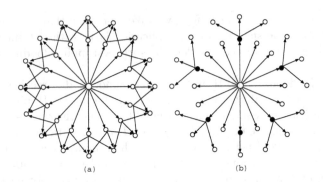

Fig. 3. Pure Flooding (a) and Diffusion Using Multiple Point Relays (b)

center represents an originator and the arrow represents packet flow. If all the neighbors of the originator retransmit packet, 2-hop-neighbors of the originator will receive same packet multiple times unnecessarily.

For meet these purposes, MPR flooding is used. MPR flooding is the scheme for selecting a reliable node as a relay node. Figure 3(b) illustrates how MPR flooding works. First of all, every node sends HELLO message to its neighbors. HELLO message contains information about the node and its neighbors. If the node receives HELLO messages from each neighbors, the node knows conditions of neighbors and existence of 2-hop-neighbors which has links to 1-hop neighbors. Then, the node selects relay nodes (MPR nodes) based on these information. In MPR flooding, every node chooses its own MPR nodes and MPR selectors.

5   Autonomous Flow Management (AFM)

Two problems such as "Dense Mobile Ad-hoc Networks" and "Desired Data Quality" are introduced in Section 3. Although SMF is designed to optimize flooding scheme, it is not enough for applications to adapt to our assumed scenarios. This section proposes "Autonomous Flow Management (AFM)" as an extension of SMF.

5.1   Concept of Autonomous Flow Management

An AFM feature is varying a size of received data dynamically according to network congestion around a received node as shown in Figure 4. When a relay receives a flooded packet, it controls flooding operations such as 1) sampling flooded data and 2) dropping flooded data based on a wireless network condition. By decreasing the size of the received data, a total costs of flooding operation can be adapted to a maximum capability of the entire MANET. In a worst case where the overhead reaches a maximum threshold, the relay node may drop the received data and stop retransmitting the data to the MANET. The threshold can be defined as a maximum congested level of the wireless interface and lower signaling level.

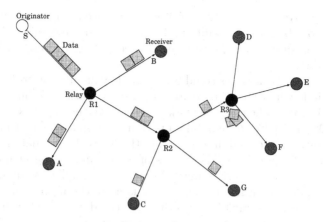

Fig. 4. Overview of Autonomous Flow Management

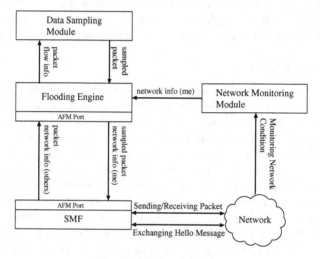

Fig. 5. AFM architecture

In AFM, flooded data is archived by intra-frame compression such as Motion-JPEG or has a hierarchical structure such as MPEG2 (SNR profile) [1] and Motion-JPEG2000 [2]. Thus, the data size can be dynamically shrank by relays. For example, in Figure 4, a relay R1 detects network congestion and sample the original data sent by the originator. Then, the R1 retransmits the sampled data to neighbor nodes (i.e. A, B and R2) so that network congestion can be avoided due to the half size of the data. This sampling operation is done by relay and not by receivers. The relay is defined as MPR node of the reliable node selection function. The relay also collects network congestion status of all the neighbor nodes by exchanging a HELLO message. The HELLO exchange is a fundamental operation of MPR and is extended for AFM to keep the network status in the HELLO message. If a relay detects network congestion at neighbor nodes, it can sample the received data for the neighbor nodes even if network condition at the relay is not observed. In addition, if a relay found the considerable network overload, it can discard all the received packet and stop retransmission of the packet (i.e. R3).

The wireless network status could be acquired by monitoring traffic and signal level on a wireless interface. Any relays operate this packet modification blindly from the originator when the network is congested and wireless resources are decreased. AFM requires to maintain a network information set. And each node manages all network related information in the network information set.

As shown in Figure 5, AFM consists of an AFM flooding engine and two following modules: A network monitoring module and a data sampling module. The network monitoring module manages the network information set and the data sampling module samples data of the flow according to the flow information. The AFM flooding engine retrieves the network information from the network monitoring module and SMF process. If a network congestion is detected, the AFM flooding engine delivers received data and flow information to the data sampling module. When sampled data is returned, the AFM engine send it to SMF process. Details of each module are described in the following sections.

5.2   Flooding Engine

AFM is designed on SMF scheme so as to inherit SMF concepts such as mechanisms of the duplicate packet detection and the reliable node selection described in Section 4.2. However, SMF is currently designed as an IP flooding scheme. In SMF, data is blindly forwarded and cannot be modified at the application layer. Therefore, AFM must be designed as an application flooding scheme although it inherits SMF concepts. To achieve the application flooding on SMF, the AFM flooding engine is operated at the application layer. Each data is carried by the UDP protocol and delivered to an AFM port. The AFM flooding engine requires an AFM header for each packet so that each relay does not need to maintain status of each flows. An AFM header contains a field of a flow identifier, a flow originator address, a flow ttl, a flow format, data sampling information. The sampling information has sampling specific commands such as supported data rate at each flow. An AFM header is appended at an originator of a flow. Relayers control the flow based on the information derived from the AFM header. A flow TTL is incremented at each relay whenever the packet is retransmitted. The AFM header can be defined as an application header.

5.3   SMF Interaction

As described in Section 4.2, SMF maintains MPR information at the IP layer if SMF is achieved at the IP layer mechanism. Note that SMF is not finalized the protocol yet and may be changed to an application flooding scheme in the future. AFM needs to access the MPR information so as to decide relay at the application layer. AFM also exchanges network condition by using HELLO messages with neighbor nodes. This information can be maintained at the MPR module of SMF, but it is also accessible by AFM (i.e. application layer).

5.4   Network Monitoring Module

The network monitoring module maintains surrounding network conditions and detects network congestion. Each node has a network information set managed by the network monitoring module. The network information set has following information.

- wireless media information
- consumed network bandwidth
- consumed network bandwidth threshold

Wireless media information are set when a node start to run the application. An interface name and a specified wireless bandwidth (e.g. 11Mbps for 802.11b) are set to the wireless media information. Consumed network bandwidth can be obtained through network monitoring. Each node measures network traffic by using promiscuous mode. The node may use the radio link quality of the wireless interface (e.g. SNR: Signal to Noise Ratios). The SNR is available at the most of device drivers of IEEE 802.11b chipsets. The consumed network bandwidth threshold is defined depending on the maximum wireless bandwidth. The network information set are periodically sent to neighbors by using HELLO message. AFM extends the HELLO message of MPR to store the network information set. Each node maintain the network information set of neighboring nodes as a neighbor node's status.

For detecting network congestion, AFM compares the consumed network bandwidth with the network bandwidth threshold. If the actual network bandwidth is over the threshold, AFM assumes a network is congested. AFM also checks the neighbor node's status. The network status of neighbor nodes is collected by listening HELLO messages from the neighbor nodes. If the number of neighbor nodes whose network bandwidth is over the network bandwidth threshold becomes particular number, AFM also assumes network congestion.

5.5   Data Sampling Module

The data sampling module is responsible to select an appropriate sampling rate according to network congestion level and to shrink the data with the picked sampling rate. After network congestion is detected, a node check an AFM header of a received packet (see Section 6.1). In the AFM header, a description of how to sample the packet can be found. If the received data can be sampled, the node decides an appropriate sampling rate and shrink the data. For example, if the data is intra-compressed format such as Motion-JPEG, some sampling rates are described in flow_opt like 1/2, 1/4, etc. By using one of these values for sampling, the relay can decrease the size of data. How to sample data depends on the data format and must be defined for each data format in the data sampling module.

6   Implementation Consideration

This section gives a guideline how to implement AMF. We will detailed the protocol and gives sample codes for the flooding engine and the retransmission control algorithm.

6.1   Flooding Engine

In AFM, all data are delivered to the application layer so that a node makes a decision how to retransmit data. For this operation, AFM uses an application header called AFM header. The format of the AFM header is shown in Figure 4.

The AFM header can be extended to store originator preference how to disseminate packets in MANET. For example, an originator may want to deliver the original contents for a few hop neighboring nodes and may allow to deliver shrank packets for other nodes. This will be our future work.

Figure 7 shows a sample code of the flooding engine for AFM. If data is received, a relay gets afm_header by get_afm_header function. Then, in the retransmission_control function, a relay decide whether to retransmit the data or not. If flow_ttl reaches flow_ttl_max, relays drops the data. If the threshold is not reached, relays increment flow_ttl and retransmits the data.

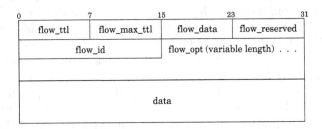

Fig. 6. afm_header format

```
/* receiving packet */
packet := recv_packet();
/* retrieving the AFM header */
afm_header := get_afm_header(packet);

/* retransmission control (see Figure 9)*/
ret := retransmission_control(packet, afm_header)

if (ret == 0) {
    /* discarding packet */
    drop_packet(packet);
} else {
    /* update the AFM header and append it to the packet */
    update_afm_header(packet, afm_header);
    /* sending the packet */
    send_packet(packet);
}
```

Fig. 7. Sample Code of the Flooding Engine

6.2   Retransmission Control Algorithm

The retransmission control algorithm is a core function of the adaptive application flooding. Each relay will process the received packet based on this algorithm.

```
struct network_info{
   char media_name[];
   u_int media_bw;
   u_int network_bw;
   u_int network_bw_th;
}
```

Fig. 8. Network Information Set

```
MPR_TH := Threshhold of MPRs

int retransmission_control(packet, afm_header) {

  /* Get the currenet network information set */
  netinfo := get_network_info();

  /* Check MPR status */
  number_of_mprs := get_congested_mpr();

  /* Condition where sampling is operated */
  if (number_of_mprs >= TH_MPRS ||
     netinfo->network_bw >= netinfo->network_bw_th) {

      /* sampling packet */
      switch(afm_header->flow_data) {
      case MOTION_JPEG:
        /* decreaseing the data size according to afm_header->flow_opt */
        newpacket := sampling_packet(packet, flow_opt);
        break;
      case MOTION_JPEG2000:
        /* decreaseing the data size according to afm_header->flow_opt */
        newpacket := sampling_packet(packet, flow_opt);
        break;
      default:
        /* dropping packet*/
        return 0;
      }
  }
  packet = newpacket;
  /* ASSERT: packet is not NULL */
  return 1;
}
```

Fig. 9. Sample Retransmission Control Algorithm

Figure 8 shows the network information set maintained by the network control module. The network information set is used by a relay to decide how retransmission is operated. The media_name stores a name of a wireless interface such as "IEEE802.11b" and the media_bw contains the specified bandwidth of the wireless interface (ex. 11Mbps for IEEE802.11b). The network control module measures an actual network bandwidth and keeps the value in the network_bw. The network_bw_th indicates a bandwidth threshold where a node treats its surrounded network is congested.

The sample code of the retransmission_control function is shown as the algorithm for the automatic application flooding in Figure 9. The retransmission_control is used in Figure 7. If the return value of this function is zero, it indicates that the flooding engine drops the packet. Otherwise, the flooding engine forwards the packet. When the retransmission_control function is called, a relay gets its network information set in netinfo value and checks the number of MPR whose network is congested by the get_congested_mpr(). This function is explained later in this Section. If the number of MPRs is over the threshold of MPRs (i.e. MPR_TH) or a network congestion occurred at the relay, the relay samples the received packet according to the data format and prepares a new packet for the sampled data. The sampling is operated with the flow_opt which has supported sampling information of received data. Otherwise, the retransmission_control function just drops the packet by returning zero.

7   Conclusion

This research proposes an application flooding scheme supporting automatic data size adaptation. In the adaptive application flooding scheme, size of flooded data is dynamically sampled depending on network congestion level. Although we introduce implementation guideline in Section 6, our on-going work is implementation and evaluation. As for evaluation, implementation based on simulator is being made to evaluate how effective our scheme is. The overhead at each relay and performance of the network congestion control will be examined in the simulation. Thus, we will examine how many packet are disposed and how many relays adapt the size of data with changing the number of originators and network size.

Acknowledgment

I would like to express my gratitude to Noriyuki Shigechika, Koshiro Mitsuya and Guillaume Valadon for many insightful remarks and discussions. I am fortuante to have Noriatsu Kudo and Kazunori Sugiura for thorough review and comments. I would like to thank all the member of WIDE project and Jun Murai Lab at Keio University for fruitful discussions.

References

1. Generic Coding of Moving Pictures and Associated Audio. ISO/IEC International Standard 13818, 1994.
2. ISO/IEC 15444-3 Motion-JPEG2000 (JPEG2000 Part 3), 2000.
3. T. Clausen and P. Jacquet. Optimized Link State Routing Protocol OLSR. Request for Comments (Experimental) 3626, Internet Engineering Task Force, October 2003.
4. P. Jacquet, V. Laouiti, P. Minet, and L. Viennot. Performance of multipoint relaying in ad hoc mobile routing protocols. Networking 2002, 2002.
5. L. Ji and M. S. Corson. Differencial Destination Multicast (DDM) Specification (work in progress, draft-ietf-manet-ddm-00.txt). Internet draft, Internet Engineering Task Force, June 2000.
6. J. Macker and SMF. Design Team. Simplified Multicast Forwarding for MANET (work in progress, draft-ietf-manet-smf-00.txt). Internet draft, Internet Engineering Task Force, June 2005.
7. M. Mauve, J. Widmer, and H. Hartenstein. A survey on position-based routing in mobile ad hoc networks. IEEE Network Magazine, 15(6):30–39., November 2001.
8. R. Ogier, F. Templin, and M. Lewis. Topology Dissemination Based on Reverse-Path Forwarding TBRPF. Request for Comments (Experimental) 3684, Internet Engineering Task Force, February 2004.
9. V. Paruchuri, A. Durresi, D. Dash, and R. Jain. Optimal flooding protocol for routing in ad-hoc networks. Technical report, Ohio State University, CS Department, 2002.
10. C. Perkins, E. Belding-Royer, and S. Das. Ad hoc On-Demand Distance Vector (AODV) Routing. Request for Comments (Experimental) 3561, Internet Engineering Task Force, July 2003.
11. Y. Yi, S. Lee, W. Su, and M. Gerla. On-Demand Multicast Routing Protocol (ODMRP) for Ad Hoc Networks (work in progress, draft-ietf-manet-odmrp-04.txt). Internet draft, Internet Engineering Task Force, November 2002.

Multicast Packet Loss Measurement and Analysis over Unidirectional Satellite Network

Mohammad Abdul Awal, Kanchana Kanchanasut, and Yasuo Tsuchimoto

Internet Education and Research Laboratory (IntERLab),
Asian Institute of Technology, P.O. Box 4, Klong Luang,
Pathumthani 12120, Thailand
awal@interlab.ait.ac.th, {kanchana, tsuchy}@ait.ac.th

Abstract. Packet loss patterns over unidirectional satellite link provide useful information for improving the performance of applications like bulk data transfer to many recipients. In this paper, we present our observations and analysis of packet loss patterns on one-to-many multicast sessions taken over an extended period of time. We collected the packet loss data on four multicast receivers in four different geographic locations. The measurements taken demonstrate possible reasons of packet loss in both sender and receiver sides including the burstiness behavior of the channel. It is found in our result that setting priority for multicast traffic in the sender side router is one of the key factors for packet loss. We found a significant amount of bursty or consecutive packet losses contributing to the overall loss percentage while the number of occurrences of bursty loss is relatively small.

1 Introduction

Satellite technology provides efficient data delivery to a large number of users by broadcasting or multicasting over a wide geographical area at a low marginal cost when come to adding additional users [18]. IP multicasting over the satellite provides massive information delivery from a single source to many receivers consuming minimum bandwidth. It extends the range of interesting internet applications to cover bulk data transfer such as movies or multimedia contents, newspapers, educational materials to remote areas reliably over unidirectional satellite link without return path.

One of the main reasons of packet loss in satellite channel is due to the fundamental characteristic of noise in the channel itself. The bit error rate (BER) of the satellite channels are 1×10^{-7} or less frequent [10]. Sometimes the channel experiences very high loss rate because of the weather condition like rain fade, fog, attenuations etc. [2]. In the satellite network, different receivers have different loss rates because of the weather effect though the receivers are situated on dispersed areas [2], [5]. Besides the channel noise, the packet loss can be caused by sender or receiver equipments or intermediate network elements such as routers, bandwidth allocation policies or satellite equipments. Routers usually try to discard multicast UDP packets more than the TCP packets if the buffer of the elements is full.

K. Cho and P. Jacquet (Eds.): AINTEC 2005, LNCS 3837, pp. 254–268, 2005.
© Springer-Verlag Berlin Heidelberg 2005

To solve the reliability issue for bulk data download application, several approaches [7], [14], [9], [11] have been adopted. One of the approaches is called digital fountain [7], [9], [11] which uses Forward Error Correction (FEC). But in this method, the sender sends redundant packets even in the absence of any loss at all which consumes network bandwidth. Furthermore FECs have proven to be effective in the absence of bursty losses [13] and there are some situations where FEC cannot be expected to solve the noise problem caused by rain fade or attenuation [10]. Another one is the Broadcast Disk approach [14], [19] which uses push based periodic data sending from the sender to the receivers until the entire data set is received by the receivers. These approaches also require redundant data transmission but the sender never know whether all the receivers have successfully received all the data or not. These protocols generate redundant data in the network but lessen the need of feedback channel hence they are applicable for Unidirectional Link (UDL).

Knowing the reasons for packet loss can reduce the packet loss rate dramatically. On the other hand, if we know the loss patterns, we may be able to redesign above protocols or get a better assumption for designing protocols for [14], [19], [4]. A good knowledge about burst patterns also helps in making good design decisions about FEC for [7], [9], [11]. Our study includes the measurement of multicast data transfer from one sender to many receivers over a satellite link in a real test-bed provided by the AI3 project [1]. The rest of the paper is organized as follows. Section 2 discusses our environments which includes the experimental test-bed, the network topology and our measurement tool description. Section 3 contains the measurement results where section 4 presents analysis and discussions of the measurement data collected. Related works on packet loss measurement on multicast traffic is presented in section 5 where a conclusion of the paper is presented in section 6.

2 Measurement Environment

2.1 Test-Bed Topology

Our experiment consists of simultaneously monitoring and recording received multicast packet transmission of AI3 satellite network [1] at four receivers at four different geographic locations from a single sender located in another location. The network topology of our experiment is shown in Fig. 1. The description of the sender and receivers are given in Table 1 including their geographic locations, IP addresses, operating systems and machine types. These machines are mainly used for a project named SOI-ASIA under the WIDE project in Japan. The machine named KEIO works as the multicast sender over unidirectional satellite link because it has the sending capable antenna. Other four machines are used as multicast data receiver as they have the receive-only (RO) antenna.

During the multicast sessions from the sender to the receivers, a packet traverses through a router in the sender side, a machine named *udl-feed* which feeds traffic to the satellite antenna for uplink, the air space from the sending antenna to the satellite and the satellite to the receive only (RO) antenna, and finally a receiving router in each receiving site forwards the packets to the receiver machine. The *sender* program

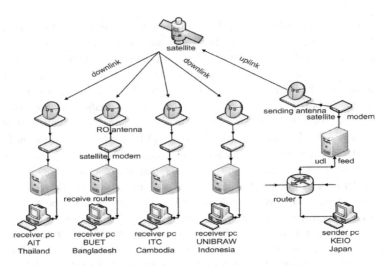

Fig. 1. Experimental topology under AI3 network

Table 1. The sender and four receiving host's descriptions

Machine Name	Location	Operating System	Machine Type
KEIO	Jun Murai's Lab, SFC Campus, Keio University, Japan	FreeBSD 4.5	Intel Pentium III 1.2GHz, 256MB RAM
AIT	Asian Institute of Technology, Thailand	Fedora Core 1 (Linux 2.4.22-1.2199)	Intel Celeron 633MHz, 128MB RAM
BUET	Bangladesh University of Engineering and Technology, Bangladesh	Fedora Core 1 (Linux 2.4.22-1.2199)	Intel Pentium IV 2.8GHz, 256MB RAM
ITC	Institute of Technology and Communication, Cambodia	Fedora Core 1 (Linux 2.4.22-1.2199)	Intel Pentium Celeron 2.4GHz, 480MB RAM
UNIBRAW	Brawijaya University, Indonesia	Fedora Core 1 (Linux 2.4.22-1.2199)	Intel Xeon 2.8GHz, 512MB RAM

sets the parameter *TTL* (time-to-live) to 3 for multicast traffic hop counts which also validates that the traffic traverse through only 3 hops. The *udl-feed* machine uses *ALTQ* (Alternate Queuing) [27] software to ensure traffic priority in the unidirectional satellite network.

2.2 Tools and Methods

Our goal was to be able to measure the application-level loss that a multicast traffic stream would experience. We wrote two application layer programs (sender program

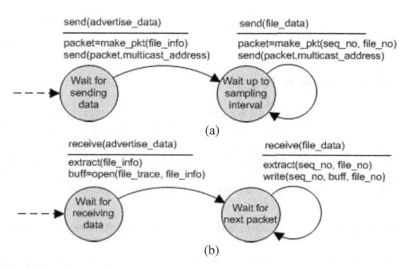

Fig. 2. Experiment tools protocol (a) sender program FSM, (b) receiver program FSM

and receiver program) and Fig. 2 shows their *finite-state machine* (FSM) definitions. The FSM in Fig. 2(a) defines the operation of the sender, while the FSM in Fig. 2(b) defines the operation of the receiver.

The sender side has two states. In the leftmost state, the sender program is waiting for data to be sent over the satellite link. The sender program creates an advertise packet (via the action make_pkt(file_info)) to inform the receivers about the file name, file size, number of packets and file number which is going to be sent to the receivers. Then, send the packet to a pre-specified multicast address (via the action send(packet, multicast_address)) over the satellite network and move to the rightmost state. In the rightmost state, the sender program is waiting for a fixed interval time to send the next data packet. The program uses the UNIX interval timer to schedule transmission of packets as if our sender can send packets in a fixed flow rate to a multicast group address. If the interval time is over, the sender program creates the data packet attaching a sequence number and a file number (via the action make_pkt(seq_no, file_no)) and sends the packet to the same multicast address over the satellite channel.

The receiver side also has two states. In the leftmost state, the receiver program is waiting to receive a packet from the same pre-specified multicast address. If the receiver program receives the advertise packet, it extracts the file information (via the action extract(file_info)); then, it opens a buffer to record packet received or lost information (via the action open(file_trace, file_info)) and moves to the rightmost state. We call this buffer a *trace file*. In the rightmost state, the receiver program is waiting to receive any data packet. If the receiver program receives a data packet, it extracts the packet sequence number and file number (via the action extract(seq_no, file_no)) and record the sequence number (via the action write(seq_no, buff, file_no)) to the corresponding trace file. During the packet transmission, we ignored the actual contents of these packets essentially considering them as periodic test packets that were sent from the sender in the multicast network.

3 Measurement Results

We measured the packet loss of multicast traffic from the sender site KEIO to other four receiving sites AIT, BUET, ITC and UNIBRAW on the unidirectional satellite link. The KEIO site ran the sender program and other four sites ran the receiver program. The experiment was basically divided into two parts where in the first part, or part I, there was no priority set up on the sending side router for our multicast traffic during 13-16 February. As a result, our traffic was treated with a lower priority.

Table 2. Examples of trace files description from the experiment performed on 15th February 2005

Flow Rate (Mbps)	Packet Size (Byte)	Loss Rate (%)				Average Bad Burst Length (number of packets)				# of Packets
		AIT	BUET	ITC	UNIB RAW	AIT	BUET	ITC	UNIB RAW	
5	1440	31.59	31.59	31.80	31.65	2.66	2.66	2.71	2.67	90,000
5	1240	31.86	31.86	31.86	31.90	2.76	2.76	2.76	2.76	90,000
5	1040	30.15	30.15	30.18	30.18	2.77	2.77	2.78	2.77	90,000
5	840	39.00	39.00	39.00	39.08	3.87	3.87	3.87	3.88	90,000
5	640	30.44	30.44	30.44	30.46	3.38	3.38	3.38	3.38	90,000
4	1440	34.89	34.89	42.87	35.51	4.29	4.29	5.19	4.30	90,000
4	1240	42.84	42.84	42.87	43.68	5.19	5.19	5.19	5.22	90,000
4	1040	40.17	40.17	40.17	40.63	4.74	4.74	4.74	4.76	90,000
4	840	41.13	41.15	41.16	41.55	5.01	5.01	5.02	5.03	90,000
4	640	0.17	0.17	0.17	0.27	3.65	3.65	3.65	3.58	90,000
3	1440	21.06	21.08	21.06	21.52	3.25	3.25	3.25	3.27	90,000
3	1240	14.28	50.77	14.28	15.03	3.41	24.13	3.41	3.44	90,000
3	1040	12.75	12.75	12.79	14.37	4.26	4.26	4.28	4.26	90,000
3	840	16.29	16.30	16.29	16.93	4.97	4.96	4.97	4.97	90,000
3	640	20.08	20.08	20.09	20.53	5.68	5.69	5.69	5.66	90,000
2	1440	1.17	1.17	1.17	1.48	2.79	2.79	2.79	2.83	90,000
2	1240	1.03	1.03	1.05	1.34	2.90	2.90	2.94	2.91	90,000
2	1040	10.16	10.16	10.16	10.39	3.12	3.12	3.12	3.12	90,000
2	840	7.71	7.71	7.71	8.11	3.68	3.68	3.68	3.70	90,000
2	640	5.64	5.65	5.68	5.88	4.41	4.41	4.45	4.40	90,000
1	1440	4.33	4.33	4.33	4.44	2.21	2.21	2.21	2.22	90,000
1	1240	4.09	4.09	4.09	4.24	2.36	2.36	2.36	2.36	90,000
1	1040	4.27	4.27	4.27	4.45	2.39	2.39	2.39	2.42	90,000
1	840	0.61	0.62	0.61	0.64	4.53	4.51	4.53	4.48	90,000
1	640	0.00	0.00	0.00	0.22	0.00	0.00	4.00	2.82	90,000

For part II, the priority rule was applied to our traffic from 27th February 2005 and later where our traffic was given the highest priority in the channel.

During the first part of our experiment our experiments consist of file transmission using different packet sizes with different flow rates as shown in Table 2. The sender sent a file consisting 90,000 packets to the receivers using multicast. There were 25 trace files (5 different packet size * 5 different bandwidth) collected which means 100 traces for 4 receivers. The experiment was repeated for four days. Table 2 shows an example of traces collected in one day with loss rates.

Here the highest loss rate appears to be 50.77% at Unibraw site. We have selected the trace files of 15th February as this shows the highest loss rate. On other days, the highest loss rate found were 12.77% in ITC, 31.13% in Unibraw and 38.16% in Unibraw on the day 13th, 14th and 16th February respectively. The trace results for Sunday 13th February show comparatively low loss rate as the day is a weekend

Table 3. Examples of trace files description from the experiment performed on 27-28 February, 1st march, 3-4 March and 9th March 2005

Flow Rate (Mbps)	Loss Rate (%)				Average Bad Burst Length (number of packets)				# of Packets (million)
	AIT	BUET	ITC	UNIB RAW	AIT	BUET	ITC	UNIB RAW	
2.000	3.30	3.47	1.51	3.31	1.30	1.37	1.38	1.30	13.40
2.000	2.17	2.42	2.41	2.18	1.20	1.34	1.34	1.20	4.60
0.500	4.65	4.74	4.68	5.24	3.02	3.07	3.05	3.09	4.00
1.000	0.83	1.02	0.95	0.85	1.05	1.29	1.21	1.05	1.00
0.250	0.87	1.05	0.99	0.90	1.03	1.23	1.16	1.03	0.42
0.125	2.12	3.08	2.13	2.13	1.04	1.51	1.04	1.04	0.24
1.000	0.76	4.53	0.84	4.57	1.07	1.00	1.18	1.00	3.00
0.250	0.82	4.78	0.90	4.84	1.07	1.06	1.17	1.06	1.20
0.125	0.78	4.55	0.85	4.58	1.06	1.06	1.16	1.06	0.70
1.000	0.64	1.82	0.64	0.74	1.04	3.10	1.04	1.05	1.30
0.500	0.65	1.83	0.65	0.74	1.03	3.00	1.03	1.04	0.91
0.250	0.66	1.83	0.66	0.75	1.04	3.03	1.04	1.05	0.55
0.125	0.65	1.83	0.65	0.75	1.06	3.07	1.06	1.05	0.30

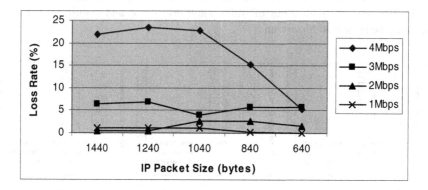

Fig. 3. Packet loss for different packet sizes using different flow rate

Table 4. Comparison of packet loss rate, before and after prioritizing the multicast traffic

Experiment	Flow Rate (Mbps)	Packet Size (Bytes)	# of Packets	Average Loss Rate (%)
Part I	1	1440	360,000	1.09
Part II	1	1440	360,000	0.08

where there was less traffics on the channel. The above graph in Fig. 3 demonstrates our observations of packet loss rates variation as against packet sizes. We have found that smaller packet sizes with higher flow rates tend to reduce the packet loss rates.

During the second part of the experiment, all our traffic from the feeder machine KEIO was given the highest priority in the feed but up to the limit of 3Mbps where the UDL channel capacity was 9 Mbps. As observed in Fig. 3, with 3 Mbps upper bound, the size of the packets has no affect on the packet loss rates; hence during this part of the experiment, we used fixed packet size of 1440 bytes. However, to observe other loss behavior such as loss occurrences and the burstiness, we extended our data collection periods in this part of the experiment. Table 3 shows an example trace file from part II.

The effect of using prioritized traffic is shown in Table 4 which compares the loss rates of the two different experimental set up for part I and part II. It can be observed that with prioritized multicast flow in part II, the loss rates become significantly lower than those in part I. This is so since we are using UDP packets when multicasting which get discarded more than the TCP packets by network elements such as routers whenever the buffer of the network elements is full.

During the whole experimental period, the sender sent 26.5 million multicast packets over the satellite network in 237 hours total. Out of which 8.91 millions non-prioritized packets were transmitted during part I. The remaining 17.59 million packets were transmitted after the prioritized routing policy was applied to our traffic. Fig. 4 shows the packet loss rate during both transmission periods. The leftmost histogram shows that more than 11% traffics were lost when the packets were not prioritized but middle histogram shows that it dramatically dropped to only less than 3% after the higher priority has been assigned to our experimental traffic. Finally, overall loss rate for both types of traffic (non-prioritized and prioritized traffic) is shown in the rightmost histogram.

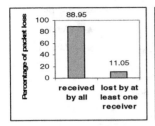 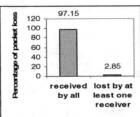 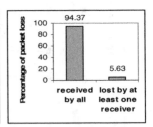

Fig. 4. Loss rate (%) during the experiment; x-axis: percentage of packet loss; left: part I non-prioritized traffic, middle: part II prioritized traffic and right: Part I + Part II combined

4 Analysis and Discussion

4.1 Occurrences of packet loss

From the loss of almost 6% in mixed traffic, we tried to investigate further about where these losses actually take place. Fig. 5 shows that out of all the lost packets, more than 85% packets were lost by all receivers in overall cases. 12.27% packets were found lost by only one of the four receivers while others could receive the packets successfully. Very small percentages of packet loss were shared by more than one receiver but not all receivers where only 0.05% of all losses shared by two receivers and 1.23% shared by three receivers.

As we know, packet loss in satellite network depends on many factors including bad weather, sun interference, channel noise, equipment (like antenna) problems, routing policy etc. Since our receivers were placed in a wide area, it is unlikely that they share the same weather condition or sun interference. To explain above observation of common losses, we further investigated on how could such a large percentage of packet loss is shared by all receivers. We collected the corresponding MRTG (Multi Router Traffic Grapher) graphs for UDL traffic maintained by AI3 [1]

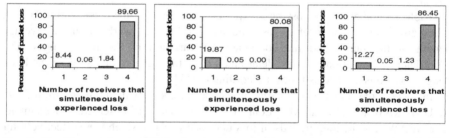

Fig. 5. Simultaneous losses (%) by the receivers; x-axis: percentage of packet loss, y-axis: number of receivers; left: part I non-prioritized traffic, middle: part II prioritized traffic and right: Part I + Part II combined

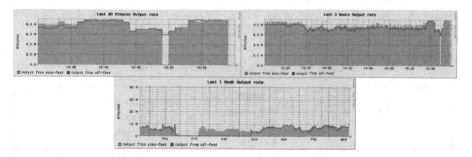

Fig. 6. Traffic on unidirectional satellite link, top: last 30 minutes output, middle: last 3 hours output, bottom: last 1 week output, Source: Traffic on UDL (http://sfc-serv.ai3.net /UDLOUT /udlout.cgi) on March 2, 2005 at 16:00:00 JST

network administrator to investigate these losses. We have found strong correlation of our packet loss data and the failures of the sender's equipment. Shown in Fig. 6 is the sender site's MRTG graph of March 2, 2005. The SOI-ASIA UDL channel bandwidth under AI3 network is 9Mbps which is almost fully occupied most of the time. The gray area in the graph is the output traffic from the equipment which transmits the traffic on the UDL. The black line represents the output from the udl-feed machine. From 1st and 2nd graph (counting from top) we see that UDL transmitter was down for 1 minute which can cause traffic loss in a huge burst size. Again the 3rd (counting from top) graph shows that the udl-feed might be down causing packet loss although the UDL transmitter is working. In summary, losses common to all receivers could be due to the equipment on sender side alone.

4.2 Burstiness Analysis

This section describes the distribution of loss lengths in each receiver. Most of the analysis presented here is based on Trace3, collected on 1st March 2005. We chose the Trace3 because it was collected for the longest time period of 24 hours 25 minutes. And another reason is, from the period of data collection of Trace3, we were able to collect the corresponding MRTG graph for UDL and SOI-ASIA network monitoring graphs.

Table 5 summarizes the loss rates experienced by each of the receivers for Trace3. They also show the number of lossy bursts seen by each workstation, the average length of the bursts and the coefficient of variation of the burst length. The Coefficient of variation c can be determined from the standard deviation of the burst lengths which is defined as follows according to [17].

In general, the coefficient of variation was seen to be very high ranging from 7.995 at UNIBRAW to 9.363 at AIT. Long burst of losses were observed. For example, UNIBRAW saw the long periods of loss upto 688 packets which were effectively for 15.136 seconds. On the other hand, BUET and ITC saw the long periods of loss upto 1468 packets which were effectively for 32.296 seconds. Fig. 7(a), 7(b), 7(c) and 7(d) show the burst distribution of Trace3 of AIT, BUET, ITC, and UNIBRAW respectively. From the trace files, we extracted the packets received in burst and packets lost in burst. We refer to the event of packets reception in burst as the good burst and packets loss in burst as the bad burst.

Table 5. Loss details of all receivers for Trace3

Receiver	Loss Rate	Num of bad Bursts	Average bad Burst Length	Coefficient of Variation	Longest good Burst length	Longest bad Burst length
AIT	4.650	61589	3.020	9.363	500466	3735
BUET	4.737	61776	3.067	8.649	208283	1460
ITC	4.679	61390	3.049	8.393	623086	1468
UNIBRAW	5.241	57836	3.090	7.995	112381	688

$$\sigma = \sqrt{\frac{1}{N}\sum_{i=1}^{N}((\ell-\bar{b})^2 *b_\ell)} \tag{1}$$

$$c = \frac{\sigma}{b} \tag{2}$$

where N = number of packets; b_ℓ = number loss bursts (number of consecutive losses) of length $\ell, \forall\ell\in\{1,2,...,N\}$; $\bar{b}$ = mean loss burst length; σ = standard deviation of the burst lengths; c = coefficient of variation of the burst lengths.

For every good burst or bad burst event we detected the sequence number from where the burst event occurred. In the following figure, we plotted the packet sequence number along the x-axis. We plotted the good burst length along the positive y-axis and the bad burst length along the negative y-axis. There were 4 million packets transmitted from sender to receivers. For some technical reason, UNIBRAW stopped receiving packets after 3.4 millions which we plotted in Fig. 7(d). If we try to relate the graph of any site (e.g. AIT in Fig. 7(a)) with our MRT graph in Fig. 8, we see that the larger good burst lengths are found when the UDL channel has less traffic, which from 6:00 am to 9:30 am. During this time, AIT experienced longest good burst length of 500466 packets. For ITC, this longest burst is even larger which consists of 623086 packets.

Most of the bad bursts occur during the period when the UDL channel is full with traffic. Sometime we see the events of bad burst with length up to 3735 packets which was more than 2% of the overall packet loss although the number of occurrences of these events is small.. The graph does not show the burst length distribution very well because of the scaling of graph, specially the bad bursts. For a better view, Fig. 7(e) shows the burst distribution for AIT of range within 800. This figure shows an interesting view that there are some periodic burst loss events for some specific time duration. Once there is burst loss occurs there is a very high probability to occur another burst loss, even of a larger size. Fig. 9(a) shows the distribution of burst lengths for AIT. It is clear that, although most of the bursts are of size less than 100, there are occasional long loss periods. Fig. 9(b) shows the percentage of total loss experienced by AIT in the various burst lengths.

The cumulative distribution functions of good bursts for every receivers of Trace3 are shown in Fig. 10(a). This graph shows the percentage of cumulative receive rate that occurred in good bursts length b or smaller. The graph shows that more than 70% packet are received in good bursts of length 100 or more. From this graph, another fact is clear that, nearly 1% packets are received by burst length 1 only. The cumulative distribution functions of bad bursts for every receiver of Trace3 are shown in Fig. 10(b). This graph shows the percentage of cumulative loss rate that occurred in bad burst length b or smaller. The graph shows that 50% packets are lost with burst length 75 or smaller. Other 50% traffics are lost by higher bad bursts of length more than 75. On the other hand, almost 25% packets are lost by bursts of length only 1. It is clear that larger bad burst events have very high contribution to the overall loss process. This is true for packet receive events also.

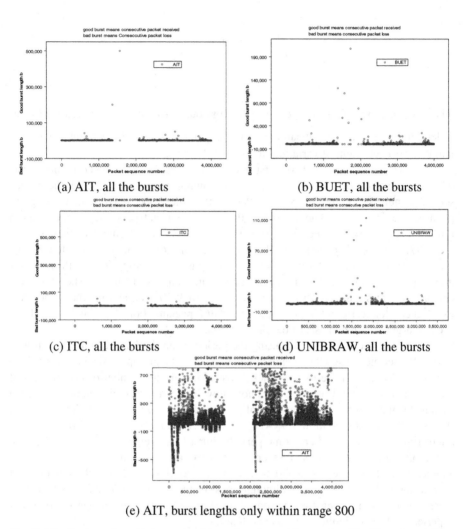

(a) AIT, all the bursts

(b) BUET, all the bursts

(c) ITC, all the bursts

(d) UNIBRAW, all the bursts

(e) AIT, burst lengths only within range 800

Fig. 7. Distribution of good burst and bad burst lengths; x-axis: good and bad burst length b, y-axis: packet sequence number; (a) Trace3.AIT, (b) Trace3.BUET, (c) Trace3.ITC, (d) Trace3.UNIBRAW, (e) trace3.AIT. Good burst lengths are in positive y-axis, bad burst lengths are in negative y-axis.

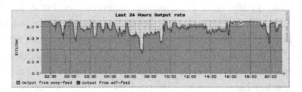

Fig. 8. Traffics on unidirectional satellite link last 24 hours output, Source: Traffic on UDL (http://sfc-serv.ai3.net/UDLOUT/udlout.cgi) on March 2, 2005 at 21:00:00 JST

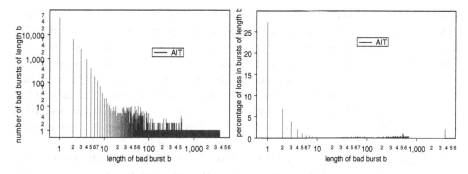

Fig. 9. (a) Distribution of bad burst lengths for Trace3.AIT; x-axis: number of bad burst of length b, y-axis: length of bad burst b; (b) Distribution of percentage of packet loss for Trace3.AIT; x-axis: percentage of packet loss in bursts of length b, y-axis: length of bad burst b

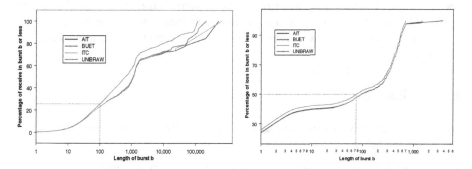

Fig. 10. Cumulative distribution function for Trace3 for (a) good bursts; x-axis: percentage of receive in burst b or less, y-axis: length of burst b logarithmic scale; and (b) bad bursts; x-axis: percentage of loss in burst b or less, y-axis: length of burst b logarithmic scale

5 Related Works

Earlier works on the measurement of multicast packet loss in the internet [20], [21], [6], [5] have noted that the number of consecutively lost packets is small. In [20], the experiment used sources to send 640 bytes of audio data in 80ms and 40ms interval. Long outages of several seconds and minutes were observed on the MBone (multicast backbone network overlaid on the Internet) [6]. A few extremely long bursts also observed lasting from a few seconds up to 3 minutes (around 2000 packets). At least one receiver saw one of these long loss bursts, in every dataset. Periodic bursts of length approximately 0.6 second or 8 consecutive packets were observed for some receivers in some of the datasets. In [21], the experiment used sources to send 640 bytes of audio data in 160ms, 80ms, 40ms and 20ms interval. The packet loss was analyzed with respect to good run lengths (consecutive packet receive) and bad run lengths (consecutive packet loss).

Although these experiments were done in terrestrial network, [5] present some measurement analysis for multicast data over satellite network. In [5], the experiment used a source to send four different size files (1MB, 10MB, 50MB, 100MB) in 118Kbps (10 packet/sec) rate to four receivers. The experiment shows less than 2% packet loss rate in almost all cases. In some cases the loss rate up to 28% was also observed. It also shows some bursty loss with two or more consecutive packets when the source sends 1500 bytes size packets in 100ms interval. The result shows almost 14.29% loss events of the overall losses occur in bursts. Experiment also showed that packet size does not have significant effect on loss rates. Experiments concluded that congestion control is needed for multicast traffics although they did not mention anything about the traffic priority policy in network elements.

From all the discussion we see that our experiment results are different from the result discussed in [5] although the test-bed is same AI^3 satellite channel. The sending algorithm in [5] always sent 10packet/sec (maximum 10,000 packets per hour) although it could send 36,000 packets in 1 hour. They collected the results for each 10,000 packets in each hour separately. Their traffic was sent in a low rate (118 Kbps) and traverses the sender side router in one hour interval. It is more logical that their traffic will cause less congestion in the router compared to our traffic because in our traffic packets were sent out continuously for longer period of time in a steady and higher rate.

6   Conclusions

After all the analysis of experimental data from satellite network, we conclude our findings in the below.

– Around 85% packet losses out of total lost occurs in the sender side network. Extra care should be given to the routing policies and network equipments to reduce this percentage. Reducing this rate would dramatically increase the packet recovery rate by satellite network receivers.
– A receiver in the satellite network may see a very large size of bad burst which can be upto 5561 packets. This single bad burst event contributes upto 10% of the total packet loss. At the same time, other receivers may see a bad burst consisting 16 packets only. It seems that, there is a very high chance for the receivers, suffering from high loss rates can recover packets from other receivers if the receivers can communicate with each other.

The bad burst lengths of more than 75 contribute heavily to the total loss, despite their small in numbers. In the same way, more than 70% packets are received by good bursts of lengths 100 or more.

Analysis of the experiments of the loss with more receivers would give more variety and widespread of loss patterns in different situations. The long bad bursts can be analyzed with inter-loss distance metric [13] to describe distance packet losses in terms of sequence numbers which would help to determine the number of packets unrecoverable by FEC. [20], [21], [6], [15], [8] shows different modeling method to validate the loss process. Similar modeling also can be applied to our results to validate the loss process in general case.

Acknowledgements

We would like to thank our colleagues of the Keio University (SFC Campus), especially Ms. Patcharee Basu and Mr. Shunsuke Fujieda, for providing us all the support. We thank the Asia Internet Interconnection Initiation (AI3) project for providing such an excellent opportunity of us to learn from a real satellite network. We are grateful to Bangladesh University of Engineering and Technology of Dhaka, Institute of Technology of Cambodia, Brawijaya University of Indonesia, and Asian Institute of Technology of Thailand for providing us multicast capable computer accounts which allows us to take the measurement. We would like to thank to two anonymous reviewers for there important comments to improve the paper.

References

1. AI3.: Asia Internet Interaction Initiative (AI3) Project. Retrieved 21 April 2005 from http://www.ai3.net/topology/.
2. Akkor, G., Hadjitheodosiou, M., and Baras, J. S.: IP Multicast via Satellite: A Survey. Technical Research Report, CSHCN, Institute for Systems Research, University of Maryland, 2002.
3. 27ALTQ.: ALTQ: Alternate Queuing for BSD UNIX. Retrieved 11 June 2004 from http://www.csl.sony.co.jp/person/kjc/kjc/software.html.
4. Awal, M. A., Tsuchimoto, Y., and Kanchanasut, K.: An Approach of Peer-Based Packet Recovery using EDbit for Unidirectional Satellite Environment. 2nd Workshop on Asia-Pacific Networking Technology and Next Generation Mobile Communication, 19th APAN Meeting, Bangkok, 2005.
5. Basu, P., and Kanchanasut, K.: Multicast Loss on Satellite UDL Study. 1st French Asian Workshop on Next Generation Internet, INRIA Sofia Antipolis, 2004. Retrived 10 October 2005 from http://www-sop.inria.fr/planete/fawngi/slides/t1_1_kanchanasut.pdf
6. Bolot, J., Crepin, H., AND Garcia, A. V.: Analysis of Audio Packet Loss in the Internet. In Proceedings of the International Worskshop on Network and Operating System Support for Digital Audio and Video (NOSSDAV) (Durham, New Hampshire, Apr. 1995), pp. 163–174.
7. J. Byers, M. Luby, M. Mitzenmacher, and A. Rege: A Digital Fountain Approach to Reliable Distribution of Bulk Data. In ACM SIGCOMM'98, Vancouver, Canada, September 1998.
8. Jiang, W. and Schulzrinne, H.: Modeling of Packet Loss and Delay and their Effect on Real-time Multimedia Service Quality. Proc. of 10th Int. Workshop Network and Operations System Support for Digital Audio and Video, 2000.
9. L. Rizzo: Effective Erasure Codes for Reliable Computer Communications Protocols. ACM Computer Communication Review, 27(2), 24 - 36, 1997.
10. M. Allman, D. Glover and L. Sanchez: Enhancing TCP Over Satellite Channels using Standard Mechanisms. RFC 2488, January 1999.
11. M.W. Koyabe, G. Fairhurst, A. Matthews, H. Cruickshank, S. Iyengar, M.P. Howarth: Satellite Reliable Multicast Transport Protocol (SAT-RMTP): A Network Tool for Multimedia File Distribution.Poster and Protocol Demonstration, Alcatel Space Industries (ASPI), 29th April 2003, Toulouse, France. Retrieved on 10 October 2005 http://geocast.netvizion.fr/download/sat-rmtp.pdf

12. Moufida Maimour: Design, Analysis and Validation of Router-Assisted Reliable Multicast Protocols in Wide Area Networks. University Claude Bernard – Lyon I, PhD Dissertation, November 2003.
13. R.Koodli and R. Ravikanth.: One-way Loss Pattern Sample Metrics. Internet Draft, IETF, June 1999.
14. S. Acharya, R. Alonso, M. Franklin, S. Zdonik: Broadcast Disks: Data Management for Asymmetric Communication Environments. Proceedings of the ACM SIGMOD Conference, San Jose, CA, 1994.
15. Sanneck, H., Carle, G., and Koodli. R.: A Framework Model for Packet Loss Metrics Based on Loss Runlengths. Proceedings of the SPIE/ACM SIGMM Multimedia Computing and Netoworking Conference, January 2000.
16. Stevens, W. R.: UNIX Network Programming, Network APIs: Sockets and XTI. Volume 1, 2nd Edition, Prentice Hall, 2002.
17. Stoica, I., Zhang H., and Eugene, T. S. N.: A Hierarchical Fair Service Curve Algorithm for Link-Sharing, Real-Time and Priority Service. Proceedings of SIGCOMM'97, September 1997.
18. Sun Z., Howarth M.P., Cruickshank H. and Iyengar S.: Networking Issues in IP Multicast over Satellite. International Journal of Satellite Communications and Networking 2003 21:489-507.
19. Wacharapol Pokavanich: Packet Delivery Scheduling for Reliable Multicasting over Unidirectional Link. Asian Institute of Technology, MS Dissertation, Thesis no. CS-00-17, 2000.
20. Yajnik, M., Kurose, J., and Towsleym D.: Packet Loss Correlation in the MBone Multicast Network. In Proceedings of the IEEE Global Internet Conference, London, UK, 1996.
21. Yajnik, M., Moon, S., Kurose, J., and Towsley, D.: Measurement and Modelling of the Temporal Dependence in Packet Loss. In IEEE INFOCOM '99, (pp. 345-352), 1999.

On a Novel Filtering Mechanism for Capacity Estimation: Extended Version

Jianping Yin, Shaohe Lv, Zhiping Cai, and Chi Liu

School of Computer, National University of Defense Technology,
Changsha Hunan, 410073, China
jpyin@nudt.edu.cn, chi.shaohe@gmail.com,
xiaocai@tom.com, 3dfx232@sohu.com

Abstract. Packet-pair has been used as one of the primary means to measure network capacity. Yet, most prior proposal tools are sensitive to network status and perform poorly in heavy-loaded network. We present a novel filtering mechanism to address the negative effects of cross traffic. After split the origin set of probe pairs into two packet sets composed of the first and second packet of all pairs respectively, we select the packets with minimum one-way delay in each set and use them to reconstruct a new pair free from interference of cross traffic, from which the final capacity estimates are derived. We show the mechanism in detail and validate it in simulations as well as Internet experiments. Preliminary results show that the proposed mechanism is feasible and robust for the heavy-loaded network, which can produce accurate estimates with relatively fewer overheads compared to similar tools. Finally, we analyze the difference in the first and second packet of probe pair in depth, which argues a novel direction of the analysis of packet-pair.

1 Introduction

Knowledge of capacity is of growing importance in various scenarios, such as overlay networks and Peer-to-Peer networks. The capacity of a link is the maximal data transmission rate that can provide in the absence of competing traffic. As to a path comprised of several store-and-forward links, its end-to-end capacity is limited by the link, referred to as narrow link, with minimum capacity. In this paper, we focus on the measurements of end-to-end capacity mainly.

The packet-pair technique is traditionally used as one of the primary mechanisms to measure the end-to-end capacity. Assuming no cross traffic interacts with probe traffic, a pair of two probes which is launched at sender, will be spaced by the transmission time of the second packet at the narrow link after traversed the path. Then the capacity is computed by $C = P/\Delta$, termed as the basic packet-pair formula, where P is the size of the second packet and Δ the inter-arrival spacing at receiver.

Due to the violation of the underlying ideal model, packet-pair estimates may be erroneous. The inflation or compression errors occur due to the difference in the effects of cross traffic experienced by the first and second packet within a pair. There has been a fair amount of research striving to address the issue, either eliminating the negative influence of cross traffic or extracting some useful information from the distorted origin data. Yet, the dependence on network characteristics, such as the size

K. Cho and P. Jacquet (Eds.): AINTEC 2005, LNCS 3837, pp. 269–281, 2005.
© Springer-Verlag Berlin Heidelberg 2005

of dominant cross packets interacted with pro-be pairs and the network load or congestion, limits their power. Especially in heavy-loaded networks where the interactions of probe and cross packets are frequent and complex, their performance drops drastically.

The whole pair is usually used as the minimal processed unit, which we believe, should be responsible for the inefficiency, since it ignores the difference in the single packet (i.e. the first or second packet within a pair) and the pair as well as between the first and second packet implicitly. But the differences can be remarkable as we show in next section. To exploit the features explicitly, we advocate a novel filtering mechanism, which exhibits a promising robust performance in terms of eliminating the adverse effects of cross traffic, even for the heavy-loaded network. The in-depth analysis of the differences argues a novel direction regarding the analysis of packet-pair. That is our ongoing work.

The rest of the paper is organized as follows. In section 2 we describe the underlying mechanism in detail and evaluate it in simulations as well as Internet experiments in section 3. The discussion about the generalization of the mechanism is presented in section 4. Finally, we present the related work and conclusions in section 5 and 6 respectively.

2   The Mechanism

We first depict the set of assumptions and notations used in this paper, which are common to most similar studies. Then the mechanism and the corresponding tool, *CapEst*, are described in detail.

2.1   Assumptions and Notations

The set of assumptions can be enumerated as below: (1) Router is store-and-forward and adopts FIFO queuing. (2) The measured path is invariant during the whole measurement episode. (3) All probes have the same size (denoted as P in this paper) and initial sending rate while forwarded or received in-order. (4) All pairs are sent back-to-back and spaced enough to avoid the mutual interference between distinct pairs. (5) Host clock resolution is enough to enable accurate timing estimation.

Consider a path of n links, we use t_k^{PA} to denote the time instance when packet PA arrives at link k. Especially, t_0^{PA} marks the sending time and t_n^{PA} the arrival time at receiver. We consider the probe packet only. Note that the one-way delay of probe PA is $OWD_{PA} = t_0^{PA} - t_0^{PA}$. The duple (FP(n), SP(n)) is used to denote the nth packet pair where n can be ignored as refers to general pair and FP or SP represents the first or second packet within the pair respectively. In addition, the intra-pair spacing of a pair at sender is $Spacing_s = t_0^{SP(k)} - t_0^{FP(k)}$ while at receiver is $Spacing_r = t_n^{SP(k)} - t_n^{FP(k)}$.

2.2   Overview of the Mechanism and Tool

The basic procedure of the mechanism can be summarized as follows. After all probe pairs traversed the measured path are obtained, the origin sample set is split into two packet sets, termed as S1 and S2, comprised the first and second packet of each pair respectively. Then the packets with minimum one-way delay in each packet set are extracted and used to reconstruct a new pair to yield the final capacity with the help of the basic packet-pair formula.

Based on the above mechanism, we design a measurement tool, called *CapEst*, whose features include: First, after a certain amount of pairs are transmitted, called a *run*, the filtering mechanism is executed effort to find the current estimate of capacity. Second, in a run all probe pairs with the same packet size are sent back-to-back at the same rate, via an UDP connection. Third, the inter-pair spacing obeys the Poisson distribution while keep large enough to avoid the mutual interference of different pair. Fourth, adopted from CapProbe [6], only after the difference between the estimates of two run with distinct packet size is within a small value, the average of the two estimates can be reported as the final estimate. Last, due to the separate process the first and second packet, *CapEst* can tolerate the partial loss of probe pair, which means that only the first or second packet is lost. Actually, no matter how seriously the loss events occur, the correct estimate can be gained as long as at least, there has a first and second packets free from distortion of cross traffic respectively.

2.3 Descriptions

We now describe the filtering mechanism in detail, including the observations on the one-way delay (such as the important time transferable property) and the re-construction procedure of the desired packet pair.

2.3.1 One-Way Delay Visiting

The one-way delay of packet *PA* after traversed a network path can be denoted by equation (1), where t_n^{PA} and t_0^{PA} indicates the arrival time at the last link, e.g. arriving the receiver, and the departure time from the sender, of packet *PA* respectively.

$$OWD_{PA} = t_n^{PA} - t_0^{PA} = \sum_{k=1}^{n}(\frac{P}{C_k} + p_k + q_k) \tag{1}$$

Note that in the right-part of equation (1), the middle term, that is the propagation latency of a link, is the time consumed in propagating signals through the link at the velocity of light, which is independent of packet *PA*. The last term is the queue latency that reflects the status of the link at the arrival time of packet *PA* and is determined by the network itself only. The first term, transmission latency, is related to the size of packet *PA* and limited by the capacity of physical link.

Thus, the propagation and transmission latency is fixed and independent of network status under the assumptions (2) and (3). Furthers, all packets with the same size, emitted at different time, have the identical one-way delay when they have the identical network experiences. This property is termed as "time transferable property of one-way delay".

In addition, if there has no queue event occurs, the relevant one-way delay only consists of propagation and transmission latency is the least one among all one-way delays of packets with the same size. This is the signature of a single packet free from disturbance of any other traffic.

Through a number of experiments, we found that the first or second packet free from interacting with cross traffic always exists within a reasonable amount of probe pairs, even when the network is heavy-loaded. The packet with minimum one-way delay in set S1 and S2, thereby, can be referred as the one without any distortion from cross traffic confidently, which is one of the fundaments of the filtering mechanism.

2.3.2 Packet Pair Reconstruction

We now focus on the relationship between one-way delay of packet and inter-packet spacing. Equation (2) denotes the difference between the inter-packet spacing at sender and at receiver of a pair.

$$Spacing_r - Spacing_s = (t_n^{SP} - t_n^{FP}) - (t_0^{SP} - t_0^{FP}) \qquad (2)$$

In addition, equation (3) indicates the difference between the one-way delay of the first and second packet within a pair.

$$OWD_{SP} - OWD_{FP} = (t_n^{SP} - t_0^{SP}) - (t_n^{FP} - t_0^{FP}) \qquad (3)$$

The obvious derivation is that the two differences are, actually, equivalent. Thus, the inter-packet spacing at receiver can be constructed from the intra-pair spacing at sender and the one-way delays of the first and second packet within a pair. The former is known and equal in every pair, while the latter can be obtained through end-to-end measurement. This is an alternative way to obtain particular intra-pair spacing at receiver rather than measuring at the endpoint directly, which is served as another one of the fundaments in the proposed scheme.

For the special case of capacity estimation where the pair should be free from interference of cross traffic, the inter-arrival spacing at receiver can be computed by the known intra-pair spacing at sender and the one-way delay of a first and second packet without any interaction with cross traffic.

Further, with the help of the time transferable property of one-way delay, all the first or second packets without any interaction with cross traffic have the same one-way delay respectively. Therefore, the inter-arrival spacing at receiver of a desired pair for capacity estimation can be gained by any one first and second packet free from interference of cross traffic.

The detail description of the reconstruction procedure is: Firstly, suppose all pairs in a run are indexed as $1,2...N$, where N is the amount of pairs in a run. We use $MinSet(FP)$ or $MinSet(SP)$ to denote the set included the indexes referred to the pairs involved the first or second packet with the minimum one-way delay respectively. Secondly, fetching any $k \in MinSet(FP)$ and $m \in MinSet(SP)$. The inter-arrival spacing of the desired pair at receiver, Δ, is computed from the one-way delay of $FP(k)$ and $SP(m)$. Equation (4) denotes the computation, where the intra-pair spacing at sender, termed as $Spacing_s$ is known and identical for all pairs under the assumption (3). Finally, the capacity can be derived from the basic packet-pair formula with the known P.

$$\Delta = Spacing_s + (OWD_{SP(m)} - OWD_{FP(k)}) \qquad (4)$$

The fact that, In general, there has no globe clock implies that there is some possible difference between the clocks of two sides of the measured path. Suppose the difference is ω ($\omega \in R$), then the new one-way delay can be represented as:

$$\text{New}\_OWD_{PA} = t_n^{PA} + \omega - t_0^{PA} = OWD_{PA} + \omega$$

Then, from Equation (4), we know the new inter-packet spacing at receiver can be computed as:

$$\text{New}\_\Delta = Spacing_s + (\text{New}\_OWD_{SP(m)} - \text{New}\_OWD_{FP(k)})$$
$$= Spacing_s + ((OWD_{SP(m)} + \omega) - (OWD_{FP(k)} + \omega))$$
$$= Spacing_s + (OWD_{SP(m)} - OWD_{FP(k)}) = \Delta$$

The conclusion is that it's no effect of the absence of globe clock on the computation of capacity in our *CapEst* scheme.

3 Experiments

We evaluate the mechanism and tool in simulations used *NS-2* [12] as well as Internet experiments. The preliminary results confirm that the filtering mechanism is effective and robust for increasing network load.

3.1 Simulations

3.1.1 Setup

The simulation topology is shown in Fig.1, where the measured path consists of $n+1$ links from Sender to Sink and the narrow link is at the center of the path. All probe pairs with the same packet size emitted back-to-back, and received at Sink through the path via a UDP connection. A number of Pareto traffic (shape parameterα=1.6) are transmitted over the connections from S_k to R_k where k=1,2…n. measured path, all connections carried cross traffic are per-hop persistent. The capacity of those links rather than narrow link is as large twice as the capacity of narrow link or end-to-end capacity of the measured path

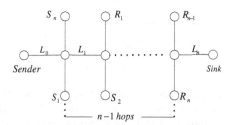

Fig. 1. The simulation setup where the measured path is from Sender to Sink consists of $n+1$ links

Every link of measured path has the same amount cross traffic as well as the identical flow type distribution, e.g. 95% TCP and 5% UDP. Further more, all traffic over all links have the same size distribution, e.g. there have three types of packet size: 1500/700/40 Bytes that consumes the 5%, 85% and 10% percentages of total amount of cross traffic respectively.

3.1.2 Results

In the first few seconds, we only conduct the probe packets. Then the minimum one-way delay of both the first and second packet can be gained to determine whether the first or second packets interact with cross traffic in follow experiments.

Fig.2 (a) and (b) plot the amount of sample pairs needed for obtaining the first *FP*, *SP* and whole pair free from interference of cross traffic in the case of *n*=5 and 11, where the end-to-end capacity is 10Mbps and 22Mbps respectively, when the size of all probe packets is 700Bytes. Note that the larger one of the amount in the case of *FP* and *SP* is the amount of packet, e.g. the first and second packet respectively, needed for the novel filtering mechanism to yield a correct estimate.

There are several observations can be drawn from the pictures. First, all the amount of probe pairs needed for obtaining a *FP, SP* or a whole pair with no interaction with

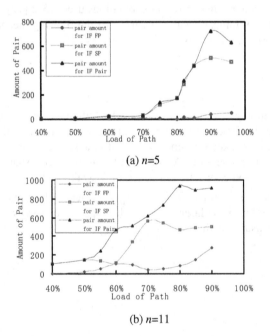

(a) *n*=5

(b) *n*=11

Fig. 2. The amount of pairs needed for obtaining a first, second packet and a whole pair free from interference of cross traffic under distinct network load, indicated by the curve "*FP*"、"*SP*" and "*pair*" respectively. The sub graph (a) and (b) refer to the case of *n*=5 and 11 respectively.

cross traffic is small and the difference among them is trivial when the load of network is light, which means that both *CapEst* and traditional methods, such as CapProbe, can acquire an accurate estimate with few overheads. But the difference among the amount for *FP*, *SP* and the whole pair become non-negligible with the increase of network load. Especially, in the case of *n*=11, the amount for obtaining a whole pair free from distortion of cross traffic is large impractical when network is heavy-loaded. This is one possible cause responses to the inefficiency of traditional methods in heavy-loaded networks. Yet, the amount for obtaining a expected *FP* and *SP* is still reasonable even the load is up to 80% in the case of *n*=11 or 96% when *n* =5.

Second, the amount of probe pairs needed for gaining a desired *FP* and *SP* becomes larger with the increase of network load. But the change trend is different: in the case of *SP*, the amount is sensitive to network load when actual load is middle, i.e. 60-70% in case of *n*=11, but becomes stable when the network load is light or heavy. In the case of *FP*, however, it exhibits a slow though consistent increasing trend with respect to network load. In addition, the amount of needed pairs for a desired first packet is less in most cases, which suggests that the second packet is more susceptible to the network dynamics. That's why we dispose the first and second packet separately. We will discuss the topics related to the difference in the first and second packet further in section 4.

3.2 Internet Measurements

3.2.1 Topology

The measured path from host in *NUDT* (*www.nudt.edu.cn*)to host in *CSU* (*www.csu.edu.cn*), as shown in Fig.3, consists of several links across the LAN and WAN (e.g. Internet) simultaneously. The first five hops are in NUDT domain and fully controlled where the narrow link, the fourth hop, is also the globe bottleneck link of path. The capacity of the path segment inside WAN, not shown in Fig.3, is 155Mbps and consistent during the period of measurements (noticed by the managers of underlying links).

The extra cross traffic are injected into the five hops in *NUDT* domain only. The injected cross traffic is self-similar. In addition, no other traffic except we injected

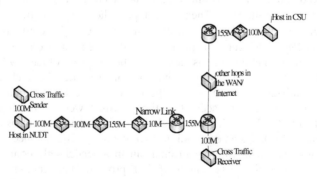

Fig. 3. The topology for Internet experiments where the first five hops of measured path are in *NUDT* domain and fully controlled

exists in the first five hops and all the five hops have the same amount of injected traffic (note that over other hops there may have real Internet traffic).

3.2.2 Results

Fig.4 shows the comparison of *CapEst* and CapProbe under various load of narrow link. We execute the corresponding estimate process 10 times and use the results with maximal occurrence likelihood as the final estimate for each certain link load. The relative error is computed by the formula $\lambda = |C_r - C_m| / C_r$, where C_r and C_m rep- resent the real and measured capacity respectively. The detailed description of CapProbe can be found in [6].

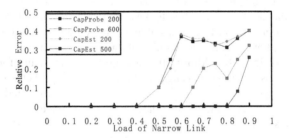

Fig. 4. Comparison of the accuracy of *CapEst* and *CapProbe*, where the curve *"tool amount"* indicate the result of the underlying *tool* when a run consists of the *amount* of pairs

From the figure, one can inspect that the estimate error is well controlled and slight for both tools when the network load is not heavy. The underlying errors of CapProbe increase rapidly, however, when the load becomes more aggressive. Particularly, after the load is up to 70%, the error of CapProbe cannot be eliminated effectively even when the amount of pairs in a run reaches 600. In opposition, *CapEst* can yield accurate estimate through no more than 500 pairs in a run even the load approaches 80%. Unfortunately, when the load is up to 85%, the estimates of both tools are inaccurate and the relative error goes up drastically. Thus, the problem of capacity estimation under much heavier load, i.e. up to 85%, is still an open issue.

In order to obtain estimate within certain relative error, the amount of pairs in a run should be altered dynamically. Then we compare the overheads of the two tools to yield the same accuracy. Fig.5 plots the ratio between the convergence time of CapProbe and *CapEst* to product estimate with the same relative error within 5%.

The figure shows that the ratio is close to 1 when the load of narrow link is light or middle; and the ratio rises up with the increase of load after the load is greater than 50%. In the load region of 60%-80%, the ratio is close to 3, which suggests that *CapEst* can product the same accurate estimate with fewer overheads in a factor of three compared to CapProbe. The relatively less cost consumed in measurement when network is heavy-loaded is an essential merit of the novel filtering mechanism.

We also test the novel filtering mechanism in a sender-only context. Similar to SProbe [17], we exploit the semantics of TCP protocol that sending a pair of SYN packets to an inactive port on a live host results in the generation of a pair of RST

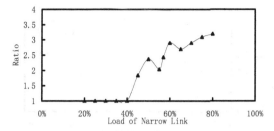

Fig. 5. The ratio of convergence time of *CapProbe* and *CapEst* with the same relative error

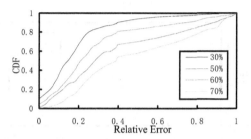

Fig. 6. The relative errors of *CapEst* in sender-only con-text under several load of narrow link indicated by the underlying curves

packets. However, we still use the pair of SYN with payload in place of a long train of regularly-sized SYNs in SProbe. Fig.6 plots the results of *CapEst* with a run consists of 500 pairs under several load of narrow link, where all probe packets have the fixed size of 700Bytes (the appended payload is 660Bytes). Compared to Fig.4, the accuracy under the sender-only context is obvious low. But the relative errors of more than half estimates are still smaller than 20% even the load is up to 50%, a considerable progress contrast to previous proposals. However, the relative errors rise up rapidly with the increase of the load of narrow link, e.g. up to 60%.

4 Discussions

In their study [1], Liu etc. examines the issue that what signals contained in packet-pair probes and pinpoints that the packet-pair probing essentially characterizes the network dynamics in terms of capacity, available bandwidth as well as cross traffic intensity. Otherwise, the signals encoded in packet-pairs are detectable and extractable only under particular conditions. In general, due to the elusory behaviors of network, it is difficult to preserve the underlying constraints to obtain metrics of interest accurately and rapidly, especially when network is heavy-loaded. We now present a novel aspect, from which the possible solution may be derived.

Adopting the same simulation setup as section 3.1, we inspect the difference between FP and SP in depth. The minimum one-way delay is removed from the total one-way delay to yield the queue latency for FP and SP respectively. Notice that the

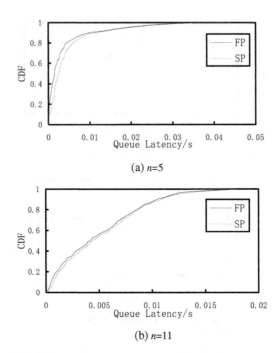

Fig. 7. The CDF of the CT queue latency experienced by the first and second packet of each pair when the network load is 70%, indicated by the curve "*FP*" and "*SP*" respectively. Sub graph (a) and (b) refer to the case of *n*=5 and 11 respectively.

minimum one-way delay of second packet already contains the queue latency caused by the corresponding first packet. So the retaining part of one-way delay is the queue latency came from cross traffic only, which term as CT latency. Fig.7 (a) and (b) show the CT latency of *FP* and *SP* when packet size is 700Bytes and the network load is 70%. Obviously, *SP* experiences more interference from cross traffic both in the case of *n*=5 and *n*=11 and the underlying interaction is more complex.

To analyze the interactions of probe and cross packets, we focus on a single link with capacity C only; the corresponding results of multiple links can be generalized simply. The duple $(AT(CP), DT(CP))$ denotes the arrival and departure time of cross packet CP and $I_{(t,s)}$ represents the total idle period in the time interval [t, s] when considering cross traffic only. As for probe packet, we use the same notation to denote its arrival and departure time.

Apparently, for FP(n), only the cross packets satisfied the condition $(AT(CP) < AT(FP(n)) < DT(CP))$, e.g. the cross packet arrive before FP(n) and depart after the arrival time of FP(n), will interact with FP(n).

For SP(n), however, the case is a bit more complex. The cross packets, similar to FP(n), satisfied the follow condition $(AT(CP) < AT(SP(n)) < DT(CP))$, will affect the process of SP(n). However, due to the effects of the first packet, the cross packet CP that arrives between FP(n) and SP(n) and departs before the arrival time of SP(n) when considering cross traffic only, will also delay the transmission of SP(n). In the case mixing the probe and cross traffic, the departure of CP can be postponed enough to delay the process of SP(n) if the experienced queue latency of CP from FP(n) is larger than the difference produced from subtracting $DT(CP)$ from the arrival time of SP(n). We use λ_1 and λ_2 to represent the latency and difference respectively. Equation (5) denotes the computation of λ_1, where $[x]^+ = \max(0, x)$.

$$\lambda_1 = [P/C - I(AT(FP), AT(FP))]^+ \tag{5}$$

A simple case is shown in Fig.8: the link is idle when FP(n) arrives; CP is the unique cross packet arrives between FP(n) and SP(n) and it departs before the arrival time of SP(n) when considering cross traffic only; however, the queue latency of CP caused by FP(n) is larger than the difference between the arrival time of SP(n) and the departure time of CP, e.g. $\lambda_1 > \lambda_2$. Thus, the actual departure time of CP, e.g. $DT(CP) + \lambda_1$, which is larger than the arrival time of SP(n), e.g. $DT(CP) + \lambda_2$.

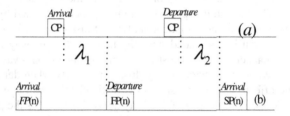

Fig. 8. An illustration of a simple interaction: (a) shows the arrival and departure time of CP when considering cross traffic only and (b) shows the cases of the nth pair. The interference incurs when $\lambda_1 > \lambda_2$.

The final observation is that the CP which had no interaction with SP(n) originally defers ultimately the transmission of SP(n) due to the adverse effects of corresponding FP(n).

The difference analyzed above argues the necessity to process the first and second packet separately. As a result, we advocate the "separating and reconstructing" procedure to characterize the packet-pair probes, which can be regarded as a generalization of the filtering mechanism described above. The basic procedure can be outlined as follows: First, the set of whole pair is separated as described in section 2. Then the corresponding analytic means are applied to the pair set and packet sets simultaneously. Finally, the estimates of interesting metrics are derived from the combination of the analysis results. Due to space limitations, more details about the generalized procedure will be presented in our future articles.

5 Related Work

There are many tools for measuring path capacity, either actively or passively. The per-hop methods [11], i.e. pathchar, are based on the observation on the equation (1) that the one-way delay is reversely proportional to link capacity. Thus, capacity can be inferred from the difference in delays of various probes with different sizes. In order to isolate the effects of cross traffic, a common filtering mechanism adopted is that only the packets with the minimum one-way delay are reserved to computing the capacity for every certain packet size. One of the drawbacks of per-hop methods is that the overheads is too large to be accepted since the probe and computation procedure (e.g. the linear regression analysis) should be repeated for all hops of measured path.

Another important category of tools for estimating end-to-end capacity is packet-pair based. Keshav [10] proposed the packet pair concept for use with Fair Queuing. Paxson [3] found that the distribution of packet-pair estimates is multi-modal, and Dovrolis etc. [2] showed that the true capacity is often different from the global modal. To address the effects of cross traffic, there have various proposal mechanisms: uses the minimum dispersion in a bunch of packet pairs; uses the global mode in the inter-arrival histograms [8] [16]; uses variable size packet pair [2]; selects the pair with the minimum sum of one-way delay to compute the capacity [6]; uses peak detection under a delay variation model [14], and so forth. The method not to filter out but to extract some useful information from the distorted origin data is also proposed in [15]. The mechanism presented in this paper, illumined by CapProbe [6], avoids the cost to executing the statistical analysis of distribution histogram.

Finally, capacity can be estimated with the help of the one-way or RTT metrics such as delay and packet loss etc. Most recent proposals are one-way based. Several researches [8] points out that the accuracy drops drastically when deploying the RTT metrics since the effects of the reverse path is difficult to be extracted or eliminated. One should be addressed in the use of one-way delay, however, is the problem of time synchronization. As discussed earlier, this problem has no effect on our scheme.

6   Conclusions

In this paper, we examine the significant difference among the first, second packet and the sample pair itself, which implies that the whole pair is commonly used as the minimal process unit should be regarded as one of the major factors contributing to the inefficiency of traditional packet-pair based measuring tools, especially in the heavy-loaded networks. Then we present an alternative scheme effort to address this issue by disposing the first and second packets separately. Preliminary simulations and Internet experiments show that our scheme is feasible and efficient. The underlying lightweight tool *CapEst* is robust for the increasing network load and exhibits promising performance in the tradeoff between accuracy and overhead. Finally we discuss the generalization of the filtering mechanism to characterize packet-pair probes in a novel aspect briefly, which is the focus of our future work.

Acknowledgements

This work is supported by National Natural Science Foundation of China under Grant No.60373023. All comments could be sent to the corresponding author Shaohe Lv (chi.shaohe@gmail.com). In addition, the authors are also very grateful to the valuable constructive comments of anonymous AINTEC reviewers. Thanks!

References

1. X. Liu, K. Ravindran and D. Loguinov. What Signals do Packet-Pair Dispersion Carry? In IEEE INFOCOM (2005).
2. C. Dovrolis, P. Ramanathan and D. Moore. Packet Dispersion Techniques and A Capacity Estimation Methodology. IEEE/ACM Transactions on Networking Dec. (2004)
3. V. Paxson. Measurement and Analysis of End-to-End Internet Dynamics. Ph.D. Dissertation, University of California, Berkeley, Apr.(1997).
4. R. Prasad, C. Dovrolis. M. Murray and Kc Claffy. Bandwidth estimation: metrics, measurement techniques, and tools. IEEE Network, November-December. (2003)
5. K. Harfoush, A. Bestavors and J. Byers. Measuring Bottleneck Bandwidth of Targeted Path Segments. In proceedings of IEEE INFOCOM, Apr.(2003).
6. R. Kapoor, L. Chen, L. Lao, M. Gerla and M. Sanadid. CapProbe: A Simple and Accurate Capacity Estimation Technique. In proceedings of SIGCOMM, Aug. (2004).
7. Kevin Lai and Mary Baker. Measuring Link Bandwidths Using a Deterministic Model of Packet Delay. In SIGCOMM (2000).
8. Kevin Lai and Mary Baker. Measuring Bandwidth. In proceedings of IEEE INFOCOM, New York, Apr.(1999).
9. K. Lai and M. Baker. Nettimer: A tool for Measuring Bottleneck Link Bandwidth. In USITS. (2001)
10. S. Keshav. A Control-theoretic Approach to Flow Control. In proceedings of SIGCOMM, Sept. 1991.
11. A. B. Downey. Using pathchar to Estimate Internet Link Characteristics. In proceedings of SIGCOMM, (1999).
12. NS version 2. Network Simulator. Http://www.isi.edu/nsnam/ns.
13. X. Liu, K. Ravindran, B. Liu and D. Loquinov. Single-Hop Probing Asymptotic in Available Bandwidth Estimation: Sample-Path Analysis. In IMC, (2004).
14. A. Paztor and D. Veitch. Active Probing Using Packet Quartets. In SIGCOMM IMW, Sep. (2002).
15. S. Katti, D. Katabi, C. Blake, E. Kohler and J. Strauss. MultiQ: Automated Detection of Multiple Bottleneck Capacities Along a Path. In IMC, Oct.(2004).
16. R. Carter and M. Crovella. Measuring Bottleneck Link Speed in Packet-Switched Networks. In ACM Performance Evaluation (1996).
17. S. Saroiu, P. Gummadi and S. Gribble. SProbe: Another Tool for Measuring Bottleneck Bandwidth. (2001).
18. K. Papagiannaki, S. Moon, C. Fraleigh, P. Thiran and C. Diot. Measurement and Analysis of Single-hop Delay on an IP Backbone Network. IEEE JSAC, Aug. (2003).
19. S. Lee, P. Sharma, S. Banerjee, S. Basu and R. Fonseca. Measuring Bandwidth between PlanetLab Nodes. In PAM, Apr. (2005).
20. R. Prasad, C. Dovrolis and B. A. Mah. The Effect of Layer-2 Store-and-forward devices on per-hop Capacity Estimation. In proceedings of IEEE INFOCOM, (2003).
21. Y. Zhang, N. Duffield, V. Paxson and S. Shenker. On the Constancy of Internet Path Properties. In ACM SIGCOMM IMW. Nov. (2001).

TCP Retransmission Monitoring and Configuration Tuning on AI³ Satellite Link

Kazuhide Koide[1], Shunsuke Fujieda[2],
Kenjiro Cho[3], and Norio Shiratori[1]

[1] Tohoku University
{koide, norio}@shiratori.riec.tohoku.ac.jp
[2] The University of Tokyo
sirokuma@k.u-tokyo.ac.jp
[3] IIJ
kjc@wide.ad.jp

Abstract. The AI³ project aims to achieve effective and fair bandwidth usage of its relatively narrow satellite links by all the partners in this project and the best possible end-to-end communication quality under limited resources. It is difficult to realize them as the operation of it is shared by all the partners with heterogeneous traffic demands and policies, as the network includes many traffic engineering(TE) facilities for traffic shaping or priority queuing in various points. In this paper we monitor the retransmission rate of TCP flows in a passive mode to obtain estimated packet loss rate of each flow and partner, and try to analyze the actual reason that affects end-to-end communications in a real-world environment. The results obtained in this paper is considered as a first step of reflecting TCP retransmission monitoring to improve operation, focusing on a certain partner, ITB, who has the largest traffic demand in AI³. The effectivity of the method is approved by the experimental configuration tuning.

1 Introduction

1.1 Overview of AI³ Project

The Asian Internet Interconnection Initiatives(AI³) project[1][2] is a research consortium for developing an international communication infrastructure in Asia. In this project, an experimental internetwork (called the AI³ network) with satellite links is deployed to directly interconnect many research groups in Asia (called AI³ partners).

The AI³ network consists of several point-to-point links (called BDL, Bi-Directional-Link) which interconnect Japan and each partner, and a wide-band UDL(Uni-Directional-Link). UDL is a broadcast link, and packets are transmitted from Japan via UDL and received by all the partners. Figure 1 shows the difference between BDL and UDL. The reason of setting UDL is 1)to meet the demand for high quality multimedia communication, and 2)to utilize the satellite link bandwidth effectively by aggregating traffic from Japan to the partners

K. Cho and P. Jacquet (Eds.): AINTEC 2005, LNCS 3837, pp. 282–295, 2005.

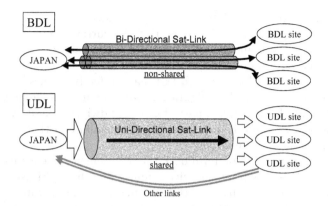

Fig. 1. Two types of satellite links: BDL and UDL

in a wideband link. For example, in multimedia communication, SOI-ASIA[3][4] sends real-time lecture video streams by IP multicast over UDL.

There are two types of partners in terms of connectivity in the AI³ network. An AI³ partner has a BDL toward Japan and an UDL receiver. On the other hand, there are partners called 'UDL partners' who do not have BDL but only UDL receiver. The AI³ partners have their own networks under its BDL (BDL site) and all the partners have networks under the UDL (UDL site). So an AI³ partner has two networks, a BDL site and an UDL site. The AI³ partners mainly use BDL for each direction of traffic (between Japan and the AI³ partners). But UDL partners have only UDL site. So traffic can be forwarded from JAPAN to UDL partners uni-directionally. The UDL partners use terrestrial links toward local ISPs for the reverse direction, from the UDL partners to Japan.

1.2 Characteristics of AI³ Network

A characteristic of the AI³ networks is the existence of many partners that have heterogeneous traffic demands, and their growing demands make the network heavily congested. UDL was established as a wideband link shared by all the partners because statistical multiplexing is more effective than distributing bandwidth to several narrow Point-to-Point links. But recently the UDL is always congested by the increased amount of traffic, and packet loss ratio is quite high in some periods.

In this situation, AI³ aims at establishing an operational scheme in order to achieve effective and fair utilization of the limited bandwidth. Here, 'effective' means to minimize underutilized bandwidth when congestion occurs. Both BDLs and UDL should be fully used under congestion. To achieve this goal, we have a "Policy Router" that enables flow-based traffic engineering in Japan. Regarding traffic from Japan to partners, when the amount of traffic on a BDL is higher than a threshold that is nearly equal to the capacity of the link, the excess traffic are distributed to UDL per micro flow.

And the meaning of 'fair' is to allocate bandwidth to the partners according to their heterogeneous traffic demands. But the current problem is to achieve fairness between flows, and between partners. There are lots of researches on fairness realization. Random Early Detection (RED)[5] is known as a pioneering work in this field. It realizes, though less than perfectly, fairness between TCP flows, but it needs detailed configuration and tuning. And it does not work between TCP and UDP flows. A part of UDL bandwidth is used by multimedia communications. So there is a link sharing problem[6] and the fairness among the partners should also be achieved.

It is quite difficult to accomplish both the requirements of effective usage and fairness service in an environment with a link shared by multicast and unicast, and individual policies and usage patterns of many partners. Traffic patterns can also be changed dynamically, because managers and users on each local network try to improve throughput with various tuning. As a result, the configurations of the AI^3 network as well as partners' have become complicated with many network components at many points.

1.3 Objective of This Research

Our objective is to establish an evaluation scheme for current situation and traffic, and to feedback to the network operation for more adequate fairness of bandwidth usage on UDL. An effective method for monitoring throughput of flows and for checking fairness among flows and partners is required for calibrations on operation.

Our goal in this paper is to achieve the first step of this objective. We are trying to achieve it by monitoring the retransmission rate of TCP flows passively, and getting information about end-to-end communication quality from the metric.

In this paper, we mainly focus on one of the AI^3 partners called *ITB*, Institute of Technology Bandung in Indonesia. ITB is connected with BDL, and spills out exceeding traffic to UDL. Also ITB utilizes UDL in parallel. Our measurement shows that there is a mismatch in quality between these two types of network utilization on UDL. ITB has the largest amount of traffic (utilizing over half of the entire bandwidth), and the behavior of ITB traffic is influential to the quality of the entire AI^3 network. In this paper, our analysis deeply focuses on ITB as a first step of achieving our objective.

A method to monitor TCP retransmission passively for measuring packet loss is also experienced in [7], but the objective is different. While [7] focuses on identifying lossy links in a large network, our work is on a certain network, especially with an UDL as a lossy link. Moreover, we focus on the packet loss rate of each TCP flow and each partner for achieving the fairness on their throughput.

In section 2, we describe the existing unfairness between networks on an AI^3 partner. In section 3, we mention the monitoring method of TCP retransmission. In section 4, we explain about our finding of a configuration that causes unfair bandwidth allocation and made an examination. In section 5, we conclude the work and discuss future works.

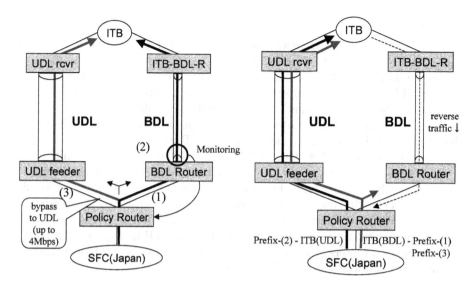

Fig. 2. Routing path from Japan to ITB

Fig. 3. The difference of routing configuration between ITB(BDL) and ITB(UDL)

2   Problem Statement

2.1   Remarkable Behavior in a Partner - ITB

ITB and Japan are interconnected by a BDL, and ITB has an UDL receiver also. On the BDL, the bandwidth from ITB to JAPAN(down link) is 1.5Mbps, and that from JAPAN to ITB(up link) is 0.5Mbps. The traffic that exceeds a threshold on up link is routed to UDL by the Policy Router. This configuration is shown in Figure 2. 1) Default route is defined as BDL router. 2) The policy router monitors the consumption of BDL, and 3) exceeding traffic are changed their next hop to the "UDL feeder" per flow up to 4Mbps.

There are various types of traffic on the UDL, so various traffic classes are configured on it. A mechanism of bandwidth limitation based on each traffic class is also employed so that it can not occupy the entire bandwidth. The bandwidth of UDL is distributed to them by HFSC (Hierarchical Fair Service Curve) on ALTQ (Alternate Queuing)[8]. The major classes are "Multicast", "Traffic of Policy Routing", and "Traffic of UDL Partners".

ITB has some blocks of IP addresses, they are (1) BDL site, (2) UDL site and (3) ITB own. Here, prefix-(1) and prefix-(3) are normally used for daily traffic. The traffic of those prefixes is, as explained above, normally routed to the BDL and policy-routed to the UDL if they exceeds the threshold. Here, we call traffic of prefix-(1) and (3) as *ITB(BDL)*. On the other hand, the traffic of prefix-(2) is directly routed to the UDL without policy routing. Here we call them as *ITB(UDL)*. On the reverse direction(from ITB to JAPAN), all traffic are carried on the BDL. Figure 3 shows those types of traffic and their route.

2.2   Arising Problem

Focusing on traffic on UDL, there is only one difference between ITB(BDL) and ITB(UDL); whether to be routed with policy or not. But recently it is observed that some users on ITB try to use prefix-(2) to use another bandwidth. Daily traffic of ITB are separated into two classes, policy routed traffic class and UDL partners' traffic class, and thus became difficult to handle the traffic.

From the operational point of view, throughput of ITB(BDL) and ITB(UDL) should be same in terms of achieving fairness. But currently, the throughput of ITB(UDL) seems higher than ITB(BDL), and the source of this difference is unknown because the situation is distinctly complex. Though those flows are passed through the same link, they are separately treated with different traffic classes. Each class includes traffic of other partners which share the same bandwidth. And the reverse direction on BDL, from ITB to Japan, is also congested. In addition to that there are some possibilities that ITB is shaping their traffic on some internal routers or some www cache servers. So an in depth investigation is required to find the source of the problem.

2.3   Metric

The reason that ITB(UDL) traffic is growing could be its higher throughput than ITB(BDL). To check that, we monitor throughputs of TCP flows on both ITB(UDL) and ITB(BDL) for comparison. From [9], it is possible to obtain a rough estimate of throughput from the packet loss rate (and RTT), by using an analytical model of TCP. But the focus of this method is limited to the bulk-transfer of TCP flow. Monitoring all throughputs of TCP flows directly is not quite effective because they are affected by factors such as the type of workloads or applications, as mentioned in [7].

One possible approach is viewing only packet loss rate of TCP flows, which is more fundamental metric than throughput. As monitoring of packet drops on all routers is unrealistic, ping from a node to another node is normally used for checking packet loss rate in large network. But ICMP based method can check only a single rate between two nodes with a specific protocol number. As our network includes some network components such as policy routing and traffic shaping on some routers, and ITB internal network may have another traffic shaping or priority queuing functions, ICMP packets and other traffic can be treated differently. Moreover, recent routers may be configured to process TCP packets and ICMP packets separately. Thus the packet loss rate monitored by ICMP does not always reflect the nature of TCP flows. We need a metric that reflects throughput of real TCP traffic.

3   TCP Retransmission Monitoring

It needs to identify the difference of throughput between flows of two subnets, ITB(BDL) and ITB(UDL), based on "monitoring TCP retransmission packets". In this section we explain the monitoring method and show the present state of retransmission statistics.

3.1   Monitoring Method

A TCP session is bidirectional, between a sender and a receiver, and contains two one-way TCP streams. Each TCP stream is unidirectional. So we focus on the TCP streams in the direction of UDL, and call them as *TCP flow*, or in short, *flow*. TCP flow, often referred to as a 'micro-flow', is defined using a 4-tuple (Source-IP, Destination-IP, Source-Port, Destination-Port).

A *retransmitted packet* has a sequence number in TCP header that is same as in a previous packet in the stream as shown in Figure 4. One packet can lead to multiple retransmitted packets. There are two types of retransmissions viz, data packet retransmission and TCP SYN retransmission. We do not distinguish between them in this paper.

The *passive monitor* captures all packets and generates flow status. The passive monitor detects retransmitted packet by matching the 4-tuple and the TCP sequence number seen in the captured packet with the previously monitored sequence numbers of the corresponding flow. If there is a match, the packet is counted as retransmitted.

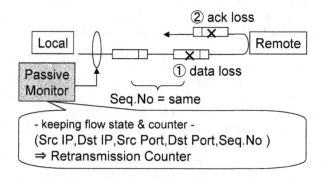

Fig. 4. Retransmission detection on passive monitoring

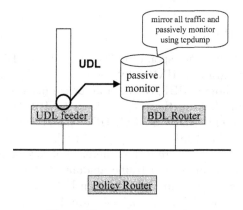

Fig. 5. The monitoring point

Bin is the time interval of analysis. A passive monitor outputs the number of retransmission packet for each flow per *bin*, and clear packet counters. But the passive monitor does not refresh flow entries including TCP sequence number records at the start of each *bin*. This is in order to detect all retransmitted packets. The length of a flow is independent of *bin*. A flow of long duration will bestride several *bins*. The flow state is cleared after the lifetime of flow elapsed from the arrival of last packet that matches the flow.

We set the analyzing interval, *bin*, to *5 minutes*. And the lifetime of flow is configured as *64 seconds*. The passive monitor is placed on the outbound interface of UDL feeder as shown in Figure 5. A port mirroring function is enabled on the interface of the switch, and copies all packets to the interface that the passive monitor connects.

3.2 Definition of Fairness/Effectiveness with Retransmission Rate

The notion of 'effectiveness' is defined using the UDL of *retransmission rate* as " when the UDL is not full, if the retransmission rate is relatively low, the usage of UDL will be called as effective".

And the notion of 'fairness' is defined using *retransmission rate* of each flows /partners. The rate is almost same for each flows, or partners, and it is called as 'fair' in the usage of UDL. In general it will be suitable for defining fairness to use an objective metric just like Jain's Index[10]. But in this paper, the examination is made about only two, ITB(BDL) and ITB(UDL). It will be sufficient to compare retransmission rate simply if a number of monitored partners is relatively small. So we use the ratio of retransmission rates, that is, *(retransmission rate of ITB(BDL))/(retransmission rate of ITB(UDL))*.

This definition may not be very correct because retransmission does not always occur due to packet loss on UDL. One such case is packet reordering. But the possibility of this is low because there is no multiple links where the delay is largely different. Another case is the loss of ACKs on the reverse path. The cumulative nature of TCP ACKs would mitigate, although not eliminate, this problem. We consider a retransmission rate as an *estimated packet loss rate*.

3.3 Monitoring Result

There are three lines that correspond to ITB(BDL), ITB(UDL), and UNIBRAW traffic each other in graphs. 'UNIBRAW' is one of "UDL partners" that only uses UDL link. We show UNIBRAW's graph for comparison with other two. It does not use satellite link on the reverse (from UNIBRAW to JAPAN) path. The reverse traffic of ITB(BDL) and ITB(UDL) pass through satellite BDL link but UNIBRAW uses local ISP's aboveground path for reverse traffic. We can evaluate the effect that comes from the difference of reverse path.

Figure 6 shows the statistics of traffic amount of these three sites, and the estimated packet loss rate comparison between ITB(BDL) and ITB(UDL). We start the monitoring from *1 May, 2005*. There are three typical set of graphs. It is clear that the flow count of ITB(UDL) is increasing over time. The traffic amount

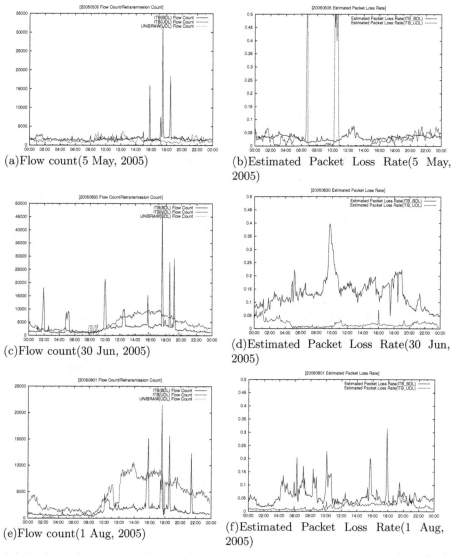

Fig. 6. Monitoring result of Flow count / Estimated packet loss rate monitoring result on some typical days

of ITB(UDL) overtakes ITB(BDL) after *18 May, 2005*. This phenomenon does not suit to the routing policy of AI^3 network. On the contrary, the estimated packet loss rate of ITB(UDL) is always lower than ITB(BDL), as shown in Figure 6-(d)(f). This means ITB(UDL)'s traffic receives a better communication environment as expected.

Figure 7-(a) shows the number of packets, (b) is the number of retransmitted packets of three types of traffic, and (c) shows the estimated packet loss rate from

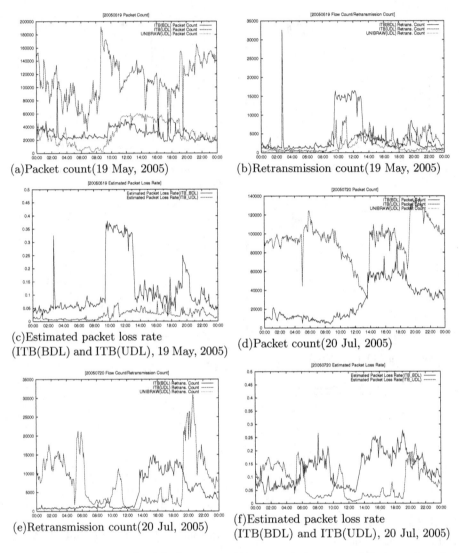

Fig. 7. Monitoring result of Packet count / Retransmission count / Estimated packet loss rate

the packet count and retransmission count, on *19 May, 2005*. This is a typical result when the number of packets in ITB(UDL) is larger than ITB(BDL) and UNIBRAW. Especially from 9:00 to 13:00 ITB(BDL) suffers from too many retransmitted packets rather than other two, and also the temporal pattern of packet loss rate is quite different from ITB(UDL) in this period. On the other hand Figure 7-(d)(e)(f), the result on *20 July, 2005*, shows the specific phenomenon. After 19:00 the traffic amount of ITB(UDL) is quite large, and the number of retransmission and the estimated packet loss rate on ITB(UDL) overtakes ITB(BDL)'s.

Figure 8 shows a deeper insight of these periods. These graphs plot the relationship of total packet count on x-axis, and retransmission count on y-axis, of each flow.

Figure 8-(a) shows the period of non-congestion, 6:05-6:10 on *19 May, 2005*. There is little difference in distribution between flows of three subnets. But in Figure 8-(b) at the period of heavy-congestion, i.e. 10:00-10:05 on *19 May, 2005*, it shows a clear difference between ITB(BDL) and the other two. This result shows that many of ITB(BDL)'s flows suffer from very high rate of retransmission, and the distribution of packet/retransmission is quite similar between flows. On the other hand, the retransmission of ITB(UDL) and UNIBRAW is relatively low, and the distribution is not much different from each other. This seems to be the effect of traffic shaping for ITB(BDL) at UDL-feeder. Figure 8-(c), between 20:10-20:20 on *20 July, 2005*, shows that the distribution of retransmission is almost same for ITB(BDL) and ITB(UDL), and the ITB(UDL) will overtake ITB(BDL) at the end (Figure 8-(d)). It seems that UDL-feeder shapes traffic also of ITB(UDL).

The conclusion from these monitoring results is that the typically estimated packet loss rate of ITB(UDL) is lower than ITB(BDL) in spite of traffic amount except some particular periods when ITB(UDL)'s traffic amount is quite large and the number of retransmission increases appreciably. From the retransmission distribution, it is thought that the traffic shaping is realized by UDL-feeder, strongly effects to those phenomena.

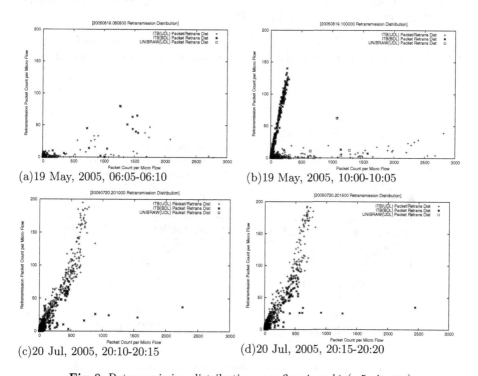

(a) 19 May, 2005, 06:05-06:10

(b) 19 May, 2005, 10:00-10:05

(c) 20 Jul, 2005, 20:10-20:15

(d) 20 Jul, 2005, 20:15-20:20

Fig. 8. Retransmission distributions per flow in a *bin*(=5minutes)

4   Experimental Operation

4.1   Hypothesis About a Cause

There are some causes that can be thought. In this section, we investigate how traffic shaping on UDL-feeder effects to the unfairness of estimated packet loss rates.

Traffic shaping is realized by HFSC(Hierarchical Fair Service Curve) on ALTQ (Alternate Queuing)[8]. The detailed configuration of traffic shaping is in Figure 9. It allows "policy routed traffic from BDL"(ITB(BDL), etc.) to use 3Mbps + 40% of residual bandwidth. On the other hand it allows "UDL prefix traffic"(ITB(UDL),UNIBRAW, etc.) to use 1Mbps + 30% of residual bandwidth.

This configuration makes each of them not to occupy UDL bandwidth. But it seems to be a cause of problem in this case, that is, it seems that other UDL partner's traffic is relatively low and only ITB(UDL) gets good throughputs in the "UDL prefix traffic" class, while the traffic of "policy routed traffic from BDL" is relatively high. It is considered that consequently this configuration is not effective for handling traffic of ITB(BDL) and ITB(UDL).

4.2   Experiment and Result

The change in experimental configuration is:

– merge ITB(UDL)'s prefix into the traffic shaping class of "policy routed traffic from BDL(same with ITB(BDL)).

The traffic of ITB(BDL) and ITB(UDL) is similarly processed in accordance with the traffic shaping at UDL-feeder. The change is enabled at 14:00 on *27 August, 2005.*

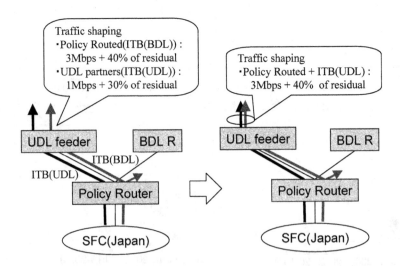

Fig. 9. The configuration of traffic shaping on UDL-feeder and the change

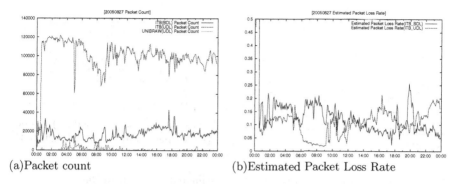

(a)Packet count   (b)Estimated Packet Loss Rate

Fig. 10. Packet count / Estimated packet loss rate monitoring result after configuration change(27 Aug, 2005)

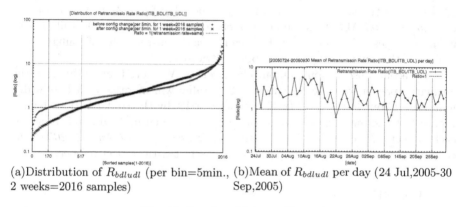

(a)Distribution of R_{bdludl} (per bin=5min., (b)Mean of R_{bdludl} per day (24 Jul,2005-30
2 weeks=2016 samples)   Sep,2005)

Fig. 11. Results of fairness(R_{bdludl})

We have to evaluate changes in fairness and effectiveness. As mentioned above, we compare the ratio of estimated packet loss rate between ITB(BDL) and ITB(UDL), R_{bdludl} =loss rate of ITB(BDL)/loss rate of ITB(UDL). If R_{bdludl} is nearly 1 it can be considered as fair. Here we calculate R_{bdludl} per *bin*, that is 5 minutes. And about the effectiveness, we evaluate changes of estimated packet loss rate. In simple, loss rate decreases, the effectiveness are thought to be increased. At first, fairness is considered. The packet count and estimated packet loss rate of *27 August, 2005* is shown in Figure 10. After the change, the estimated packet loss rate of ITB(BDL) and ITB(UDL) reverses. Figure 11-(a) shows the distribution in R_{bdludl}. We calculate R_{bdludl} of data from 2 weeks before *27 August, 2005* and from 2 weeks after. So there are 2016 data plotted in each graph. It is revealed that the number of R_{bdludl} under 1 increases from 170 to 517. This means that the estimated loss rate of ITB(UDL) start to overtake ITB(BDL) more frequently than before. It can be said that an imbalance in fairness between ITB(BDL) and ITB(UDL) is comparatively reduced.

But from Figure 11-(a) it can be claimed that there are still many cases where R_{bdludl} is too high. It means that ITB(BDL) suffers from quite high packet loss

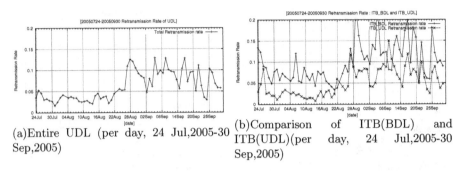

(a)Entire UDL (per day, 24 Jul,2005-30 Sep,2005)

(b)Comparison of ITB(BDL) and ITB(UDL)(per day, 24 Jul,2005-30 Sep,2005)

Fig. 12. Results of effectiveness(Estimated Packet Loss Rate)

compare to ITB(UDL) in many periods, similar to the previous case. Figure 11-(b) shows the statistics of mean R_{bdludl} per day from *24 July, 2005* to *30 September, 2005*. Middle is the point of changing configuration. Overall pattern in fairness does not improve largely. It can be said that the effect of configuration change in this examination 'exists' but 'restrictive'.

The Second observation is about effectiveness. We note that in this case the estimated packet loss rate in entire ITB's traffic will increase as a bandwidth limitation about ITB's total traffic becomes strict by the configuration change. Figure 12-(a) shows the estimated packet loss rate in entire UDL, and (b) shows each subnets. As expected, each of them increases after *27 August, 2005*. The configuration change in this paper has a bad effect with respect to effectiveness.

4.3   Consideration

We assume the effect of traffic shaping for this experiment. But the result of this experiment is restrictive in improvement of fairness. So other possible causes should be considered about the high packet loss rate of ITB(BDL). Figure 13 shows a sample (*08 September, 2005*) of the packet count and the estimated packet loss rate after the configuration change. During this period when the

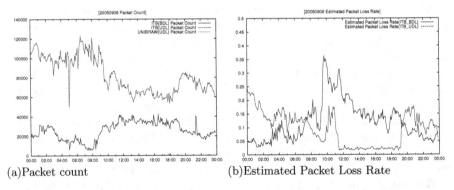

(a)Packet count

(b)Estimated Packet Loss Rate

Fig. 13. Packet count / Estimated packet loss rate monitoring result in 08 Sep, 2005

packet count of ITB(BDL) is nearly 40,000, the increase in rate of packet loss seems quite large irrespective of the decrease in packet loss rate of ITB(UDL). One of expected causes is the existence of a network component that realizes traffic shaping, independently of UDL-feeder that may exist inside ITB. Another possible cause is the heavily loaded www cache server in ITB.

The experiment done in this paper is not sufficient enough to fully resolve the problem related to ITB. Further investigation and examination is needed in this regard to get a clear view to overcome this problem.

5   Conclusion

We monitored the retransmission rate of TCP flows passively on UDL in AI^3 network for several months, and we got information about quality of each flows and partners. We found that there is unfairness in retransmission rate between traffic of ITB(BDL) and ITB(UDL). Next we considered about affects of traffic shaping on UDL-feeder and experimentally changed the configuration, then considerable improvement in fairness was achieved as a result. More thorough investigation is required but we made sure that our method helps improvement of operation, by experimental network tuning and analysis of the results.

Our objective is to establish an evaluation scheme for current situation and traffic, and offer feedback to the network operation for more adequate fairness of bandwidth usage on UDL. Developing (semi-)automatic configuration calibration system for operation would be one of our future works.

References

1. AI^3 Project http://www.ai3.net/
2. Suguru Yamaguchi, and Jun Murai, "Asian Internet Interconnection Initiatives," In Proceedings of INET'96, Jun. 1996.
3. SOI Asia Project http://www.soi.wide.ad.jp/soi-asia/
4. Shoko Mikawa, Keiko Okawa and Jun Murai, "Establishment of a lecture environment using Internet Technology over satellite communication in Asian Countries," In Proceedings of SAINT 2003 Workshop, Jan. 2003.
5. Sally Floyd, and Van Jacobson, "Random Early Detection Gateways for Congestion Avoidance," IEEE/ACM Transactions on Networking, Volume 1, Issue 4, Page:397–413, Aug. 1993.
6. Sally Floyd, and Van Jacobson, "Link-sharing and Resource Management Models for Packet Networks," IEEE/ACM Transactions on Networking, Volume 3, Issue 4, Page:365–386, Aug. 1995.
7. V.N.Padmanabhan, L.Qiu, and H.J.Wang, "Server-based Inference of Internet Link Lossiness," In Proceedings of IEEE INFOCOM 2003.
8. Kenjiro Cho, "The Design and Implementation of the ALTQ Traffic Management System," Ph.D thesis, Keio University, January 2001.
9. J.Padhye, V.Firoiu, D.Towsley, and J.Kurose, "Modeling TCP Throughput: A Simple Model and its Empirical Validation," In Proceedings of ACM SIGCOMM'98, August 1998.
10. R.Jain, "The Art of Computer Systems Performance Analysis," John Wiley and Sons, 1991.

An Analytical Evaluation of Autocorrelations in TCP Traffic

Georgios Rodolakis and Philippe Jacquet

INRIA and Ecole Polytechnique,
France
{Georges.Rodolakis, Philippe.Jacquet}@inria.fr

Abstract. This paper addresses the problem of characterizing analytically the autocorrelation structure of TCP traffic. We show that, under simple models, a single TCP connection generates traffic with an exponentially decreasing autocorrelation function. However, several TCP connections sharing a given link can generate traffic with long term dependencies, under the condition that the distribution of their round trip delays is heavy tailed.

1   Introduction

It is a well known observation that, in many cases, traffic in the internet displays long term dependencies [1]. In this paper, we argue that one of the possible causes of this property is the combination of the widely used over the internet TCP protocol [2] and heavy tailed round trip delays. We are able to gain more insight in TCP traffic correlations by extending previous analytic derivations of TCP performance [3, 4, 5].

By definition, a stationary process has long term dependencies when its autocorrelation function is non-summable. More precisely, let $I(t)$ be the intensity of traffic at time t, and $C(x)$ the covariance of $I(t)$ and $I(t + x)$ when t varies. Function $C(x)$ corresponds to the autocovariance of the traffic intensity and does not depend on t because the process is stationary:

$$C(x) = E[I(t)I(t + x)] - (E[I(t)])^2$$

We will say that the traffic has long term dependencies when $C(x) \sim Bx^{-\beta}$, with $0 < \beta < 1$.

One of the many undesirable effects of long term dependencies stands in the loss rates in the buffers. Long term dependent traffics generate loss rates that are inverse power functions of the buffer size, contrary to the exponential function of the buffer size in Poisson models. Consequently, the increase in buffer size is prohibitively expensive to significantly reduce the packet loss rate.

The rest of the paper is organized as follows.

In Section 2, we present a model for multi-user TCP and we summarize the analytic derivation of TCP performance. We then use the fact that, according to the previous analysis, the remaining space in the buffer tends to be exponentially

K. Cho and P. Jacquet (Eds.): AINTEC 2005, LNCS 3837, pp. 296–306, 2005.

distributed in order to derive a simplified model of a single TCP connection with fixed packet loss probability.

Based on this model, in Section 3, we characterize the autocorrelation function of a single TCP connection via an estimate of the second eigenvalue of the Markov transition matrix.

In Section 4, we investigate the autocorrelation function of several independent TCP connections with heavy tailed round-trip delays. We argue that heavy tailed round trip delays result in heavy tailed traffic autocorrelation and we present a comparison between analytic and simulated results.

2 TCP models

2.1 Multi-user TCP

We consider a large number N of parallel TCP connections towards a bottleneck router with a finite buffer of capacity B connected to a slow network interface with service rate μ (see figure 1). In particular, the network is divided into a local loop with relatively low speed and a backbone with high throughput, hence we assume that all congestion occurs in the bottleneck buffer. A detailed description and analysis of this model can be found in [3, 6].

We assume that the transmitted files are infinite. We also assume that all packets in a TCP window arrive in a single batch. The round trip time (RTT) has a fixed component $\frac{B}{\mu}$ (the maximum sojourn time in the buffer) and a random exponential delay of mean $\frac{N}{\lambda}$, assumed to be much smaller than $\frac{B}{\mu}$.

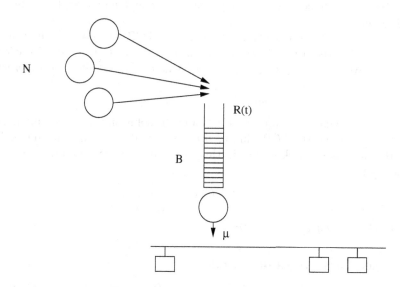

Fig. 1. N TCP connections towards the same bottleneck buffer

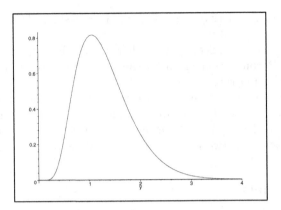

Fig. 2. Limiting function $g(x)$ of the window size distribution

In such a model the buffer is expected to be full most of the time. Indeed, if we call $R(t)$ the available room in the buffer at time t, we have $R(t) = O(1)$ when $N \to \infty$. This result is derived from the analysis in [3, 6], where it is shown that, when the system is stationary, the probability that the available room on top of the buffer $R(t)$ exceeds x has the expression:

$$P(R(t) \geq x) \to \exp(-ax)$$

for some constant $a > 0$, which corresponds to the probability of the buffer being full.

This observation is used in the following section to derive a simplified model where there is a fixed packet loss probability a and moreover the losses are independent.

We denote $w(y)$ the density function of the TCP window size distribution when the system is in steady state. When $a \to 0$ and the round trip delay is large, we have $w(y) = \sqrt{a}g(\sqrt{a}y) + O(a)$, where $g(y)$ satisfies the differential equation:

$$yg(y) + g'(y) = 4yg(2y)$$

This equation can be explicitly solved and the solution is depicted in figure 2. In passing, note that the TCP window sizes vary greatly although all connections share the same bottleneck buffer. In this case, we have the asymptotic estimate for parameter a:

$$\sqrt{a} = (1 + O(\sqrt{a})) \frac{g^*(2)}{\frac{B}{N} + \frac{\mu}{\lambda}}$$

with $g^*(2) = \int_0^\infty yg(y)dy \simeq 1.3098$.

2.2   Fixed Loss Probability Model

We now derive a simplified model of a single TCP connection, in order to analyze the autocorrelation function of TCP traffic. We assume that the packet loss probability is constant and equal to a, and that the losses are independent. This is justified by the model of the previous section where a is constant. For simplicity, we turn to a discrete time model, where the unit is the RTT (we assume that we collect all buffer changes within one RTT), which does not vary much since it is composed of a large fixed delay and a much smaller random processing time. Hence, we can model the window size adaptation with a Markov chain. A similar model has been studied via simulation in [5].

We study the behavior of a TCP connection transferring an infinite file, during the congestion avoidance phase. If there is a packet loss in a window transmission, the retransmission of packets resumes from the first lost packet but with a window size divided by two: $W = \lfloor W/2 \rfloor$, otherwise the window size increases by one: $W = W + 1$. We note that whether there are more losses or not after the first loss in a window is of no importance in our model.

We deduce the following probabilities for $a \ll 1$:

$$Pr(\text{no loss in a window}) = (1 - a)^W \approx e^{-aW} \tag{1}$$
$$Pr(\text{loss in a window}) = 1 - (1 - a)^W \approx 1 - e^{-aW} \tag{2}$$

We now define the TCP discrete time Markov chain. The states are the window sizes and the transition probabilities can be calculated from (1) and (2). The resulting Markov chain is depicted in figure 3. In reality, the receiver announces a maximum window size, which means that the Markov chain is finite. As the tendency in the Internet is towards increasing this value, we will ignore it in the analysis. We suppose that it corresponds to a very large window size, which is never achieved in practice.

We denote π_i the stationary probability of state i. In steady state, the following equations hold when $a \ll 1$:

$$\pi_1 = a\pi_1 + (1 - e^{-2a})\pi_2 + (1 - e^{-3a})\pi_3 \tag{3}$$
$$\pi_k = e^{-a(k-1)}\pi_{k-1} + (1 - e^{-a(2k+1)})\pi_{2k+1} + (1 - e^{-a2k})\pi_{2k}, k \geq 2 \tag{4}$$

Let $\mathbf{P}$ be the transition matrix of the TCP Markov chain. It is easy to see that $\mathbf{P}$ is stochastic, irreducible and aperiodic. Thus, according to the Perron-Frobenius theorem, its dominant eigenvalue is 1 with corresponding right eigen-

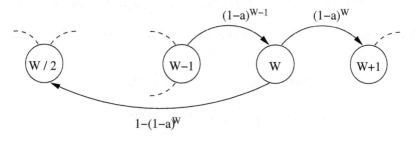

Fig. 3. Marcovian TCP model

vector $\boldsymbol{\pi}$, which gives the stationary distribution of the Markov chain. We can also deduce that $\lim_{n \to \infty} \mathbf{P}^n = \boldsymbol{\pi} \cdot \mathbf{1}$, where $\mathbf{1}$ is a line vector of ones.

The matrix $\mathbf{P}$ is of the following form:

$$\mathbf{P} = \begin{pmatrix} 1-e^{-a} & 1-e^{-2a} & 1-e^{-3a} & 0 & \cdots \\ e^{-a} & 0 & 0 & 1-e^{-4a} & \\ 0 & e^{-2a} & 0 & 0 & \\ & 0 & e^{-3a} & 0 & \\ & & 0 & e^{-4a} & \cdots \\ \vdots & & & \vdots & \ddots \end{pmatrix} \tag{5}$$

If the initial distribution of windows is $\boldsymbol{\pi}(0)$, then after n RTT's: $\boldsymbol{\pi}(n) = \mathbf{P}^n \boldsymbol{\pi}(0)$. From the spectral decomposition of the matrix, we get that the convergence rate is exponential:

$$\|\boldsymbol{\pi}(n) - \boldsymbol{\pi}\| = O(\rho^n) \tag{6}$$

for every ρ such that $|\lambda_2| < \rho < 1$, where λ_2 is the second largest eigenvalue of $\mathbf{P}$.

3   Autocorrelation Function of a Single TCP Connection

3.1   Analysis Under the Fixed Loss Probability Model

The traffic intensity $I(t)$ at time t is a function of the present window size $W(t)$ and the round trip time, which we denote as D:

$$I(t) = \frac{W(t)}{D} \text{ packets/second} \tag{7}$$

We calculate the traffic autocovariance function $C(x)$ at time t:

$$C(x) = \mathrm{cov}[I(t), I(t+x)] = E[I(t)I(t+x)] - (E[I(t)])^2 \tag{8}$$

As we are dealing with a stationary process, $C(x)$ does not depend on time t, so we can fix $t = 0$. If we express the time in RTT multiples n, the first term of (8) becomes:

$$E[I(t)I(t+nD)] = E[I(0)I(nD)] = \frac{1}{D^2} E[W(0)W(nD)] \tag{9}$$

We continue by supposing that the initial window size is k and averaging on all k:

$$E[I(0)I(nD)] = \frac{1}{D^2} \underset{k}{E}[E(W(nD)k|W(0) = k)]$$

$$= \frac{1}{D^2} \sum_k k E(W(nD)|W(0) = k)\boldsymbol{\pi}(0)_k \tag{10}$$

where $\boldsymbol{\pi}(n)_i$ is the probability of the window being of size i after n RTT's.

We want to calculate the autocovariance of a system that has reached equilibrium, so we can assume that the initial distribution $\pi(0)$ is the stationary distribution π:

$$E[I(0)I(nD)] = \frac{1}{D^2} \sum_k k E(W(nD)|W(0) = k)\pi_k$$

$$= \frac{1}{D^2} \sum_k k \sum_l l \, (\mathbf{P}^n \mathbf{1}_k)_l \pi_k \qquad (11)$$

where $\mathbf{1}_k$ is a column vector with all entries being 0 except for entry k which is 1.

If we define $\mathbf{p}$ as a column vector such that $\mathbf{p}_i = i\pi_i$, and $\mathbf{u}$ as a line vector such that $\mathbf{u}_i = i$, then:

$$E[I(0)I(nD)] = \frac{1}{D^2} \sum_l l \, (\mathbf{P}^n \mathbf{p})_l = \frac{1}{D^2} \mathbf{u}(\mathbf{P}^n \mathbf{p}) \qquad (12)$$

For $n \to \infty$ we have:

$$\frac{1}{D^2} \mathbf{u}(\lim_{n\to\infty} \mathbf{P}^n \mathbf{p}) = \frac{1}{D^2} \sum_l l \left(\lim_{n\to\infty} \mathbf{P}^n \mathbf{p} \right)_l = \frac{1}{D^2} \sum_l l \, (\pi \cdot \mathbf{1} \cdot \mathbf{p})_l$$

$$= \frac{1}{D^2} \sum_l l(\pi_l \sum_k k\pi_k) = (E[I(0)])^2 \qquad (13)$$

Combining equations (8), (12) and (13) we get the autocovariance function:

$$C(nD) = E[I(0)I(nD)] - (E[I(0)])^2$$

$$= \frac{1}{D^2} \mathbf{u} \left(\mathbf{P}^n \mathbf{p} - \lim_{n\to\infty} \mathbf{P}^n \mathbf{p} \right) \qquad (14)$$

From the spectral decomposition of $\mathbf{P}$, as in (6), we conclude that the autocovariance function decreases exponentially with rate $O(\rho^n)$ for all ρ such that $|\lambda_2| < \rho < 1$. By normalizing we obtain the autocorrelation function, which is also $O(\rho^n)$.

The next step in our analysis is to calculate the eigenvalues of $\mathbf{P}$ for a large scale of error rates. The calculations are made by fixing a maximal window size of 1000 packets, thus truncating the matrix $\mathbf{P}$. The probability of reaching the maximum window size is close to 0 for the error rates we consider, so ignoring larger values does not affect significantly our results.

In the equivalent continuous model of TCP, we scaled window sizes by a factor $\frac{1}{\sqrt{a}}$ to obtain the limit distribution. We expect to find a similar factor in the autocovariance function. This observation leads us to approximate the spectral gap of the TCP Markov chain by expression $C\sqrt{a}$, where C is a constant. In figure 4, we compare the calculated spectral gap values, for different error rates a, and the values corresponding to the proposed approximation for $C = 1.6$. In all our calculations, the second eigenvalue is real, of multiplicity 1, which means that the autocovariance is $O(\lambda_2{}^n)$.

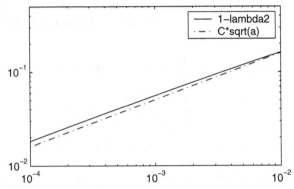

Fig. 4. Spectral gap of the TCP Markov chain for different error rates a

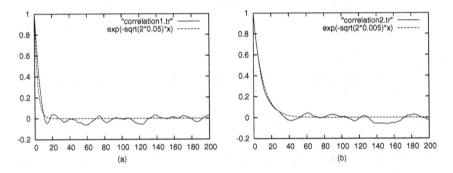

Fig. 5. TCP traffic autocorrelations for error rates (a) $a = 0.01$, (b) $a = 0.005$. The time unit is the RTT.

More precisely, we have the following first order approximation:

$$C(nD) = \frac{A}{D^2}\lambda_2{}^n \approx \frac{A}{D^2}e^{-C\sqrt{a}n} \tag{15}$$

where A is another constant.

Observe that a very small error rate results in a second eigenvalue close to 1, meaning that the autocorrelation function decreases very slowly (although exponentially).

3.2 Simulations of a Single TCP Connection

To verify that our simplified model can predict the autocorrelation of a real TCP connection, we conducted a number of simulations with the network simulator ns2 [7]. In the simulations, the RTT is mainly caused by the link delays and the error rate is constant and due to a loss agent. By measuring the number of packets transmitted during an interval equal to the RTT, we obtain the traffic intensity $I(t)$. We then calculate the covariance of $I(t)$ and $I(t+x)$ when t varies and we normalize to obtain the traffic autocorrelation.

We present the results for two models with different error rates, compared to the calculations of the previous section. The duration of these simulations is 1000 seconds of simulated time, and we start making measurements after waiting for the system to stabilize for 100 seconds. In figure 5 we draw the autocorrelation for error rates $a = 0.01$ and $a = 0.005$. The time unit is the round trip delay, which in this case is equal to 100ms.

The oscillations are due to the finite duration of the simulations.

4   Long Term Dependencies

In this section we will show that long term dependencies can arise from heavy tailed round trip delays. The remaining question is to figure out whether or not the round trip delays are heavy tailed. Recently it has been discovered that the internet topology contains numerous heavy tailed features. Among them are the router degree, router reachability degree and the length of paths inside the internet [8, 9, 10]. In [11] there is evidence that the RTT distribution is also heavy tailed, *i.e.*, the RTT complementary cumulative density function $P(RTT > x)$ corresponds to a power law with exponent approximately 1.5. Equivalently, the rank of the RTTs follows a power law of exponent 2/3.

In the first subsection we show that a link shared by several TCP connections with round trip times with a heavy tailed distribution generates long term dependence. In the second subsection we provide simulations to compare with the theoretical results.

4.1   Autocorrelation of Several TCP Connections with Heavy Tailed Round Trip Delays

We consider that the link is shared by several TCP connections with different round trip delays, so that the round trip delays distribution is heavy tailed (see figure 6). To simplify, we assume there is an infinite sequence of TCP connections and the connection with sequence number i has a round trip time equal to $D_i = Di^\beta$ for $\beta > 0$ and D fixed, while the error rate a remains the same. Of course the analysis can also be carried out with all parameters varying. Since the TCP connections are assumed to be independent, the autocovariance function of the aggregated traffic is equal to the sum of the autocovariance functions of the individual connections:

$$C(x) = \sum_{i=1}^{\infty} C_i(x) \qquad (16)$$

It comes from (15) and the RTT distribution that:

$$C(x) \approx \frac{A}{D^2} \sum_{i=1}^{\infty} i^{-2\beta} \exp(-C\frac{x}{D}i^{-\beta}) \qquad (17)$$

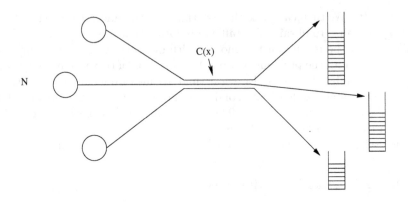

Fig. 6. Several TCP connections sharing the same link

Function $C(x)$ is a harmonic sum generated from function $\exp(-C\frac{x}{D})$. The asymptotic analysis of such sums can be easily performed with the use of the Mellin transform [12]. In particular, we need to determine the definition domain and the singularity set of the Mellin transform $C^*(s)$ of function $C(x)$, defined for appropriate complex numbers s by:

$$C^*(s) = \int_0^\infty x^{s-1}C(x)dx \tag{18}$$

From (17) we obtain (*cf.* [12]):

$$C^*(s) = \frac{A}{D^2}\left(\frac{C}{D}\right)^{-s}\zeta((2-s)\beta)\,\Gamma(s) \tag{19}$$

where $\zeta(s) = \sum_{i=1}^\infty i^{-s}$ is Euler's *zeta* function and $\Gamma(s)$ is Euler's *Gamma* function ($\Gamma(s) = \int_0^\infty x^{s-1}e^{-x}dx$).

The function $C^*(s)$ converges for all s such that $\Gamma(s)$ and $\zeta((2-s)\beta)$ converge. Quantity $\zeta((2-s)\beta)$ has a simple pole at $s = 2 - \frac{1}{\beta}$. Therefore $C^*(s)$ is defined on the strip $0 < \Re(s) < 2 - \frac{1}{\beta}$. The classical results on the Mellin transform state that there exist B and $\varepsilon > 0$ such that:

$$C(x) = Bx^{\frac{1}{\beta}-2}(1 + O(x^{-\varepsilon})) \tag{20}$$

when $x \to \infty$ [12]. This implies that the traffic autocorrelation function decays following a power law with exponent $\frac{1}{\beta} - 2$, and there are long term dependencies when $\frac{1}{2} \le \beta \le 1$.

However, if the number of TCP connections is finite, we expect to observe a heavy tailed behavior for a finite time scale, which is a multiple of the largest RTT in the system. In the next section, we will see that, even for a small number of connections, this upper bound can be quite large.

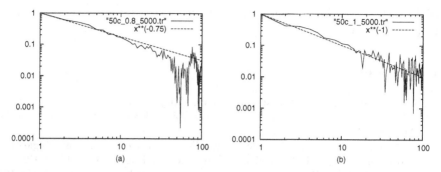

Fig. 7. Traffic autocorrelations of 50 TCP connections with heavy tailed RTT distributions (a) $D_i = 40i^{0.8}$, (b) $D_i = 40i$. The time unit is 500ms.

4.2   Simulations of Several TCP Connections

We have simulated with ns2 the case of many TCP connections with heavy tailed RTT's sharing the same link. The simulation models 50 clients downloading very large files from different servers. The traffic of all connections traverses a shared link of capacity 100 Mbps. Each client is connected to the shared link via a slow link at 10 Mbps and, at the other end, each server is connected via a private link at 1 Gbps. The fixed propagation delays in these private links are chosen to follow a power law. The traffic measurements are made on the shared link in intervals of 500 ms. The packet loss rate is $a = 0.001$ for each connection. The packet size is 1KB and the maximum window size is 1000 packets.

In figure 7 we draw the autocorrelation functions in log-log scale. The dashed line shows the theoretical power law decrease. The RTT distributions are of the form $D_i = 40i^{\beta}$, with $\beta = 0.8$ in (a) and $\beta = 1$ in (b). According to the theoretical analysis, the expected power law exponents for the traffic autocorrelations are $\frac{1}{\beta} - 2$, that is -0.75 and -1, respectively.

The duration of the simulations is 5000s and the measurements start after a stabilization period of 100s.

The fluctuations in figure 7 are due to the finite duration of the simulations and the fact that there are RTT's which are larger than the time unit. The exact characterization of TCP autocorrelation in a finer time scale is a subject of further research.

5   Conclusion

We have shown via analytic means that TCP traffic from a single connection cannot generate long term dependencies. However, several connections with heavy tailed round trip delays can generate long term dependencies. We argue that such a distribution is plausible in the internet and we point to supporting experimental evidence. Finally, although the analysis was performed in the asymptotic case where the number N of parallel TCP flows tends to infinity, the simulations of the system show a good agreement with the analytical results, even when the number N is rather small.

References

1. K. Park, W. Willinger, Self-similar traffics, Wiley, 2000.
2. V. Jacobson, Congestion avoidance and control, in: Proc. of ACM SIGCOMM '88, August 1988.
3. C. Adjih, P. Jacquet and N. Vvedenskaya, Performance evaluation of a single queue under multi-user TCP/IP version 2, INRIA Research report RR-4478, 2002.
4. F. Baccelli, D. McDonald and J. Reynier, A mean field model for multiple TCP connections through a buffer implementing RED, Perform. Eval. 49(1/4): 77-97 (2002).
5. D.R. Figueiredo, B. Liu, V. Misra, and D. Towsley, On the autocorrelation structure of TCP traffic, Computer Networks Journal, Special Issue on Advances in Modeling and Engineering of Long-Range Dependent Traffic, 2002.
6. C. Adjih, P. Jacquet, G. Rodolakis and N. Vvedenskaya, Performance of multiple TCP flows: an anlytical aproach, INRIA Research report RR-5417, 2004.
7. UCB/LBNL/VINT Network Simulator - ns2, http://www.isi.edu/nsnam/ns, 2001.
8. T. Bu, D. Towsley, On distinguishing between internet power law topology generators, INFOCOM 2002.
9. J. Chuang, M. Sirbu, Pricing multicast communication: a cost based approach, INET 1998.
10. C. Adjih, L. Georgiadis, P. Jacquet, W. Szpankowski, Is the internet fractal: the multicast power law revisited, SODA 2002.
11. A. Broido, E. Basic and K.C. Claffy, Invariance of the Internet RTT spectrum. Global RTT analysis, ICIR, August 2002, http://www.caida.org/ broido /rtt/rtt.html
12. P. Flajolet, X. Gourdon, and P. Dumas, Mellin transforms and asymptotics: Harmonic sums, Theoretical Computer Science 144, 1–2 (June 1995), 3–58.

Author Index

Lecture Notes in Computer Science

For information about Vols. 1–3720

please contact your bookseller or Springer